PHYSICS FOR DIAGNOSTIC RADIOLOGY, 2ND EDITION

This book is due

Medical Science Series

PHYSICS FOR DIAGNOSTIC RADIOLOGY

2ND EDITION

P P Dendy

Addenbrooke's NHS Trust, Cambridge, UK

B Heaton

University of Aberdeen, UK

Published in association with the
Institute of Physics and Engineering in Medicine

Institute of Physics Publishing
Bristol and Philadelphia

British Library Cataloguing-in-Publication Data

A catalogue record for this book is available from the British Library.

ISBN 0 7503 0590 8 (hbk)
ISBN 0 7503 0591 6 (pbk)

Library of Congress Cataloging-in-Publication Data are available

Series Editors:
   R F Mould, Croydon, UK
   C G Orton, Karmanos Cancer Institute and Wayne State University,
   Detroit, USA
   J A E Spaan, University of Amsterdam, The Netherlands
   J G Webster, University of Wisconsin-Madison, USA

Published by Institute of Physics Publishing, wholly owned by The Institute of Physics, London

Institute of Physics Publishing, Dirac House, Temple Back, Bristol BS1 6BE, UK

US Office: Institute of Physics Publishing, The Public Ledger Building, Suite 1035, 150 South Independence Mall West, Philadelphia, PA 19106, USA

Typeset in TEX using the IOP Bookmaker Macros
Printed in Great Britain by Bookcraft, Bath

The Medical Science Series is the official book series of the International Federation for Medical and Biological Engineering (IFMBE) and the International Organization for Medical Physics (IOMP).

IFMBE

The IFMBE was established in 1959 to provide medical and biological engineering with an international presence. The Federation has a long history of encouraging and promoting international cooperation and collaboration in the use of technology for improving the health and life quality of man.

The IFMBE is an organization that is mostly an affiliation of national societies. Transnational organizations can also obtain membership. At present there are 42 national members, and one transnational member with a total membership in excess of 15 000. An observer category is provided to give personal status to groups or organizations considering formal affiliation.

Objectives

- To reflect the interests and initiatives of the affiliated organizations.
- To generate and disseminate information of interest to the medical and biological engineering community and international organizations.
- To provide an international forum for the exchange of ideas and concepts.
- To encourage and foster research and application of medical and biological engineering knowledge and techniques in support of life quality and cost-effective health care.
- To stimulate international cooperation and collaboration on medical and biological engineering matters.
- To encourage educational programmes which develop scientific and technical expertise in medical and biological engineering.

Activities

The IFMBE has published the journal *Medical and Biological Engineering and Computing* for over 34 years. A new journal *Cellular Engineering* was established in 1996 in order to stimulate this emerging field in biomedical engineering. In *IFMBE News* members are kept informed of the developments in the Federation. *Clinical Engineering Update* is a publication of our division of Clinical Engineering. The Federation also has a division for Technology Assessment in Health Care.

Every three years, the IFMBE holds a World Congress on Medical Physics and Biomedical Engineering, organized in cooperation with the IOMP and the IUPESM. In addition, annual, milestone, regional conferences are organized in different regions of the world, such as the Asia Pacific, Baltic, Mediterranean, African and South American regions.

The administrative council of the IFMBE meets once or twice a year and is the steering body for the IFMBE. The council is subject to the rulings of the General Assembly which meets every three years.

For further information on the activities of the IFMBE, please contact Jos A E Spaan, Professor of Medical Physics, Academic Medical Centre, University of Amsterdam, PO Box 22660, Meibergdreef 9, 1105 AZ, Amsterdam, The Netherlands. Tel: 31 (0) 20 566 5200. Fax: 31 (0) 20 691 7233. Email: IFMBE@amc.uva.nl. WWW: http://vub.vub.ac.be/~ifmbe.

IOMP

The IOMP was founded in 1963. The membership includes 64 national societies, two international organizations and 12 000 individuals. Membership of IOMP consists of individual members of the Adhering National Organizations. Two other forms of membership are available, namely Affiliated Regional Organization and Corporate Members. The IOMP is administered by a Council, which consists of delegates from each of the Adhering National Organization; regular meetings of Council are held every three years at the International Conference on Medical Physics (ICMP). The Officers of the Council are the President, the Vice-President and the Secretary-General. IOMP committees include: developing countries, education and training; nominating; and publications.

Objectives

• To organize international cooperation in medical physics in all its aspects, especially in developing countries.
• To encourage and advise on the formation of national organizations of medical physics in those countries which lack such organizations.

Activities

Official publications of the IOMP are *Physiological Measurement*, *Physics in Medicine and Biology* and the *Medical Science Series*, all published by Institute of Physics Publishing. The IOMP publishes a bulletin *Medical Physics World* twice a year.

Two Council meetings and one General Assembly are held every three years at the ICMP. The most recent ICMPs were held in Kyoto, Japan (1991), Rio de Janeiro, Brazil (1994) and Nice, France (1997). The next conference is scheduled for Chicago, USA (2000). These conferences are normally held in collaboration with the IFMBE to form the World Congress on Medical Physics and Biomedical Engineering. The IOMP also sponsors occasional international conferences, workshops and courses.

For further information contact: Hans Svensson, PhD, DSc, Professor, Radiation Physics Department, University Hospital, 90185 Umeå, Sweden. Tel: (46) 90 785 3891. Fax: (46) 90 785 1588. Email: Hans.Svensson@radfys.umu.se.

LIST OF CONTRIBUTORS

P P Dendy
Department of Medical Physics and Clinical Engineering
Addenbrooke's NHS Trust
Cambridge
UK

K E Goldstone (Chapter 6)
East Anglian Regional Radiation Protection Service
Addenbrooke's NHS Trust
Cambridge
UK

B Heaton
Department of Bio-Medical Physics and Bio-Engineering
University of Aberdeen
Forresterhill
Aberdeen
UK

P C Jackson (Chapter 14)
Department of Medical Physics and Bioengineering
Southampton University Hospitals NHS Trust
Southampton
UK

E A Moore (Chapter 14)
Department of Medical Physics and Bioengineering
Southampton University Hospitals NHS Trust
Southampton
UK

(Now at Lysholm Radiological Department
National Hospital for Neurology and Neurosurgery
London
UK)

T A Whittingham (Chapter 13)
Regional Medical Physics Department
Newcastle General Hospital
Newcastle upon Tyne
UK

ACKNOWLEDGMENTS

We are grateful to many persons for constructive comments on the content of this book. In particular we wish to thank Mr R W Barber, Mr D Goodman, Miss D Siddle, Professor A K Dixon, Dr R Harrison, Ms J P Wade, Mr I Wright, Professor P F Sharp, Ms R Kestelman, Mr D Parry-Jones, Professor G M Roberts and Dr J Eatough.

Thanks are also due to Mrs D Scott, Mrs S Turner, Mrs L Ross and Mrs A Dendy for their infinite patience in typing and retyping several drafts of the manuscript.

Figures 2.19, 2.20 and 2.24 are reproduced by permission of Picker International Limited. Figure 2.23 is reproduced by permission of Siemens Medical Engineering. Figure 3.7 is reproduced by permission of the British Institute of Radiology. Figures 6.1 and 11.14 are reprinted by permission of Wiley-Liss Inc., a subsidiary of John Wiley and Sons Inc. (Whyte G N 1959 *Principles of Radiation Dosimetry* and Russell L B and Russell W L 1954 An analysis of the changing radiation response of the developing mouse embryo *J. Cell Comp. Physiol.* 43 (Suppl. 1), 103). Figure 6.19 is reproduced by permission of Science and Technology Letters. Figure 6.20 ©European Communities, 1993. Reproduced by permission of the publishers, the Office for Official Publications of the European Communities, 2, Rue Mercier, L-2985 Luxembourg. Figure 8.3(*a*) is reproduced by permission of Nuclear Associates. Figure 8.4 is reproduced by permission of The Royal Society (F W Campbell 1980 *Phil. Trans. R. Soc. Lond.* B290, 5–9, figure 1). Figures 8.12, 8.21 and 9.6 are reproduced by permission of Churchill Livingstone. Figure 8.22 is reproduced by permission of the International Commission on Radiation Units and Measurements (ICRU). Figures 8.23 and 11.12 are reproduced by permission of the Radiological Society of North America (1986 *Radiology* 158, 21–6 and 1979 *Radiology* 131, 589–97). Figure 9.17 is reproduced by permission of Nuclear Technology Publishing.

CONTENTS

INTRODUCTION

INTRODUCTION TO THE FIRST EDITION

The past 20 years have seen a rapid development in the range of imaging techniques available to the diagnostic radiologist and over the same period a marked increase in the level of sophistication of imaging equipment. These changes have inevitably led to changes in the background knowledge of physics that a radiologist is expected to acquire. For example, the concept of linking an image intensifier to a television camera in order to produce digitized images introduces many new ideas; many radiology departments now provide a nuclear medicine service, and of course the technique of magnetic resonance imaging is quite new. All of these subjects, and several others, require a knowledge and understanding of additional physical principles.

Fortunately, as equipment has become more sophisticated, it has also become more reliable. Furthermore the base-line knowledge of physics possessed by radiologists in training has risen. Thus the requirement to teach fundamental current electricity and electronics at this level has diminished considerably.

For all these reasons the content of a physics course for radiologists has changed quite appreciably and this book has been written to move with the times and to cover that physics now thought to be most relevant as reflected in the current syllabus of the Royal College of Radiologists (London).

The book has been designed as a comprehensive, compact primer, for aspiring radiologists, but to achieve this a number of hard decisions have had to be taken. First, it has been assumed that the reader will have studied physics to the standard required for university entrance (with some allowance for the knowledge being a little rusty), and is familiar with the terminology of such a physics syllabus. SI units have been adopted, with alternatives given where the reader might be confused.

Second, the subject matter is very interactive and this can lead to a tendency to repetition. In the interests of economy of space, most concepts are only considered once. Occasionally this requires a simple knowledge of material discussed later in the book. For example, the idea of filtration, which is dealt with in chapter 2, requires a knowledge of beam attenuation, not discussed in detail until chapter 3. Therefore it has been assumed that the reader will already possess a rudimentary knowledge of the subject or will be prepared to return to more difficult aspects at a second reading or during revision. The text is extensively cross-referenced and analogies have been drawn wherever possible.

Finally, for reasons both of ease of learning and compactness, lateral thinking has been encouraged where possible. For example, the ideas of exponential decay developed in chapter 1 when discussing radioactivity are later applied to attenuation of x-ray beams and the measurement of relaxation times in magnetic resonance imaging with little further explanation.

INTRODUCTION TO SECOND EDITION

All the above remains just as relevant as it was in 1987. However, there have been a number of important changes since that time as a result of a further 10 years development in diagnostic imaging.

The changes are reflected in two important ways. First it has been necessary to revise substantially and update parts of the book. Major changes have been made to recognize the much greater importance attached to patient dose, the increasingly widespread use of digital radiography, the importance of both patient dose and image quality in mammography, a greater awareness of the need to protect staff and of legislation and the associated emphasis on good quality assurance procedures. Finally the chapters on ultrasound and magnetic resonance imaging have been completely rewritten to bring them up to date.

The second change relates to the anticipated readership. The book remains primarily a text for radiologists in training, especially for Part 1 of the Fellowship examination of the Royal College of Radiologists in the UK and follows closely the relevant parts of the Guidelines for Training in General Radiology recently issued by the European Association of Radiology. However, the first edition was also received favourably by other professional groups, for example many radiographers. Since that time radiography training has become university based and radiographers frequently combine their vocational training with honours degrees and sometimes higher degrees. Thus we hope that the analytical approach to radiology adopted here will be of value in this wider approach to training. Finally the book will be of value to medical physicists who teach radiologists and radiographers and parts may also appeal to younger medical physicists in training.

The material is presented in a logical order. After a review of the fundamental principles of radiation physics and radioactivity in chapter 1, chapter 2 deals with the production of x-rays. All aspects of this topic have been considered here including x-ray tube rating and quality control measurements. Chapters 3 and 4, dealing respectively with the physics of the interaction of radiation with matter and the image receptor, provide the necessary basic information for consideration in chapter 5 of the formation of the radiological image and the factors controlling image quality, discussed in qualitative terms. Chapter 6 on radiation measurement and doses to patients has been contributed jointly with one of our guest authors.

Particular emphasis has been placed on those aspects of the subject covered in less detail in some other books used by radiologists. Thus, in chapter 7 the physics of nuclear medicine is considered in some detail. This reflects the increasing involvement of radiologists in such work. The idea of digitized information is also introduced here for the first time. Quantification, a relatively new concept in diagnostic radiology, is the theme of chapter 8, where both quantification of image quality via the production of digitized images and quantification of imager performance are discussed. This chapter also deals with an increasingly important concept, the need to assess, in objective terms, the diagnostic value of a particular imaging investigation. A number of special radiographic techniques are drawn together in chapter 9. For example, having introduced some of the ideas of digital images in the two previous chapters, digital radiology is considered in detail here. All aspects of tomographic imaging are considered in chapter 10.

Chapters 11 and 12 are seen as two of the most important chapters in the book since they bring together, from a wide variety of sources, most of the information all staff working in an x-ray department should know in radiobiology and radiological protection. The material presented here is completely up to date in respect of the UK legislation.

Chapters 13 and 14 have been completely rewritten by guest authors to reflect the rapid advances in ultrasound and magnetic resonance imaging whilst still presenting the material at the appropriate level for the readership.

The detail given in the book is thought to be appropriate to a teaching course some 50 hours in length. A list of relevant references, both text books and review articles, has been provided for further reading at the end of each chapter, where a set of questions is also given to exercise the mind of the reader. Finally a comprehensive selection of multiple choice questions is provided at the back of the book.

This text will provide both a starting point and a firm basis for revision for any modern course of physics designed to prepare radiologists for Fellowship examinations and for other professionals in the radiology department wishing to study to a comparable standard.

CHAPTER 1

FUNDAMENTALS OF RADIATION PHYSICS AND RADIOACTIVITY

1.1. STRUCTURE OF THE ATOM

All matter is made up of atoms, each of which has an essentially similar structure. All atoms are formed of small, dense, positively charged nuclei, typically some 10^{-14} m in diameter, orbited at much larger distances (about 10^{-10} m) by negatively charged, very light particles. The atom as a whole is electrically neutral. Note that because matter consists mainly of empty space, radiation may penetrate many atoms before a collision results.

The positive charge in the nucleus consists of a number of **protons** each of which has a charge of 1.6×10^{-19} coulombs (C) and a mass of 1.7×10^{-27} kilograms (kg). The negative charges are **electrons**. An electron carries the same numerical charge as the proton, but of opposite sign. However, an electron has only about 1/2000th the mass of the proton (9×10^{-31} kg). Each element is characterized by a specific number of protons, and an equal number of orbital electrons. This is called the **atomic number** and is normally denoted by the symbol Z. For example, Z for aluminium is 13, whereas for lead $Z = 82$.

Electrons are most likely to be at fairly well defined distances from the nucleus and are described as being in 'shells' around the nucleus (figure 1.1). More important than the distance of the electron from the nucleus is the electrostatic force binding the electron to the nucleus, or the amount of energy the electron would have to be given to escape from the field of the nucleus. This is equal to the amount of energy a free electron will lose when it is captured by the electrostatic field of a nucleus. It is possible to think in terms of an energy 'well' that becomes deeper as the electron is trapped in shells closer and closer to the nucleus.

The unit in which electron energies are measured is the **electron volt** (eV)—this is the energy one electron would gain if it were accelerated through 1 volt of potential difference. One thousand electron volts is a kilo electron volt (keV) and one million electron volts is a mega electron volt (MeV). Some typical electron shell energies are shown in figure 1.2. Note that

(1) if a free electron is assumed to have zero energy, all electrons within atoms have negative energy—i.e. they are bound to the nucleus and must be given energy to escape;
(2) the energy levels are not equally spaced and the difference between the K shell and the L shell is much bigger than any of the other differences.

1.2. NUCLEAR STABILITY AND INSTABILITY

If a large number of protons were forced together in a nucleus they would immediately explode owing to electrostatic repulsion. Very short-range attractive forces are therefore required within the nucleus for stability, and these are provided by **neutrons**, uncharged particles with a mass almost identical to that of the proton.

3

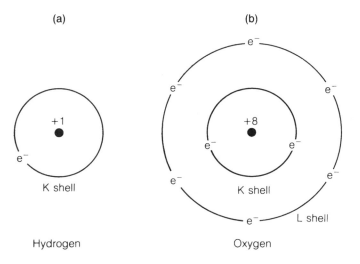

Figure 1.1. *Examples of atomic structure. (a) Hydrogen with one K shell electron. (b) Oxygen with two K shell electrons and six L shell electrons.*

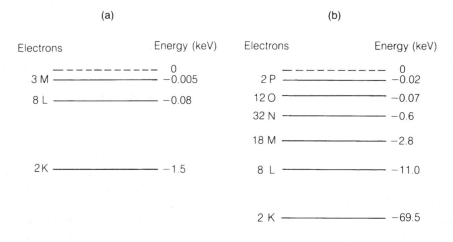

Figure 1.2. *Typical electron energy levels. (a) Aluminium (Z = 13). (b) Tungsten (Z = 74).*

The total number of protons and neutrons, collectively referred to as **nucleons**, within the nucleus is called the **mass number**, usually given the symbol A. Each particular combination of Z and A defines a **nuclide**. One notation used to describe a nuclide is $^A_Z X$. The number of protons Z defines the element X, so for hydrogen $Z = 1$, for oxygen $Z = 8$ etc, but the number of neutrons is variable. Therefore an alternative, and generally simpler, notation that carries all necessary information is X–A. The notation $^A_Z X$ will only be used for equations where it is important to check that the number of protons and the number of nucleons balance.

Nuclides that have the same number of protons but different numbers of neutrons are known as isotopes. Thus O-16, the most abundant isotope of oxygen, has eight protons (by definition) and eight neutrons. O-17 is the isotope of oxygen which has eight protons and nine neutrons. Since isotopes have the same number of protons and hence when neutral the same number of orbital electrons, they have the same chemical properties.

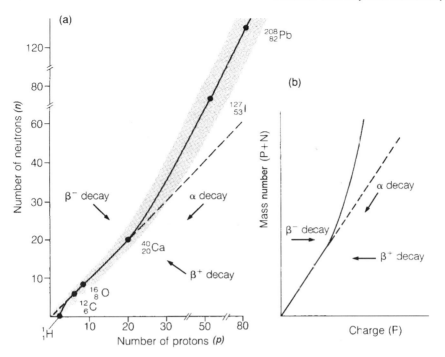

Figure 1.3. *Graphs showing the relationship between number of neutrons and number of protons for the most abundant stable elements. (a) Number of neutrons plotted against number of protons. The dashed line is at 45°. The cross-hatched area shows the range of values for which the nucleus is likely to be stable. (b) Total number of nucleons (neutrons and protons) plotted against number of protons. On each graph the changes associated with β^+, β^- and α decay are shown.*

The number of neutrons required to stabilize a given number of protons lies within fairly narrow limits and figure 1.3(a) shows a plot of these numbers. Note that for many elements of biological importance the number of neutrons is equal to the number of protons, but the most abundant form of hydrogen, which has one proton but no neutrons, is an important exception. At higher atomic numbers the number of neutrons begins to increase faster than the number of protons—lead, for example, has 126 neutrons but only 82 protons.

An alternative way to display the data is to plot the sum of neutrons and protons against the number of protons (figure 1.3(b)). This is essentially a plot of nuclear mass against nuclear charge (or the total charge on the orbiting electrons). This concept will be useful when considering the interaction of ionizing radiation with matter, and in section 3.4.3 the near-constancy of mass/charge (A/Z is close to 2) for most of the biological range of elements will be considered in more detail.

If the ratio of neutrons to protons is outside narrow limits, the nuclide is radioactive or a **radionuclide**. For example H-1 (normal hydrogen) is stable, H-2 (deuterium) is also stable, but H-3 (tritium) is radioactive. A nuclide may be radioactive because it has too many or too few neutrons.

A simple way to make radioactive nuclei is to bombard a stable element with a flux of neutrons in a reactor. For example radioactive phosphorus may be made by the reaction shown below

$$^{31}_{15}P + ^{1}_{0}n = ^{32}_{15}P + ^{0}_{0}\gamma$$

(the emission of a gamma ray as part of this reaction will be discussed later). However, this method of production results in a radionuclide that is mixed with the stable isotope since the number of protons in the nucleus has not changed and not all the P-31 is converted to P-32. Radionuclides that are 'carrier free' can be produced by bombarding with charged particles such as protons or deuterons, in a cyclotron; for example, if sulphur is bombarded with protons,

$$^{34}_{16}S + ^{1}_{1}p = ^{34}_{17}Cl + ^{1}_{0}n.$$

The radioactive product is now a different element and thus may be separated by chemical methods.

The **activity** of a source is a measure of its rate of decay or the number of disintegrations per second. It is measured in becquerels (Bq) where 1 Bq is equal to one disintegration per second. The becquerel has replaced the older unit of the curie (Ci), but since the latter is still encountered frequently in text books and published papers, it is important to know the conversion factor

$$1 \text{ Ci} = 3.7 \times 10^{10} \text{ Bq.}$$

Hence

$$1 \text{ mCi (millicurie)} = 3.7 \times 10^{7} \text{ Bq (37 megabecqerels or MBq)}$$
$$1 \text{ } \mu\text{Ci (microcurie)} = 3.7 \times 10^{4} \text{ Bq (37 kilobecquerels or kBq).}$$

1.3.   RADIOACTIVE CONCENTRATION AND SPECIFIC ACTIVITY

These two concepts are frequently confused.

Radioactive concentration

This relates to the amount of radioactivity per unit volume. Hence it will be expressed in Bq ml^{-1}. It is important to consider the radioactive concentration when giving a bolus injection. If one wishes to inject a large activity of technetium-99m (Tc-99m) in a small volume, perhaps for a dynamic nuclear medicine investigation, it is preferable to elute a 'new' molybdenum–technetium generator when the yield might be 8 GBq (200 mCi) in a 10 ml eluate [0.8 GBq ml^{-1} (20 mCi ml^{-1})] rather than an old generator when the yield might be only about 2 GBq (50 mCi) [0.2 GBq ml^{-1} (5 mCi ml^{-1})]. For a fuller discussion of the production of Tc-99m and its use in nuclear medicine see section 1.7 and chapter 7.

Specific activity

This relates to the proportion of nuclei of the element of interest that are actually labelled. Non-radioactive material, for example iodine-127 (I-127) in a sample of I-125 may be present as a result of the preparation procedure or may have been added as carrier. The unit for the total number of atoms or nuclei present is the mole so the proportion that are radioactive or the **specific activity** can be expressed in Bq mol^{-1} or Bq kg^{-1}. The specific activity of a preparation should always be checked since it determines the total amount of the element being administered. Modern radiopharmaceuticals generally have a very high specific activity so the total amount of the element administered is very small, and problems such as iodine sensitivity do not normally arise in diagnostic nuclear medicine.

1.4.   RADIOACTIVE DECAY PROCESSES

Three types of radioactive decay that result in the emission of charged particles will be considered at this stage.

β^- decay

A negative β particle is an electron. Its emission is actually a very complex process but it will suffice here to think of a change **in the nucleus** in which a neutron is converted into a proton. The particles are emitted with a range of energies. Note that although the process results in emission of electrons, it is a **nuclear** process and has nothing to do with the orbiting electrons.

The mass number of the nucleus remains unchanged but its charge increases by one, thus this change is favoured by nuclides which have too many neutrons.

β^+ decay

A positive β particle, or positron, is the anti-particle to an electron, having the same mass and equal but opposite charge. Again, its precise mode of production is complex but it can be thought of as being released when a proton in the nucleus is converted to a neutron. Note that a positron can only exist while it has kinetic energy. When it comes to rest it spontaneously combines with an electron.

The mass number of the nucleus again remains unchanged but its charge decreases by one, thus this change is favoured by nuclides which have too many protons.

α decay

An α particle is a helium nucleus, thus it comprises two protons and two neutrons. After α emission, the charge is reduced by two units and the mass number by four units.

The effects of β^-, β^+ and α decay are shown in figure 1.3(*a*) and figure 1.3(*b*). Note that emission of α particles only occurs for the higher atomic number nuclides.

1.5. EXPONENTIAL DECAY

Although it is possible to predict from the number of protons and neutrons in the nucleus which type of decay might occur, it is not possible to predict how fast decay will occur. One might imagine that nuclides that were furthest from the stability line would decay fastest. This is not so and the factors which determine the rate of decay are beyond the scope of this book.

However, all radioactive decay processes do obey a very important rule. This states that the only variables affecting the number of nuclei ΔN decaying in a short interval of time Δt are the number of unstable nuclei present N and the time interval Δt. Hence

$$\Delta N \propto N \Delta t.$$

If the time interval is very short, the equation becomes

$$\mathrm{d}N = -kN\mathrm{d}t$$

where the constant of proportionality k is characteristic of the radionuclide, known as its **decay constant** or **transformation constant**, and the negative sign has been introduced to show that, mathematically, the number of radioactive nuclei actually decreases with elapsed time.

The equation may be integrated to give the well known exponential relationship

$$N = N_0 \exp(-kt)$$

where N_0 is the number of unstable nuclei present at $t = 0$. Since the activity of a source, A, is equal to the number of disintegrations per second,

$$A = \frac{\mathrm{d}N}{\mathrm{d}t} = -kN = kN_0 \exp(-kt)$$

when $t = 0$,

$$\left(\frac{dN}{dt}\right) = kN_0 = A_0,$$

so

$$A = A_0 \exp(-kt). \tag{1.1}$$

Thus the activity also decreases exponentially.

1.6. HALF-LIFE

An important concept is the **half-life** or the time ($T_{1/2}$) after which the activity has decayed to half its original value.

   If A is set equal to $A_0/2$ in equation (1.1),

$$\tfrac{1}{2} = \exp(-kT_{1/2}) \qquad \text{or} \qquad kT_{1/2} = \ln 2.$$

Hence

$$A = A_0 \exp\left(\frac{-\ln 2t}{T_{1/2}}\right) = A_0 \exp\left(\frac{-0.693t}{T_{1/2}}\right) \tag{1.2}$$

since $\ln 2 = 0.693$. Equally,

$$N = N_0 \exp\left(\frac{-0.693t}{T_{1/2}}\right).$$

Two extremely important properties of exponential decay must be remembered.

(1)   The idea of half-life may be applied from any starting point in time. Whatever the activity at a given time, after one half-life the activity will have been halved.
(2)   The activity never becomes zero, since there are many millions of radioactive nuclei present, so their number can always be halved to give a residue of radioactivity.

   Clearly, if the value of $T_{1/2}$ is known, and the activity or rate of decay is known at one time, the rate of decay may be found at any later time by solving equation (1.2) above. However, the activity may also be found, with sufficient accuracy, by a simple graphical method. Proceed as follows.

(1)   Use the y-axis to represent activity and the x-axis to represent time.
(2)   Mark the x-axis in equal units of half-lives.
(3)   Assume the activity at $t = 0$ is 1. Hence the first point on the graph is (0,1).
(4)   Now apply the half-life rule. After one half-life, the activity is $\frac{1}{2}$, so the next point on the graph is $(1, \frac{1}{2})$.
(5)   Apply the half-life rule again to obtain the point $(2, \frac{1}{4})$ and successively $(3, \frac{1}{8})(4, \frac{1}{16})(5, \frac{1}{32})$.

   See figure 1.4(a).
   Note that, so far, the graph is quite general without consideration of any particular nuclide, half-life or activity. To answer a specific problem, it is now only necessary to relabel the axes with the given data, e.g. 'The activity of an oral dose of I-131 is 90 MBq at 12 noon on Tuesday 4 October. If the half-life of I-131 is 8 days, when will the activity be 36 MBq?'. Figure 1.4(b) shows the same axes as figure 1.4(a) relabelled to answer this specific problem. This quickly yields the answer of $10\frac{1}{2}$ days, i.e. at 12 midnight on 14 October.[1]

---

[1] To solve this problem using equation (1.2):

$$36 = 90 \exp(-(\ln 2 \cdot t)/8) \quad \text{where } t \text{ is the required time in days.}$$

$$\ln\left(\frac{90}{36}\right) = \left(\frac{\ln 2 \cdot t}{8}\right) \quad \text{from which } t = 10.6 \text{ days.}$$

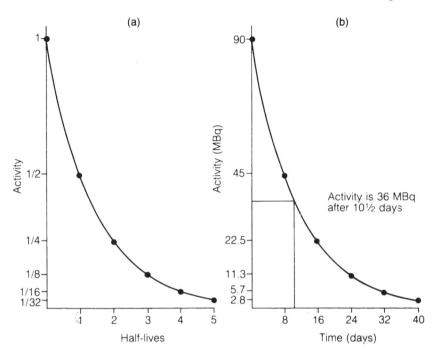

Figure 1.4. *Simple graphical method for solving any problem where the behaviour is exponential. (a) A basic curve that may be used to describe any exponential process. (b) The same curve used to solve the specific problem on radioactive decay set in the text.*

This graphical approach may be applied to any problem that can be described in terms of simple exponential decay.

1.7. SECULAR AND TRANSIENT EQUILIBRIUM

As already explained, radioactive decay is a process by which the nucleus attempts to achieve stability. It is not always successful at the first attempt and further decay processes may be necessary. For example, four decay schemes occur in nature each of which involves a long sequence of decay processes, terminating finally in stable lead-208 (Pb-208).

In such a sequence the nuclide which decays is frequently called the **parent** and its decay product the **daughter**. If both the parent and daughter nuclides are radioactive, and the parent has a longer half-life than the daughter, the rate of decay of the daughter is determined not only by its own half-life but also by the rate at which it is produced. As a first approximation, assume that the activity of the parent remains constant, or is constantly replenished so that the rate of production of the daughter remains constant. If none of the daughter is present initially, its rate of production will at first exceed its rate of decay and equilibrium will be reached when the rate of production is just equal to the rate of decay (figure 1.5(*a*)).

The curve is of the exponential type so the activity never actually reaches equilibrium. The rate of approach to equilibrium depends on the half-life of the daughter and after 10 half-lives the activity will be

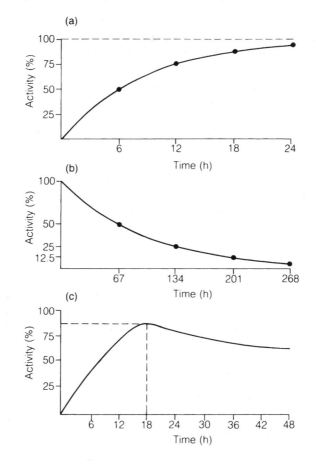

Figure 1.5. *(a) Increase in activity of a daughter when the activity of the parent is assumed to be constant. Generation of Tc-99m from Mo-99 has been taken as a specific example, but with the assumption that the supply of Mo-99 is constantly replenished. (b) The decay curve for Mo-99 which has a half-life of 67 h. (c) Increase in activity of a daughter when the activity of the parent is decreasing. Generation of Tc-99m from Mo-99 has been taken as a specific example. Curves (a) and (b) are multiplied to give the resultant activity of Tc-99m.*

within 0.1% of equilibrium.[2]

The equilibrium activity is governed by the activity of the parent.

Two practical situations should be distinguished.

(1)   The half-life of the parent is much longer than the half-life of the daughter; for example, radium-226, which has a half-life of 1620 years, decays to radon gas which has a half-life of 3.82 days. For most

---

[2] Mathematically, the shape of figure 1.5(a) is given by

$$N = N_{max}[1 - \exp(-\ln 2 \cdot t)/T_{1/2}]$$

where $T_{1/2}$ is now the half-life of the daughter radionuclide. Thus after n half-lives

$$N = N_{max}[1 - \exp(-n \ln 2)] = N_{max}[1 - (\tfrac{1}{2})^n].$$

practical purposes the activity of the radon gas reaches a constant value, only changing very slowly as the radium decays. This is known as **secular equilibrium**.

(2)   The half-life of the parent is not much longer than that of the daughter. The most important example for radiologists arises in diagnostic nuclear medicine and is molybdenum-99 (Mo-99) which has a half-life of 67 h before decaying to technetium-99m (Tc-99m) which has a half-life of 6 h. Now the growth curve for Tc-99m when the Mo-99 activity is assumed constant (figure 1.5(a)) must be multiplied by the decay curve for Mo-99 (figure 1.5(b)). The resultant (figure 1.5(c)) shows that an actual maximum of Tc-99m activity is reached after about 18 h. By the time the 10 half-lives (60 h) required for Tc-99m to come to equilibrium with Mo-99 have elapsed, the activity of Mo-99 has fallen to half its original value.

This is known as **transient equilibrium** because although the Tc-99m is in equilibrium with the Mo-99, the activity of the Tc-99m is not constant. It explains why the amount of activity that can be eluted from an Mo–Tc generator (see section 1.3 and chapter 7) is much higher when the generator is first delivered than it is a week later.

1.8.   BIOLOGICAL AND EFFECTIVE HALF-LIFE

When a radionuclide is administered, either orally or by injection, in addition to the reduction of activity with time due to the physical process of decay, activity is also lost from the body as a result of biological processes. Generally speaking, these processes also show exponential behaviour so the concentration of substance C remaining at time t after injection is given by

$$C = C_0 \exp(-(\ln 2) \cdot t / T_{\frac{1}{2}\text{biol}})$$

(cf. equation (1.2)), where $T_{1/2}$ is the biological half-life.

When physical and biological processes are combined, the overall loss is the product of two exponential terms and the activity at any time after injection is given by

$$A = A_0 \exp(-(\ln 2) \cdot t / T_{\frac{1}{2}\text{phys}}) \cdot \exp(-(\ln 2) \cdot t / T_{\frac{1}{2}\text{biol}})$$
$$= A_0 \exp[-(\ln 2) \cdot t (1 / T_{\frac{1}{2}\text{phys}} + 1 / T_{\frac{1}{2}\text{biol}})].$$

To find the effective half-life $T_{\frac{1}{2}\text{eff}}$, set

$$A = A_0 \exp(-0.693t) / T_{\frac{1}{2}\text{eff}}).$$

Hence, by inspection,

$$1 / T_{\frac{1}{2}\text{eff}} = 1 / T_{\frac{1}{2}\text{phys}} + 1 / T_{\frac{1}{2}\text{biol}}.$$

Note that if $T_{\frac{1}{2}\text{phys}}$ is much shorter than $T_{\frac{1}{2}\text{biol}}$, the latter may be neglected, and vice versa. For example if $T_{\frac{1}{2}\text{phys}} = 1$ h and $T_{\frac{1}{2}\text{biol}} = 20$ h,

$$1 / T_{\frac{1}{2}\text{eff}} = 1 + 1/20 = 1.05$$

and $T_{\frac{1}{2}\text{eff}} = 0.95$ h or almost the same as $T_{\frac{1}{2}\text{phys}}$.

1.9.   GAMMA RADIATION

Some radionuclides emit radioactive particles to gain stability. Normally, in addition to the particle, the nucleus also has to emit some energy, which it does in the form of gamma radiation. Note that emission of gamma rays as a mechanism for losing energy is very general and, as shown in section 1.2, may also occur when radionuclides are produced.

Although the emission of the particle and the gamma ray are, strictly speaking, separate processes, they normally occur very close together in time. However, some nuclides enter a metastable state after emitting the particle and emit their gamma ray some time later. When the two processes are separated in time in this way, the second stage is known as an **isomeric transition**. An important example in nuclear medicine is technetium-99m (the 'm' stands for metastable) which has a half-life of 6 h. This is long enough for it to be separated from the parent molybdenum-99 and the decay is then by gamma ray emission only which is particularly suitable for *in vivo* investigations (see chapter 7).

Just as electrons in shells around the nucleus occupy well defined energy levels, there are also well defined energy levels in the nucleus. Since gamma rays represent transitions between these levels, they are monoenergetic. However gamma rays with more than one well defined energy may be emitted by the same nuclide, for example indium-111 emits gamma rays at 163 keV and 247 keV.

1.10.   X-RAYS AND GAMMA RAYS AS FORMS OF ELECTROMAGNETIC RADIATION

The propagation of energy by simultaneous vibration of electric and magnetic fields is known as electromagnetic (EM) radiation. Unlike sound, which is produced by the vibration of molecules and therefore requires a medium for propagation (see chapter 13), EM radiation can travel through a vacuum. However, like sound, EM radiation exhibits many wavelike properties such as reflection, refraction, diffraction and interference and is frequently characterized by its wavelength. EM waves can vary in wavelength from 10^{-13} to 10^3 m and different parts of the EM spectrum are recognized by different names (see table 1.1).

Table 1.1. *The different parts of the electromagnetic spectrum classified in terms of wavelength, frequency and quantum energy.*

	Radio waves	Infra-red	Visible light	Ultra-violet	X-rays and gamma rays
Wavelength (m)	10^3–10^{-2}	10^{-4}–10^{-6}	5×10^{-7}	5×10^{-8}	10^{-9}–10^{-13}
Frequency (Hz)	3×10^5–3×10^{10}	3×10^{12}–3×10^{14}	6×10^{14}	6×10^{15}	3×10^{17}–3×10^{21}
Quantum energy (eV)	10^{-9}–10^{-4}	10^{-2}–1	2	20	10^3–10^7

X-rays and gamma rays are both part of the EM spectrum and an 80 keV X-ray is identical to, and hence indistinguishable from, an 80 keV gamma ray. In order to appreciate the reason for the apparent confusion, it is necessary to consider briefly the origin of the discoveries of X-rays and gamma rays. As already noted, gamma rays were discovered as a type of radiation emitted by radioactive materials. They were clearly different from alpha rays and from beta rays, so they were given the name gamma rays. X-rays were discovered in quite a different way as 'emission from high energy machines of radiations that caused certain materials, such as barium platino-cyanide to fluoresce'. It was some time before the identity of X-rays produced by machines and gamma rays produced by radioactive materials was confirmed.

For a number of years, X-rays produced by machines were of lower energy than gamma rays, but with the development of linear accelerators and other high energy machines, this distinction is no longer useful.

No distinction between X-rays and gamma rays is totally self-consistent, but it is reasonable to describe gamma rays as the radiation emitted as a result of nuclear interactions, and X-rays as the radiation emitted as a result of events outside the nucleus. For example, one method by which nuclides with too few neutrons may approach stability is by **K-electron capture**. This mode of radioactive decay has not yet been discussed. The nucleus 'steals' an electron from the K shell to neutralize one of its protons. The K shell vacancy is filled by electrons from outer shells and the associated characteristic radiations are referred to as X-rays, not gamma rays, even though they result from radioactive decay.

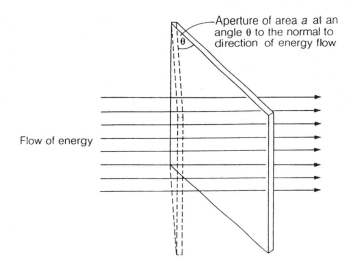

Figure 1.6. *A simple representation of the meaning of intensity.*

An important concept is the **intensity** of a beam of x- or gamma rays. This is defined as the amount of energy crossing unit area placed normal to the beam in unit time. In figure 1.6, if the total amount of radiant energy passing though the aperture of area a in time t is E, the intensity $I = E/(a(\cos\theta) \cdot t)$ where $a\cos\theta$ is the cross sectional area of the aperture normal to the beam. If a is in m^2, t in s and E in joules, the units of I will be J m^{-2} s^{-1}.

1.11. QUANTUM PROPERTIES OF RADIATION

As well as showing the properties of waves, short wavelength EM radiation, such as x- and gamma rays, can sometimes show particle-like properties. Each particle, or **photon**, is in fact a small packet of energy and the size of the energy packet (ε) is related to frequency (f) and wavelength (λ) by the fundamental equations

$$\varepsilon = hf = hc/\lambda$$

where h is the **Planck constant** and c is the speed of electromagnetic waves.

Taking $c = 3 \times 10^8$ m s^{-1} and $h = 6.6 \times 10^{-34}$ J s

$$\varepsilon \text{ (in joules)} = \frac{2 \times 10^{-25}}{\lambda \text{ (in metres)}}.$$

Thus the smaller the value of λ, the larger the value of the energy packet. For a typical diagnostic X-ray wavelength of 2×10^{-11} m, the value of ε in joules for a single photon is inconveniently small, so the electron volt, a unit of energy that has already been introduced, is used where 1 eV $= 1.6 \times 10^{-19}$ J. A wavelength of 2×10^{-11} m corresponds to a photon energy of 62 keV.

1.12. INVERSE SQUARE LAW

Before considering the interaction of radiation with matter, one important law that all radiations obey under carefully defined conditions will be introduced. This is the **inverse square law** which states that for a point source, and in the absence of attenuation, the intensity of a beam of radiation will decrease as the inverse of the square of the distance from that source.

The law is essentially just a statement of conservation of energy, since if the rate at which energy is emitted as radiation is E, the energy will spread out in all directions and the amount crossing unit area per second at radius r, $I_r = E/4\pi r^2$ (figure 1.7). Similarly the intensity crossing unit area at radius R, $I_R = E/4\pi R^2$. Thus the intensity is decreasing as $1/(\text{radius})^2$.

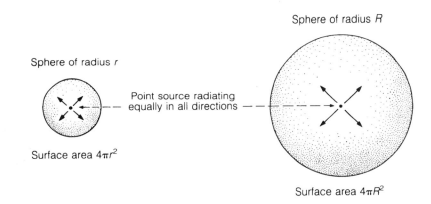

Figure 1.7. *A diagram showing the principle of the inverse square law.*

1.13.   INTERACTION OF RADIATION WITH MATTER

As a simple model of the interaction of radiation with matter, consider the radiation as a stream of fast moving particles (alphas, betas or photons) and the medium as an array of nuclei each with a shell of electrons around it (figure 1.8). As the particle tries to penetrate the medium, it will collide with atoms. Sometimes it will transfer energy of excitation during a collision. This type of interaction will be considered in more detail in chapter 3. The energy is quickly dissipated as heat. Occasionally, the interaction will be so violent that one of the outermost electrons will be torn away from the nucleus to which it was bound and become free. **Ionization** has occurred because an ion pair has been created. Sometimes, as in interaction C, the electron thus released has enough energy to cause further ionizations and a cluster of ions is produced.

The amount of energy required to create an ion pair is about 34 eV. Charged particles of interest in medicine invariably possess this amount of energy. For EM radiation, a quantum of x- or gamma rays always has more than 34 eV but a quantum of, say, ultra-violet or visible light does not. Hence the EM spectrum may be divided into ionizing and non-ionizing radiations.

The above, very simple model may also be used to predict how easily different types of radiation will be attenuated by different types of material. Clearly, as far as the stopping material is concerned, a high density of large nuclei (i.e. high atomic number) will be most effective for causing many collisions. Thus gases are poor stopping materials, but lead ($Z = 82$) is excellent and, if there is a special reason for compact shielding, even depleted uranium ($Z = 92$) is sometimes used.

With regard to the bombarding particles, size (or mass) is again important and since the particle is moving through a highly charged region, interaction is much more probable if the particle itself is charged and therefore likely to come under the influence of the strong electric fields associated with the electron and nucleus. Since x- and gamma ray quanta are uncharged and have zero rest mass, they are difficult to stop and lead may be required to cause appreciable attenuation.

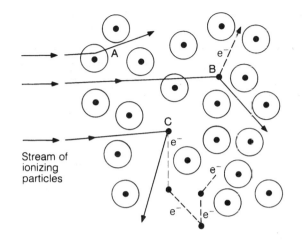

Figure 1.8. *Simple model of the interaction of radiation with matter. Interaction A causes excitation, interaction B causes ionization and interaction C causes multiple ionizations. At each ionization an electron is released from the nucleus to which it was bound. Recall the comment in section 1.1 that matter consists mostly of empty space. Hence the chance of a collision is much smaller in practice than this diagram suggests.*

The β^- particle has mass and is charged so it is stopped more easily—a few mm of low atomic number materials such as Perspex will usually suffice. Since it will be shown in chapter 3 that the mechanism of energy dissipation by x- and gamma rays is via secondary electron formation, a table of electron ranges in soft tissue will be helpful (table 1.2).

Table 1.2. *Approximate ranges of electrons in soft tissue.*

Electron energy (keV)	Approximate range (mm)
20	0.01
40	0.03
100	0.14
400	1.3

Protons and alpha particles are even more massive than β^- particles and are charged, so they are stopped easily. Alpha particles, for example, are so easily stopped, even by a sheet of paper, that great care must be taken when attempting to detect them to ensure that the detector has a thin enough window to allow them to enter the counting chamber. Neutrons are more penetrating because, although of comparable mass to the proton, they are uncharged.

One final remark should be made regarding the ranges of radiations. Charged particles eventually become trapped in the high electric fields around nuclei and have a finite range. Beams of x- or gamma rays are stopped by random processes, and as shown in chapter 3 are attenuated exponentially. This process has many features in common with radioactive decay. For example the rate of attenuation by a particular material is predictable but the radiation does not have a finite range.

1.14.   LINEAR ENERGY TRANSFER

Beams of ionizing radiation are frequently characterized in terms of their linear energy transfer (LET). This is a measure of the rate at which energy is transferred to the medium and hence of the density of ionization along the track of the radiation. Although a difficult concept to apply rigorously, it will suffice here to use a simple definition, namely that LET is the energy transferred to the medium per unit track length. It follows from this definition that radiations which are easily stopped will have a high LET, those which are penetrating will have a low LET. Some examples are given in table 1.3.

Table 1.3. *Approximate values of linear energy transfer for different types of radiation.*

Radiation	LET (keV μm^{-1})
1 MeV γ rays	0.5
100 kVp X-rays	6
20 keV β particles	10
5 MeV neutrons	20
5 MeV α particles	50

1.15.   SUMMARY OF ENERGY CHANGES IN RADIOLOGICAL PHYSICS

Energy cannot be created or destroyed but can only be converted from one form to another. Therefore it is important to summarize the different forms in which energy may appear. Remember that **work** is really just another word for energy—stating that body A does work on body B means that energy is transferred from body A to body B.

Mechanical energy

This can take two well known forms.

(1)   **Kinetic energy**, $\frac{1}{2}mv^2$, where m is the mass of the body and v its velocity.
(2)   **Potential energy**, mgh, where g is the gravitational acceleration and h the height of the body above the ground.

Kinetic energy is more relevant than potential energy in the physics of X-ray production and the behaviour of X-rays.

Electrical energy

When an electron, charge e, is accelerated through a potential difference V, it acquires energy eV. Thus if there are n electrons they acquire total energy neV. Note:

(1)   Current (i) is rate of flow of charge. Thus $i = ne/t$ where t is the time. Hence, rearranging, an alternative expression for the energy in a beam of electrons is Vit.
(2)   Just as Vit is the amount of energy gained by electrons as they accelerate through a potential difference V, it is also the amount of energy lost by electrons (usually as heat) when they fall through a potential difference of V, for example when travelling through a wire that has resistance R.
(3)   If the resistor is 'ohmic', that is to say it obeys **Ohm's law**, then $V = iR$ and alternative expressions for the heat dissipated are V^2/R or i^2R. Note, however, that many of the resistors encountered in the technology of X-ray production are non-ohmic.

Heat energy

When working with X-rays, most forms of energy are eventually degraded to heat and when a body of mass m and specific heat capacity s receives energy E and converts it into heat, the rise in temperature ΔT will be given by

$$E = ms\,\Delta T.$$

Excitation and ionization energy

Electrons are bound in energy levels around the nucleus of the atom. If they acquire energy of excitation they may jump into a higher energy level. Sometimes the energy may be enough for the electrons to escape completely (ionization). Note that if this occurs the electron may also acquire some kinetic energy in addition to the energy required to cause ionization.

Radiation energy

Radiation represents a flow of energy. This is usually expressed in terms of beam intensity I such that $I = E/(at)$ where E is the total energy passing through an area a placed normal to the beam, in time t.

Quantum energy

X- and gamma radiation frequently behave as exceedingly small energy packets. The energy of one quantum is hf where h is the Planck constant and f the frequency of the radiation. The energy of one quantum is so small that the joule is an inconveniently large unit so the electron volt is introduced where $1\text{ eV} = 1.6 \times 10^{-19}$ J.

Mass energy

As a result of Einstein's work on relativity, it has become apparent that mass is just an alternative form of energy. If a small amount of matter, mass m, is converted into energy, the energy released $E = mc^2$ where c is the speed of electromagnetic waves. This change is encountered most frequently in radioactive decay processes. Careful calculation, to about one part in a million, shows that the total mass of the products is slightly less than the total mass of the starting materials, the residual mass having been converted to energy according to the above equation. Annihilation of positrons (see section 3.4.4) is another good example of the equivalence of mass and energy.

Summary

As an example of the importance of conservation of energy in diagnostic radiology, consider the energy changes in the production and attenuation of X-rays and registration on photographic film. First, electrical energy is converted into the kinetic energy of the electrons in the X-ray tube. When the electrons hit the anode, their kinetic energy is destroyed. The majority is converted into heat, a little into X-rays. As the X-rays penetrate the body some of their energy is absorbed, more in bone than in soft tissue, and causes ionization before eventually being converted into heat. Finally the X-rays which strike the intensifying screen cause excitation and the emission of visible light quanta and these lower energy quanta stimulate the physico-chemical processes in photographic film leading eventually to blackening.

FURTHER READING

Aird E G A 1988 *Basic Physics for Medical Imaging* (Oxford: Heinemann)
Chackett K F 1981 *Radionuclide Technology—an Introduction to Quantitative Nuclear Medicine* (New York: Van Nostrand Reinhold)
Gifford D 1984 *A Handbook of Physics for Radiologists and Radiographers* (New York: Wiley)
Johns H E and Cunningham J R 1983 *The Physics of Radiology* 4th edn (Springfield, IL: Thomas)
Meredith W J and Massey J B 1977 *Fundamental Physics of Radiology* 3rd edn (Bristol: Wright)
Wilks R J 1987 *Principles of Radiological Physics* 2nd edn (Edinburgh: Churchill Livingstone)

EXERCISES

1   Describe in simple terms the structure of the atom and explain what is meant by atomic number, atomic weight and radionuclide.
2   What is meant by the binding energy of an atomic nucleus? Define the unit in which it is normally expressed and indicate the order of magnitude involved.
3   Describe the different ways in which radioactive disintegration can occur.
4   What is meant by the decay scheme of a radionuclide and radioactive equilibrium?
5   What is a radionuclide generator?
6   A radiopharmaceutical has a physical half-life of 6 h and a biological half-life of 20 h. How long will it take for the activity in the patient to fall to 25% of that injected?
7   The decay constant of iodine-123 is 1.34×10^{-5} s^{-1}. What is its half-life and how long will it take for the radionuclide to decay to one-tenth of its original activity?
8   Investigate whether the values of radiation intensity given below decrease exponentially with time.

Intensity (J m^{-2} s^{-1})	100	70	50	33	25	20	10	6.7	5.0	4.0
Time (s)	0	1.0	2.0	3.0	4.0	5.0	10.0	15.0	20.0	25.0

9   A radionuclide A decays into a nuclide B which has an atomic number one less than that of A. What types of radiation might be emitted either directly or indirectly in the disintegration process? Indicate briefly how they are produced.
10   Give typical values for the ranges of α particles and β^- particles in soft tissue. Why is the concept of range not applicable to gamma rays?
11   For an unknown sample of radioactive material explain how it would be possible to determine by simple experiment (a) the types of radiation emitted, (b) the half-life.
12   State the inverse square law for a beam of radiation and give the conditions under which it will apply exactly.
13   A surface is irradiated uniformly with a monochromatic beam of X-rays of wavelength 2×10^{-11} m. If 20 quanta fall on each square cm of the surface per second, what is the intensity of the radiation at the surface? (Use data given in section 1.11.)
14   Place the following components in order of power of dissipation: (a) a fluorescent light; (b) an X-ray tube; (c) an electric fire; (d) a pocket calculator; (e) an electric iron.

CHAPTER 2

PRODUCTION OF X-RAYS

2.1. INTRODUCTION

When electrons are accelerated to energies in excess of 5 keV and are then directed onto a target surface, X-rays may be emitted. The X-rays originate principally from rapid deceleration of the electrons when they interact with the nucleus of the target atoms. These X-rays are known as '*bremsstrahlung*' or braking radiation.

The essential features of a simple X-ray tube are shown in figure 2.1 and comprise:

(1) a heated metal filament to provide a copious supply of electrons by thermionic emission and to act as a cathode;
(2) an evacuated chamber across which a potential difference can be applied;
(3) a metal anode (the target) with a high efficiency for conversion of electron energy into X-ray photons;
(4) a thinner window in the chamber wall that will be transparent to most of the X-rays.

In this chapter the mechanisms of X-ray production will be considered in detail and the main components of a modern X-ray tube will be described. Physical factors affecting the design and performance of X-ray sets will be discussed along with implications for taking good radiographs.

2.2. THE X-RAY SPECTRUM

If the accelerating voltage across the X-ray tube shown in figure 2.1 were about 100 kV, the spectrum of radiation that would be used for radiology might be something like that shown in figure 2.2. The various features of this spectrum will now be discussed.

2.2.1. *The continuous spectrum*

When a fast-moving electron strikes the anode, several things may happen. The most common is that the electron will suffer a minor elastic interaction with an orbital electron as depicted at A in figure 2.3. This will result in the transfer to the target of a small amount of energy which will appear eventually as heat. At diagnostic energies, at least 99% of the electron energy is converted into heat and the dissipation of this heat is a major technical problem that will be considered in section 2.5.

Occasionally, an electron will come close to the nucleus of a target atom, where it will suffer a much more violent change of direction because the charge and mass of the nucleus are much greater than those of an electron (example B). The electron does not penetrate the nucleus because the energy barrier presented by these positive charges in the nucleus is far in excess of the electron energy. This results in the electron being deviated around it. The interaction results in a change in kinetic energy of the electron and the emission

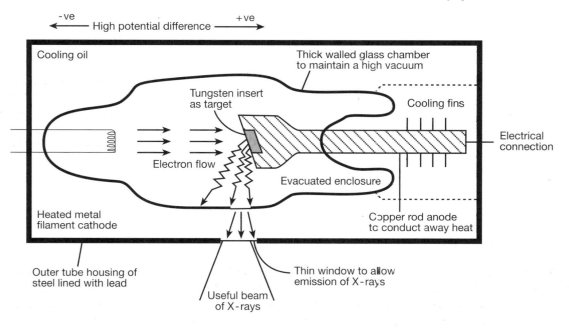

Figure 2.1. *Essential features of a simple, stationary anode X-ray tube. Note that for simplicity only X-ray photons passing through the window are shown. In practice they are emitted in all directions.*

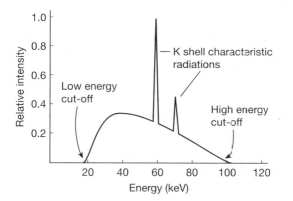

Figure 2.2. *Spectrum of radiation incident on a patient from an X-ray tube operating at 100 kVp using a tungsten target and 2.5 mm aluminium filtration.*

of electromagnetic radiation that is in the X-ray range of the spectrum. The amount of energy lost by the electron in such an interaction is very variable and hence the energy given to the X-ray photon can take a wide range of values. Note that X-ray emission may occur after two or three earlier slight deviations (example C). Therefore not all emissions occur from the surface of the anode. This factor is important when the spatial distribution of X-ray emission is considered.

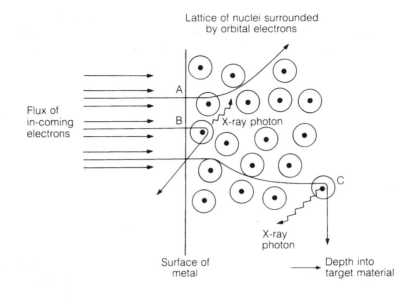

Figure 2.3. *Schematic representation of the interaction of electrons with matter. (A) Interaction resulting in the generation of low energy electromagnetic radiations (infra-red, visible, ultra-violet and very soft X-rays). All these are rapidly converted into heat. (B) Interaction resulting in the production of an X-ray . (C) Production of an X-ray after previous interactions that resulted only in heat generation.*

2.2.2.   *The low and high energy cut-off*

These parts of the spectrum are simply explained. Low energy electromagnetic radiations are easily attenuated and below a certain energy they are so heavily attenuated—by the materials of the anode, by the cooling/insulating oil, by the window of the X-ray tube and by any added filtration—that the intensity emerging is negligible. X-ray attenuation is discussed in detail in chapter 3.

The high energy cut-off occurs because an electron may, very occasionally, lose all its energy to X-ray production (section 2.2.1). Hence for any given electron energy, i.e. that determined by the accelerating voltage across the X-ray tube, there is a well defined maximum X-ray energy equal to the energy of a single electron. This corresponds to a minimum X-ray wavelength. Note that it is not possible, by quantum theory, for the energy of several electrons to be stored up in the anode to produce a jumbo-sized X-ray quantum.

It is useful to calculate the electron velocity, the maximum X-ray photon energy and the minimum X-ray photon wavelength associated with a given tube kilovoltage. To avoid complications associated with relativistic effects, a tube operating at only 30 kV is considered.

The energy of each electron is given by the product of its charge (e coulombs) and the accelerating voltage (V volts)

$$eV = 1.6 \times 10^{-19} (\text{C}) \times 3 \times 10^4 (\text{V})$$
$$= 4.8 \times 10^{-15} \text{ J (or } 3 \times 10^4 \text{ electron volts, i.e. 30 keV).}$$

Note the distinction between an accelerating voltage, measured in kV, and the electron energy after passing from the cathode to the anode, measured in keV.

The electron velocity can be obtained from the fact that its kinetic energy is $\frac{1}{2} m_e v_e^2$ where m_e is the mass of the electron and v_e its velocity. Hence

$$\tfrac{1}{2} m_e v_e^2 = 4.8 \times 10^{-15} \text{ J.}$$

Since $m_e = 9 \times 10^{-31}$ kg, v_e is approximately 10^8 m s^{-1} which is one third the speed of light. It can be seen from this example that relativistic effects are important even at quite low tube kilovoltages.

From above, the maximum X-ray photon energy ε is 4.8×10^{-15} J and the minimum wavelength is obtained by substitution in:

$$\varepsilon = hf = hc/\lambda$$

(h is the Planck constant, c the speed of light and λ the wavelength of the resulting X-ray). Hence

$$\lambda = hc/\varepsilon = \frac{6.6 \times 10^{-34} \times 3 \times 10^8}{4.8 \times 10^{-15}} = 4.1 \times 10^{-11} \text{ m (or 0.041 nm)}.$$

Note that calculations giving the maximum X-ray photon energy and minimum wavelength are valid even when the electrons travel at relativistic speeds.

2.2.3. Shape of the continuous spectrum

A detailed treatment of the continuous spectrum is beyond the scope of this book, but the following approach is helpful since it involves some other important features of the X-ray production process. First, imagine a very thin anode, and consider the production of X-rays, not the X-rays that finally emerge. It may be shown by theoretical arguments that the intensity of X-rays produced will be constant up to a maximum X-ray energy determined by the energy of the electrons (see figure 2.4(a)).

A thick anode may now be thought of as composed of a large number of thin layers. Each will produce a similar distribution to that shown in figure 2.4(a), but the maximum photon energy will gradually be reduced because the incident electrons lose energy as they penetrate the anode material. Thus the composite picture for X-ray production might be as shown in figure 2.4(b).

However, before the X-rays emerge, the intensity distribution will be modified in two ways. First, X-rays produced deep in the anode will be attenuated in reaching the surface of the anode and secondly X-rays will be attenuated in penetrating the window of the X-ray tube. Both processes reduce the intensity of the low energy radiation more than that of the higher energies so the result is the solid curve in figure 2.4(c).

In the absence of further filtration (see section 3.8) the X-ray energy corresponding to maximum intensity will be about one third of the highest energy X-ray photons.

2.2.4. Line or characteristic spectra

Superimposed on the continuous spectrum there may be a set of line spectra which result from an incoming electron interacting with a bound orbital electron in the target. If the incoming electron has sufficient energy to overcome the binding energy, it can remove the bound electron creating a vacancy in the shell. The probability of this happening is greatest for the innermost shells. This vacant energy level is then filled by an electron from a higher energy level falling into it and the excess energy is emitted as an X-ray . Thus, if for example the vacancy is created in the K shell, it may be filled by an electron falling from either the L shell, the M shell or outer shells. Even a free electron may fill the vacancy but the most likely transition is from the L shell.

As discussed in section 1.1, orbital electrons must occupy well defined energy levels and these energy levels are different for different elements. Thus the X-ray photon emitted when an electron moves from one energy level to another has an energy equal to the difference between the two energy levels in that atom and hence is characteristic of that element.

Reference to figure 1.2(b) shows that the K series of lines for tungsten ($Z = 74$) will range from 58.5 keV (for a transition from the L shell to the K shell) to 69.5 keV (if a free electron fills the K shell vacancy). Transitions to the L shell are of no practical importance in diagnostic radiology since the maximum energy change for tungsten is 11 keV and photons of this energy are absorbed before they leave the tube.

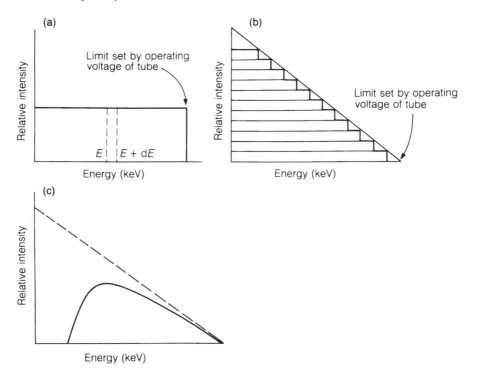

Figure 2.4. *A simplified explanation of the shape of the continuous X-ray spectrum. (a) Production of X-rays from a very thin anode. Note that the intensity of the beam in the small range E to E + dE will be equal to the number of photons per square metre per second multiplied by the photon energy. Fewer high energy photons are produced but their energy is higher and the product is constant. (b) Production of X-rays from a thicker anode treated as a series of thin anodes. (c) X-ray emission (solid line) compared with X-ray production (dotted line).*

Lower atomic number elements produce characteristic X-rays at lower energies. The K shell radiations from molybdenum ($Z = 42$) at circa 19 keV are important in mammography (see section 9.6). Note that characteristic radiation cannot be produced unless the operating kV of the X-ray tube is high enough to give the accelerating electrons sufficient energy to remove the relevant bound electrons from the anode target atoms.

2.2.5. *Factors affecting the X-ray spectrum*

If the spectrum changes in such a manner that its shape remains unaltered, i.e. the intensity or number of photons at every photon energy changes by the same factor, there has been a change in radiation **quantity**. If on the other hand, the intensities change such that the shape of the spectrum also changes, there has been a change in radiation **quality** (the penetrating power of the X-ray beam). A number of factors that affect the X-ray spectrum may be considered.

Tube current, I_c

This determines the number of electrons striking the anode. Thus the energy emitted as an X-ray photon or exposure E is proportional to tube current, but only the quantity of X-rays is affected ($E \propto I_c$).

Time of exposure

This again determines the number of electrons striking the anode so exposure is proportional to time but only the quantity of X-rays is affected ($E \propto t$).

Applied voltage

If other tube operating conditions are kept constant, the exposure increases approximately as the square of tube kilovoltage ($E \propto kV^2$). Two factors contribute to this increase. First the electrons have more energy to lose when they hit the target. Second, as shown in table 2.1, the efficiency of conversion of electrons into X-rays rather than into heat also increases with tube kilovoltage by a small amount over the range of diagnostic kilovoltage (the change associated with a large increase in kV is shown to emphasize the effect). Thus both the flux of X-ray photons and their mean energy increases.

    Furthermore, increasing the tube kilovoltage also alters the radiation **quality** since the high energy cut-off has now increased. Note that the position of any characteristic lines will not change.

Table 2.1. *Efficiency of conversion of electron energy into X-rays as a function of tube kilovoltage.*

Tube kilovoltage (kV)	Heat (%)	X-rays (%)
60	99.5	0.5
200	99	1.0
4000	60	40

Waveform of applied voltage

So far it has been assumed that the X-ray tube is operating from direct current, whereas in practice it operates from alternating current (figure 2.5(a)). Since in figure 2.1 one end of the X-ray tube must act as a 'cathode' and the other end as an 'anode', no current flows when an alternating potential is applied during the half cycle when the cathode is positive with respect to the anode. Half wave rectification (figure 2.5(b)) may be achieved by inserting a rectifier in the anode circuit (see section 2.3.4), but since X-rays are only emitted for half the cycle output is poor. Improved output can be achieved by full wave rectification obtained by using a simple bridge circuit. However, the tube is still not emitting X-rays all the time (figure 2.5(c)). Furthermore, the majority of X-rays are emitted at a kilovoltage below the peak value (kVp).

    A more constant voltage will improve the quality of the radiation and this can be achieved by using a three phase supply. The X-ray tube is now driven by three separate voltage supplies, each of which has been fully rectified. The three supplies are 60° out of phase and switching circuits ensure that each supply only drives the X-ray tube when the voltage is near to peak value. The resultant voltage profile (figure 2.5(d)) shows only about 15% variation. If the cathode supply is also three phase and is arranged to be 30° out of phase with the anode supply, a 12 peak generator is possible and fluctuations can be reduced to about 5%. Alternatively with modern technology a single phase supply at 50 Hz can be increased in frequency to the point where the output, once rectification and smoothing have taken place, is essentially a DC output.

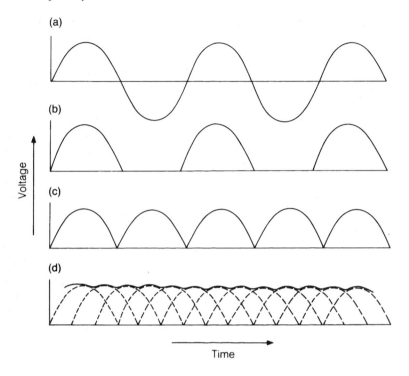

Figure 2.5. *Examples of different voltage wave forms. (a) Mains supply. (b) Half wave rectification. (c) Full wave rectification using a bridge rectifier. (d) Three phase supply (with rectification).*

This subject is considered in more detail in section 2.3.4. In future, in accordance with standard practice, operating voltages will be expressed in kVp to emphasize that the peak voltage with respect to time is being given.

Filtration

This also has a marked effect on both the quantity and quality of the X-ray beam, not only reducing the overall output but also reducing the proportion of low energy photons. Special filters (K edge filters) can be used to create a window of transmitted X-ray energies and thus reduce the number of both high and low energy photons. The effect of beam filtration is considered in detail in section 3.8.

Anode material

Choice of anode material affects the efficiency of X-ray production (see section 2.3.2) and the characteristic spectrum.

2.3. COMPONENTS OF THE X-RAY TUBE

2.3.1. The cathode

The cathode is constructed as a coiled wire filament of reasonably high resistance R so that for a given filament heating current I_F (typically in the region of 5 A), effective ohmic heating ($I_F^2 R$) and minimum heat losses will occur. A metal is chosen for the cathode that will give a copious supply of electrons by thermionic emission at temperatures where there is very little evaporation of metal atoms into the vacuum (e.g. tungsten with a melting point of 3370 °C).

Between exposures, the filament is kept warm in a stand-by mode because, although its resistance may be typically 5 Ω at 2000 K, at room temperature it falls to about 0.1 Ω. Thus a large current would be required to heat the filament rapidly from room temperature to its working temperature. Surrounding the filament is a cloud of negatively charged electrons, commonly called the **space charge**. The number of electrons in the space charge tends to a self-limiting constant value dependent on the filament temperature.

For reasons related primarily to geometrical unsharpness in the image, a small target for electron bombardment on the anode is essential. However, unless special steps are taken, the random thermally induced velocities and mutual repulsion of the electrons leaving the cathode will cause a broad beam to strike the anode. Therefore the filament is surrounded by a metal cup, normally maintained at the same potential as the filament. This cup provides an electric field which exercises a focusing action on the electrons to produce a small target area on the anode of the required size.

Most diagnostic X-ray tubes have a dual filament assembly, each filament having its own focusing cup, so as to produce two spots of different sizes. Note that spot size does vary somewhat with tube current and tube kilovoltage since the focusing action cannot be readily adjusted to compensate for variations in the mutual electrostatic repulsion between electrons when either their density or energy changes. The effective or apparent size of the focal spot on the anode is smaller than the actual focal spot because of the anode angle. The smaller the anode angle the smaller the apparent focal spot size (see section 2.4).

2.3.2. *The anode material*

The material chosen for the anode should satisfy a number of requirements. It should have:

(1) A high conversion efficiency for electrons into X-rays. High atomic numbers are favoured since the X-ray intensity is proportional to Z. At 100 keV, lead ($Z = 82$) converts 1% of the energy into X-rays but aluminium ($Z = 13$) converts only 0.1%.
(2) A high melting point so that the large amount of heat released causes minimal damage to the anode.
(3) A high conductivity so that the heat is removed rapidly.
(4) A low vapour pressure, even at very high temperatures, so that atoms are not boiled off from the anode.
(5) Suitable mechanical properties for anode construction.

In stationary anodes the target area is pure tungsten (W) ($Z = 74$, melting point 3370 °C) set in a metal of higher conductivity such as copper. Originally, rotating anodes were made of pure tungsten. However, at the high temperatures generated in the rotating anode (see section 2.3.3), deep cracks developed at the point of impact of the electrons. The deleterious effects of damaging the target in this way are discussed in sections 2.3.5 and 2.4. The addition of 5–10% rhenium (Rh) ($Z = 75$, melting point 3170 °C) greatly reduced the cracking by increasing the ductility of tungsten at high temperatures. The wear resistant rhenium alloy in the focal spot path ensures minimal aging, thus high and constant exposure values for a long life. However, pure W/Rh anodes would be extremely expensive so molybdenum is now chosen as the base metal. Molybdenum ($Z = 42$, melting point 2620 °C) stores twice as much heat, weight for weight, as tungsten, but the anode volume is now greater because molybdenum has a smaller density than tungsten. As shown in figure 2.6(*a*) only a thin layer of W/Rh is used, to prevent distortion that might arise from the differences in thermal expansion of the different metals. Some manufacturers also now additionally use a high volume graphite disc plate. The black graphite and large surface area permit effective cooling of the anode.

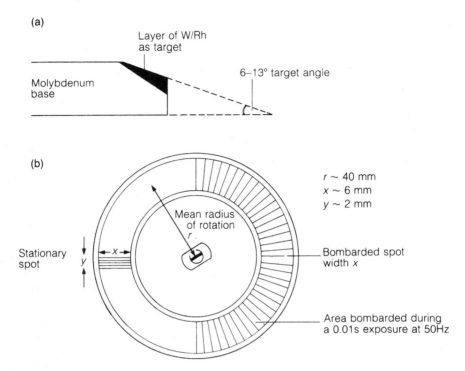

(a)

Layer of W/Rh
as target

Molybdenum
base

6–13° target angle

(b)

Mean radius
of rotation
r

$r \sim$ 40 mm
$x \sim$ 6 mm
$y \sim$ 2 mm

Stationary
spot

Bombarded spot
width *x*

Area bombarded during
a 0.01s exposure at 50Hz

Figure 2.6. *(a) Detail of the target area on a modern rotating anode. (b) Principle of the rotating anode showing the area bombarded in a 0.01 s exposure at 50 Hz.*

2.3.3.   Anode design

The two principal requirements of anode design are first to make adequate arrangements for dissipation of the large quantity of heat generated and second to ensure a good spatial distribution of X-rays. Design features related primarily to heat dissipation are discussed below; the spatial distribution of X-rays is considered later.

Stationary anode

This form of anode is rarely used in a modern X-ray department but is still used in dental units. Tungsten in the form of a small disc about 1 mm thick and 1 cm in diameter is embedded in a large block of copper. As shown in figure 2.1 the copper protrudes through the tube envelope into the surrounding oil. Heat is transferred from the tungsten to copper by conduction and thence to the oil by convection. The cooling fins assist the convection process. The oil transfers this heat to the X-ray tube shield by conduction and it is eventually removed by air in the X-ray room by convection.

Rotating anode

This is the main form of anode used in diagnostic X-ray units. The principle of the rotating anode is very simple (figure 2.6(*b*)). If the required focal spot size on the target is, say 2 mm × 2 mm, for an anode angled to the beam at about 16° the dimensions of the area actually bombarded by electrons are about 6 mm × 2 mm (see section 2.4). The area over which heat is dissipated can be increased by arranging for the tungsten target to be an annulus of material which rotates rapidly. It may be shown that, if the exposure time is long enough for the anode to rotate at least once,

$$\frac{\text{effective area for heat absorption with rotation anode}}{\text{effective area for heat absorption with stationary anode}} = \frac{2\pi r x}{y x}.$$

From the diagram this is an improvement of about $6 \times 40/2 = 120$-fold and the heat input can be increased considerably (although not by a factor of 120).

Rotation rates range from 3000 rpm (the 50 Hz mains supply) to 10 000 rpm, ensuring that the anode rotates several times during even the shortest exposure. However, this does create some problems with respect to the type of mounting and cooling mechanism. Adequate electrical contact is maintained via bearings on which the anode rotates, but the area of contact is quite insufficient for adequate heat conduction. Lubrication of these bearings cannot be by oil or grease because of the vacuum required in the X-ray tube and is now commonly performed by a dry metallic lubricant such as silver. Since the anode is in an evacuated tube, there are no heat losses by convection. The initial mode of heat transfer from the anode to the cooling oil must therefore be radiation at a rate proportional to (anode temperature)^4 − (oil temperature)^4. With a rotating anode, heat loss by conduction is actually minimized since it might result in over-heating of the bearings. Thus the rotating anode is mounted on a thin rod of low conductivity material such as molybdenum. Care must be taken that the length and method of support of this rod ensure that the anode remains stable when rotating.

Some manufacturers have now produced X-ray tubes with a metal casing instead of glass. This has several advantages one of which is that the anode can be supported on a shaft with bearings at both ends allowing much larger anodes to be installed. These have much better thermal properties allowing higher tube currents to be used. Some tubes now also have direct cooling by an oil supply into the anode allowing much higher work loads.

2.3.4. *Electrical circuits*

Only brief details will be given of the essential electrical components of an X-ray generator. For a fuller explanation see the book by Meredith and Massey (1977), but the reader should be aware that the workings of a modern generator are very much more complicated than as described there.

The transformer

This provides a method of converting high alternating currents at low potential difference to low alternating currents at high potential difference.

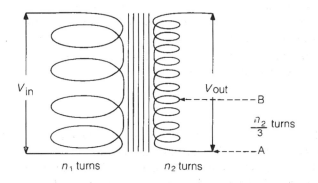

Figure 2.7. *Essential features of a simple transformer.*

It consists of two coils of wire which are electrically insulated from one another but are wound on the same soft iron former (figure 2.7). If no energy is lost in the transformer

$$V_{out}/V_{in} = n_2/n_1$$

where n_1 and n_2 are the number of turns on the primary and secondary coils respectively.
Note:

(1) By making a connection at different points, different output voltages may be obtained. For example the alternating potential difference across AB is given by

$$V_{AB} = V_{out}/3.$$

(2) Since, in the ideal case, all power is transferred from the input circuit to the output circuit

$$V_{in}I_{in} = V_{out}I_{out}.$$

Hence if V_{out} and I_{out} to the X-ray set are 100 kVp and 50 mA respectively, since V_{in} will be the 240 V AC supply,

$$I_{in} = \frac{100 \times 10^3 \times 50 \times 10^{-3}}{240} = 20 \text{ A}$$

so input currents are very high. (Hence the requirement for special 30 A hospital circuits when using non-condenser discharge mobile X-ray units.)

(3) Power loss occurs in all transformers, and the amount depends on working conditions, especially I_{in}. Hence V_{out} and I_{out} also vary and auxiliary electrical circuits are required to stabilize outputs from an X-ray set.

(4) The efficiency of a transformer increases as the frequency of operation rises and consequently its size decreases.

Figure 2.11 (see later) also shows, on the extreme left, an autotransformer. An autotransformer comprises one winding only and works on the principle of self-induction. Since the primary and secondary circuits are in contact, it cannot transform high voltages or step up from low to high voltages. However, it does give a variable secondary output on the low voltage side of the transformer and hence controls kV directly.

Generation of different voltage wave forms

As explained in section 2.2.5, the alternating potential must be rectified before it is applied to an X-ray set. The X-ray tube can act as its own rectifier (self-rectification) since it will only pass current when the anode is positive and the cathode is negative. However, this is a very inefficient method of X-ray production, because if the anode becomes hot, it will start to release electrons by thermionic emission. These electrons will be accelerated towards the cathode filament during the half cycle when the cathode is positive and will damage the tube. Thus the voltage supply is rectified independently.

If a gas-filled diode valve or a solid-state p–n junction diode rectifier is placed in the anode circuit, half wave rectification (figure 2.5(*b*)) is obtained. The gas-filled diode is a simplified X-ray tube comprising a heated cathode filament and an anode in an evacuated enclosure. Electrons may only flow from cathode to anode but the diode differs from the X-ray tube in that it is designed so that only a small proportion of the electrons boiled off the cathode travel to the anode. In terms of figure 2.17(*a*) (see later) the diode operates on the rapidly rising portion of the curve, whereas the X-ray tube operates on the near-saturation portion.

The design and mode of operation of a p–n junction diode will be considered in section 6.10 when its use as a radiation detector is discussed. It has many advantages over the gas-filled diode as a rectifier including its small size, long working lifetime and robustness. It is also easy to manufacture in bulk, inexpensive, requires no filament heating circuit and has a low heat dissipation and a fast response time. For rectification silicon rectifiers have a number of advantages over selenium, including a negligible forward voltage drop and a very high reverse resistance resulting in negligible reverse current flow. They can also withstand high reverse bias voltages so only a few hundred silicon rectifiers are required rather than a few thousand if made of selenium, and they can work up to 200 °C if required.

The essential features of a full wave rectified supply are shown in figure 2.8.

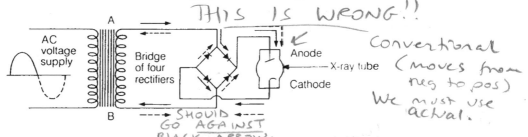

Figure 2.8. *Essential features of a full wave rectified supply. Solid and dotted arrows show that irrespective of whether A or B is at a positive potential the current always flows through the X-ray tube in the same direction. Note that conventionally this is in the opposite direction to electron flow.*

Medium frequency generators

In these generators a frequency converter or 'inverter'—a combination of a rectifier and inverting rectifier—is used to convert an alternating current (AC) of one frequency into an AC at another frequency. Figure 2.9 summarizes the stages in this process.

(1)   Rectification of the line AC voltage u_1 at frequency f_1.
(2)   This provides a direct current (DC) voltage u_0 (after smoothing).
(3)   Rapid chopping of this DC voltage by the inverter to provide an alternating voltage u_2 at a much higher frequency f_2, typically 6000–7000 Hz.

A high voltage transformer now transforms this to a higher voltage u_3 and, after again being rectified and smoothed, this voltage is fed to the X-ray tube.

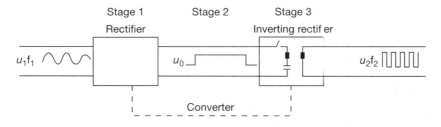

Figure 2.9. *Schematic representation of frequency converter.*

Advantages of the medium frequency generator are:

(1)   The output is comparable with a three phase, 12 peak generator.
(2)   At this high frequency there is much less ripple on the smoothed voltage than when using even a conventional three phase supply, hence a higher yield of X-rays.

(3)   The transformer equation, u/fnA = constant where n is the number of turns on the transformer and A its cross sectional area, shows that if f is increased by a factor of 100, nA may be reduced by a similar amount. Since the efficiency has been improved the transformer is much smaller, perhaps one tenth the size of a three phase 12 peak generator.

(4)   The high voltage is switched on and off, and its level may be regulated even during exposure, under feed-back control of the inverter. The rise time of the tube voltage can be less than 2 ms.

(5)   The tube current is more stable at the higher frequency f_2 and independent of the voltage.

(6)   The precision of the exposure timer can be improved.

Action of smoothing capacitors

A capacitor in parallel with the X-ray unit will help to smooth out any variations in applied potential (figure 2.10).

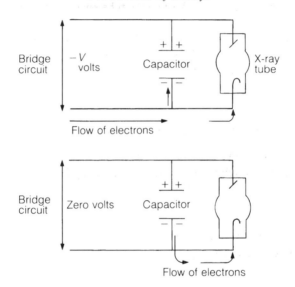

Figure 2.10. *Illustration of the use of a capacitor for voltage smoothing.*

Consider for example the full wave rectified supply shown in figure 2.5(*c*). When electrons are flowing from the bridge circuit, some of them flow onto the capacitor plates and are stored there. When the potential across the bridge circuit falls to zero, electrons flow from the capacitor to maintain the current through the X-ray tube.

Tube kilovoltage and tube current meters

These are essential components of the circuit and are shown in relation to other components in figure 2.11. Note that the voltmeter is placed in the primary circuit so that a reading may be obtained before the exposure key is closed. There are two ammeters. A_F measures the filament supply current (I_F) which may be adjusted to give the required thermionic emission before exposure starts. The actual tube current flowing during exposure (I_c) is measured by ammeter A_c.

2.3.5.   The tube envelope and housing

The envelope

The envelope is generally of thick walled glass and must be constructed under very clean conditions to a high precision so as to provide adequate insulation between the cathode and anode. It also provides a vacuum seal to the metallic components that protrude through it. Great care must be taken at the manufacturing stage to achieve a very high level of vacuum before the tube is finally sealed. If residual gas molecules are bombarded by electrons, the electrons may be scattered and strike the walls of the glass envelope, thereby causing reactions that result in release of gas from the glass and further reduction of the vacuum. Metal walled tubes have ceramic insulation between the tube and the anode and cathode connections.

The presence of atoms or molecules of gas or vapour in the vacuum, whatever their origin, is likely to have a deleterious effect on the performance of the tube. For example, metal evaporation from the anode can cause a conducting film across the glass envelope, thereby distorting the pattern of charge across the tube. This can change the output characteristics since it is assumed that the flow of electrons from cathode to anode will be influenced by the repulsive effect of a static layer of charge on the tube envelope. If this charge is not static, the electrons in the beam are not repelled by the tube envelope and deviate to it. This diversion of current may significantly reduce tube output.

Both residual gas and anode evaporation cause a form of tube instability which may occasionally be detected during screening as a kick on the milliammeter as discharges take place. In the extreme case, the tube goes 'soft' and arcs over during an exposure.

The tube housing

This has various functions which may be summarized as follows:

(1)   shields against stray X-rays because it is lined with lead;
(2)   provides an X-ray window—which filters out some low energy X-rays;
(3)   contains the anode rotation power source;
(4)   provides high voltage terminals;
(5)   insulates the high voltage;
(6)   allows precise mounting of the X-ray tube envelope;
(7)   provides a means for mounting the X-ray tube;

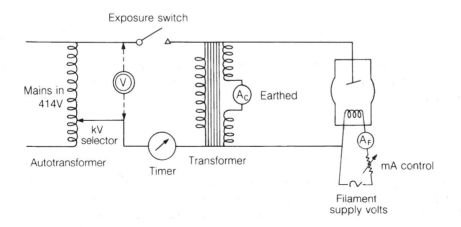

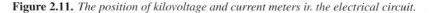

Figure 2.11. *The position of kilovoltage and current meters in the electrical circuit.*

(8)   provides a reference and attaching surface for X-ray beam collimation devices;
(9)   contains the cooling oil.

The advantages of filling this housing with oil are:

(a)   high voltage insulation;
(b)   effective convection of heat from the X-ray inset tube;
(c)   since the oil expands, an expansion diaphragm can be arranged to operate a switch, preventing further exposures, when the oil reaches its maximum safe temperature.

2.3.6.   Switching and timing mechanisms

Most switching takes place in the primary circuit of the high voltage transformer where although the currents are high the voltage is low. Switching in the high voltage secondary circuit is only undertaken if very short exposures are required or exposures must be taken in quick succession.

Primary switches

These are now almost all based on a thyristor which is a solid-state-controlled rectifier turned on and off by a logic pulse. A small positive logic pulse allows the thyristor to avalanche and a large current to flow which operates a relay closing the primary circuit. The switching is very rapid and is suitable for most radiographic exposures. The circuit used to drive the device is however quite complicated and beyond the scope of this book.

Switching the high voltage side of the transformer can be undertaken in two ways. The large electrical power that has to be accommodated (up to 100 kW) precludes the use of solid-state devices and triode valves have to be used. As an alternative, some more sophisticated tubes have a third electrode, a grid, in front of the cathode. This grid is used to switch the tube on and off very rapidly by changing its voltage from negative to just positive relative to the cathode. When negative even though the full kVp is applied across the tube, electrons cannot move from the cathode to the anode.

Timing mechanisms

Although a variety of timing mechanisms has been used in the past, only the two that are most widely used will be discussed.

The electronic timer   If a capacitor C is charged to a fixed potential V_0, either positive or negative, and then placed in series with a resistor R, the rate of discharge of the capacitor depends on the values of C and R. A family of curves for fixed C and variable R is shown in figure 2.12. Note that a large resistance reduces the rate of flow of charge so the rate of fall of V is slower.

These curves may be used as the basis for a timer if a switching device is arranged to operate when the potential across C reaches say V_s.

The photo-timer   The weakness of the electronic timer, and the other older timers that predetermine the exposure, is that a change in any factor which affects the amount of radiation actually reaching the film, notably patient thickness, will alter the amount of film blackening. The skill of the radiographer in estimating the thickness of the patient and choosing the correct exposure is therefore of great importance. In the photo-timer the exposure is linked more directly to the amount of radiation reaching the film. This is known as **automatic exposure control** or AEC.

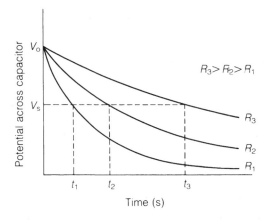

Figure 2.12. *Curves showing the rate of discharge of a capacitor through resistors of different resistance. The time taken to reach V_s, when the switching mechanism would operate, depends on the value of R.*

One design places small ionization chamber monitors in the cassette tray system between the patient and the film–screen combination. The amount of radiation required to produce a given degree of film blackening with a given film–screen combination under standard development conditions is known, so when the ion chamber indicates that this amount of radiation has been received, the exposure is terminated. This type of exposure control does not need to be 'set' prior to each exposure, but some freedom of adjustment is provided to allow for minor variations in film blackening if required. Adjustment will also be required if screens of different sensitivity are used.

As an alternative to ion chambers, photomultiplier tubes (see section 6.8) may be used. Some have the disadvantage however of being X-ray opaque because they use small photoelectric cells so they must be placed behind the cassette, where the X-ray intensity is low, and special radiolucent cassettes must be used. Others use phosphor-coated lucite which can be placed in front of the cassette as it does not attenuate the X-ray beam. The light produced is internally reflected to the side where the photomultiplier can view it. The energy response of some (most) phosphors can be quite different to that of the film and some form of compensation must be built into the circuitry by using software control.

A weakness with some types of photo-timer is that the ion chamber or photomultiplier tube only monitors the radiation reaching a small part of the film and this may not be representative of the radiation reaching the rest of the film. This problem can be partially overcome by using several small ion chambers, usually three, and controlling the exposure with the one that is closest to the region of greatest interest on the resulting X-ray film.

Photo-timers must be adjusted to the film/screen combination speed being used. There must also be a back-up exposure timer so that should the AEC fail the patient exposure, although longer than necessary, will be terminated without intervention by the operator. It should be noted that this back-up time is often set at the thermal limit of the tube and can result in a large radiation dose to the patient.

For a fuller treatment of timing mechanisms the reader should consult references given at the end of the chapter.

2.3.7. *Electrical safety features*

A number of features of the design of X-ray sets are primarily for safety and will be summarized briefly.

The tube housing

As already indicated, this provides a totally enclosing metallic shield that can be firmly earthed thereby contributing to electrical safety.

High tension cables

High tension cables are constructed so that they can operate up to potentials of at least 150 kV. Since the outermost casing metal braid of the cable must be at earth potential for safety, a construction of multiple coaxial layers of rubber and other insulators must be used to provide adequate resistance between the innermost conducting core and the outside to prevent current flow across the cable.

It is essential that high tension cables are not twisted or distorted in any other manner that might result in breakdown of the insulation. They must not be load bearing.

Electrical circuits

These are designed in such a way that the control panel and all meters on the control panel are at earth potential. Nevertheless, it is important to appreciate that many parts of the equipment are at very high potential and the following simple precautions should be observed.

(1)   Ensure that equipment is installed and maintained regularly by competent technicians.
(2)   Record and report to the service engineer any evidence of excessive mechanical wear, especially to electrical cables, plugs and sockets.
(3)   Similarly, report any equipment malfunction.
(4)   Adopt all other safety procedures that are standard when working with electrical equipment.

2.4.   SPATIAL DISTRIBUTION OF X-RAYS

When 40 keV electrons strike a thin metal target, the directions in which X-rays are emitted are as shown in figure 2.13(*a*). Most X-rays are emitted at angles between 45° and 90° to the direction of electron travel. The more energetic X-rays travel in a more forward direction (smaller value of θ). It follows that if the mean X-ray energy is increased by increasing the energy of the electrons, the lobes are tilted in the direction of the electron flow.

When electrons strike a thick metal target, the situation is more complicated because X-ray production may occur from the surface or it may occur at depth in the target. Also, the spatial distribution of X-rays will now depend on the angle presented by the anode to the incoming electron beam. Consider the anode shown in figure 2.13(*b*) and angled at 60° to the beam. X-rays produced in the direction B are much more heavily attenuated than those produced in the direction A because they travel further through anode material. This is clearly a disadvantage since a primary objective of good X-ray tube design is to ensure that the field of view is uniformly exposed to radiation. Only if this is achieved can variations in film blackening be attributed to variations in scatter and absorption within the patient. Variation in intensity across the field is one factor to consider when deciding the angle at which the anode surface is inclined to the vertical (figure 2.14).

Note the following points:

(1)   The radiation intensity reaching the film is still not quite uniform, being maximum near the centre of the field of view. This is due to

(a)   an inverse square law effect—radiation reaching the edges of the field has to travel further, and
(b)   a small obliquity effect—beams travelling through the patient at a slight angle must traverse a greater thickness of the patient and are thus more attenuated.

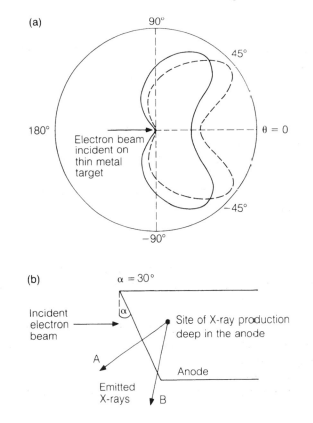

Figure 2.13. *(a) Approximate spatial distribution of X-rays generated from a thin metal target bombarded with 40 keV electrons (solid line). This figure is known as a polar diagram: the distance of the curve from the origin represents the relative intensity of X-rays emitted in that direction. The polar diagram that might be obtained with 100 keV electrons is shown dotted. (b) The effect of self-absorption within the target on X-ray production from a thick anode.*

Neither of these factors is normally of great practical importance.

(2)   The anode angle selected does not remove the asymmetry completely and this is known as the **heel effect**. The effect of X-ray absorption in the target, which results in a bigger exposure at A than at B, is more important than asymmetry in X-ray production, which would favour a bigger exposure at B.

(3)   Some compensation for the heel effect can be achieved by tilting the filter. The left-hand edge of the beam will pass through a smaller thickness of filter than the right-hand edge. This modification is being used in some mammography tubes, but *not* in normal diagnostic units.

(4)   No such asymmetry exists in a direction normal to that of the incident electron beam so if careful comparison of the blackening on the two sides of the film is essential the patient should be positioned accordingly although care must be taken balancing a tall patient at right angles to the table.

(5)   The shape of the exposure profile is critically dependent on the quality of the anode surface. If the latter is pitted owing to over-heating by bombarding electrons, much greater differences in exposure may ensue.

(6)   An angle of about 13–16° is frequently chosen and this has one further benefit. One linear dimension of the effective spot for the production of X-rays is less than the dimension of the irradiated area by a factor

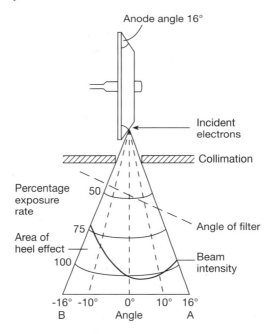

Figure 2.14. *Variation of X-ray intensity across the field of view for a typical anode target angle.*

equal to sin α. Sin 13° is about 0.2, so angling the anode in this way allows the focusing requirement on the electron beam to be relaxed whilst ensuring a good focal spot for X-ray production (figure 2.15). If a very small focal spot ($\sim$0.3 mm) is required, an angle of only 6° may be used. Note that with a small anode angle, the heel effect greatly restricts the field size. This can only be compensated by increasing the focus–film distance. For example if the minimum acceptable variation in optical density across the field of view is 0.2, for a film size of 43 cm × 35 cm the minimum focus–film distance increases from about 110 cm for a 16° angle to 150 cm for a 12° angle. There is a consequent loss of intensity at the film due to the inverse square law. Some X-ray tubes now use anodes with two angles so that the best angle for the focal spot size chosen can be used.

Even with a well designed anode, a certain amount of **extrafocal radiation** arises from regions of the anode outwith the focal spot. These X-rays may be the result of poorly collimated electrons but are more usually the consequence of secondary electrons bouncing off the target and striking the anode elsewhere. Note that extrafocal radiation is not scattered radiation. Extrafocal X-rays may contribute as much as 15–20% of the total output exposure of the tube but are of lower average energy. Many of them will fall outside the area defined by the light beam diaphragm and under extreme conditions may cast a shadow of the patient (figure 2.16).

Over the region of interest, the extrafocal radiation creates a uniform low level X-ray intensity. This contributes to the reduction in contrast produced by scattered radiation (see section 5.5). Since it is no more than 20% of the scattered radiation it plays a small part in the total reduction effect in conventional radiography. However, it is a potentially serious problem in image intensifier fluoroscopy because if unattenuated radiation reaches the input screen of the image intensifier the very bright fluorescent areas reduce contrast by light scattering and light conduction effects. This effect of extrafocal radiation in digital radiology probably merits investigation.

With rotating anode tubes extrafocal radiation is generated on the anode plate—i.e. in a strip perpen-

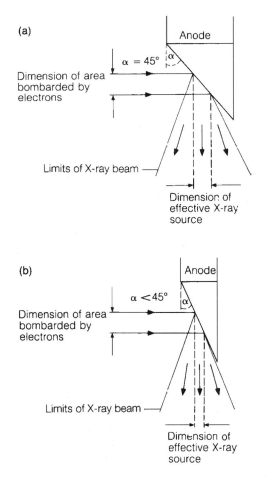

Figure 2.15. *The effect of the anode angle on the effective focal spot size for X-ray production. (a) If the anode is angled at 45°, the effective spot for X-ray production is equal to the target bombarded by electrons. (b) If the anode is angled at less than 45° to the vertical the effective spot for X-ray production is less than the bombarded area. Note that in all cases these areas are measured normal to the X-ray beam. The actual area of the impact of electrons on the anode will be greater for reasons explained in the text.*

dicular to the tube axis. So the effect is most clearly visible on edges of the radiograph that are parallel to the axis of the X-ray tube. In the metal-cased tubes referred to earlier in section 2.3.3 the case is at earth potential and attracts off-focus electrons. The amount of off-focus radiation produced is reduced as the metal case is of a low atomic number and produces only a few low energy X-rays.

2.5. RATING OF AN X-RAY TUBE

2.5.1. *Introduction*

The production of a good radiograph depends on the correct choice of tube kVp, current, exposure time and focal spot size. In many situations a theoretical optimum would be to use a point source of X-rays to minimize geometrical blurring (see section 5.9.1), and a very short exposure time, say 1 ms, to eliminate

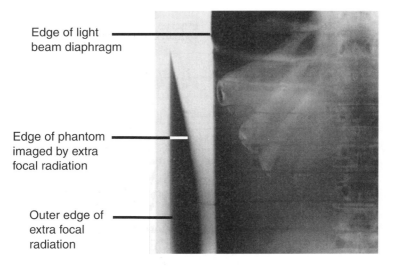

Edge of light
beam diaphragm

Edge of phantom
imaged by extra
focal radiation

Outer edge of
extra focal
radiation

Figure 2.16. *Radiograph showing the effect of extrafocal radiation. The field of view defined by the light beam diaphragm is shown on the right but the outer edge of the 'patient' (a phantom in this instance) is also radiographed on the left by the extrafocal radiation.*

movement blurring (see section 5.9.2). However, these conditions would place impossible demands on the power requirement of the set. For example an exposure of 50 mAs would require a current of 50 A. Even if this current could be achieved, the amount of heat generated in such a small target area in such a short time would cause the anode to melt. This condition must be avoided by increasing the focal spot size or the exposure time, generally in practice the latter. Furthermore, during prolonged exposures, for example in fluoroscopy, a secondary limitation may be placed on the total amount of heat generated in the tube and shield.

Thus the design of an X-ray tube places both electrical and thermal constraints on its performance and these are frequently expressed in the form of **rating** charts, which recommend *maximum* operating conditions to ensure a reasonably long tube life when used in equipment that is properly designed, installed, calibrated and operated. Note that lower exposure factors should be used whenever possible to maximize tube life.

2.5.2. *Electrical rating*

Electrical limits are not normally a problem for a modern X-ray set but are summarized here for reference.

Maximum voltage

This will be determined by the design, especially the insulation, of the set and the cables. It is normally assumed that the high voltage transformer is centre grounded (see figure 2.11), i.e. that the voltages between each high voltage tube terminal and ground are equal. A realistic upper limit is 150 kVp.

Maximum tube current

This is determined primarily by the filament current. Very approximately, the tube current (I_C) will be about one tenth of the filament current (I_F). In other words only about one tenth of the electrons passing through the filament coil are 'boiled off' from it. A modern X-ray tube may be designed to operate with a tube current of up to 1000 mA but under normal conditions it will be less than half this value. The lifetime of the tube

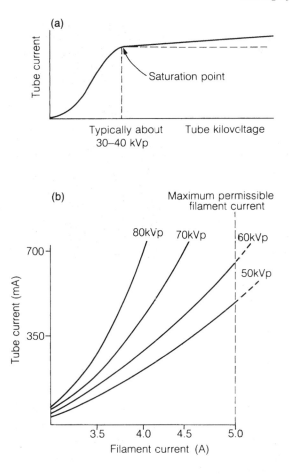

Figure 2.17. *(a) The effect of increasing tube kilovoltage on the tube current for a fixed filament current. (b) A family of curves relating tube current to filament current for different applied voltages.*

can be significantly extended by a small reduction in current. The lifetime of a filament operating at 4.3 A is about ten times that of one operating at 4.8 A.

If the voltage is increased at fixed filament current, the tube current will change as shown in figure 2.17(*a*). At low voltages, the tube current increases as the kV is increased because more and more electrons from the space charge around the cathode are being attracted to the anode. In theory, the tube current should plateau when the voltage is large enough to attract all electrons to the anode. In practice there is always a cloud of electrons (the space charge) around the cathode and as the potential difference is increased, a few more electrons are attracted to the anode. The result is that, as the tube kilovoltage is increased, the maximum tube current attainable also increases. Hence a typical family of curves relating tube current to filament current might be as in figure 2.17(*b*). Modern X-ray tubes contain several compensating circuits one of which stabilizes the tube current against the effect of changes in voltage.

Maximum power

This is the product of tube current and voltage, but is not a practical limitation.

2.5.3.   *Thermal rating—considerations at short exposures*

When electrons strike the anode of a diagnostic tube, 99% of their energy is converted into heat. If this heat cannot be adequately dissipated, the anode temperature may quickly rise to a value at which damage occurs due to excessive evaporation, or the anode may melt which is even worse. The amount of heat the anode can absorb before this happens is governed by its thermal rating.

For exposure times between 0.02 s and 10 s the primary thermal consideration is that the area over which the electrons strike the anode should not overheat. This is achieved by dissipating the heat over the anode surface as much as possible. The factors that determine heat dissipation will now be considered.

Effect of cooling

It is important to appreciate that when the maximum heat capacity of a system is reached, any attempt to achieve acceptable exposure factors by increasing the exposure time is dependent on the fact that during a protracted exposure some cooling of the anode occurs. Consider the extreme case of a tube operating at its anode thermal rating limit for a given exposure. If the exposure time is doubled in an attempt to increase film blackening then, in the absence of cooling, the tube current must be halved. This is because at a given kVp the energy deposited in the anode is directly proportional to the product of the current and the period of exposure. However, in the presence of cooling, longer exposure times do permit greater power dissipation as shown in figure 2.18.

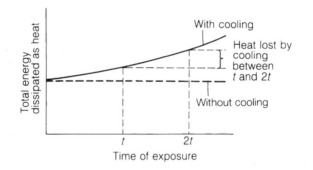

Figure 2.18. *Total energy dissipated as heat for different exposure times with and without cooling.*

Target spot size

For fixed kVp and exposure time, the maximum permitted current increases with target spot size because the heat is absorbed over a larger area. For very small spots (~0.3 mm) the maximum current is approximately proportional to the area of the spot since this determines the volume in which heat is generated. For larger spots (~2 mm) the maximum current is more nearly proportional to the perimeter of the spot since the rate at which heat is conducted away becomes the most important consideration.

The larger focal spot, although allowing short exposure times, increases geometrical unsharpness (section 5.9.1). A recent technical development allows the mixing of contributions from the two focal spots so as to optimize sharpness and rating performance.

Anode design

The main features that determine the instantaneous rating of a rotating anode are

(1)   its radius, which will determine the circumference of the circle on which the electrons fall,
(2)   its rate of rotation and
(3)   for a fixed target spot size, the anode angle.

Rating curves showing the maximum permissible tube current for different exposure times for anodes of different design are shown in figure 2.19. A small anode angle and rapid rotation give the highest rating but note that the differences between the curves become progressively less as the exposure time is extended.

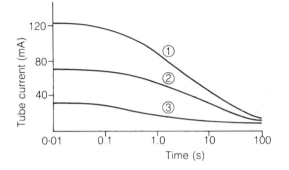

Figure 2.19. *Historical rating curves showing the maximum permissible tube current at different exposure times for anodes of different design. Each tube is operating at 100 kVp three phase with a 0.3 mm focal spot. (1) Type PX 410 4 inch diameter anode with a 10° target angle and 150 Hz stator. (2) Type PX 410 4 inch diameter anode with a 10° target angle and 50 Hz stator. (3) Type PX 410 4 inch diameter anode with a 15° target angle and 50 Hz stator. Curves (1) and (2) show the effect of increasing the speed of rotation of the anode. Curves (2) and (3) show the effect of changing the target angle. From Waters G 1968/69 J. Soc. X-ray Tech.* **5**.

Note also that since for the first complete rotation of the anode surface electrons are falling on unheated metal, the curve is initially almost horizontal. The maximum permissible tube current for a stationary anode operating under similar conditions would be much lower.

Tube kilovoltage

As the kVp increases, the maximum permissible tube current for a fixed exposure time decreases (figure 2.20). This is self-evident if a given power dissipation is not to be exceeded.

Such a rating chart may readily be used to determine whether a given set of exposure conditions is admissible with a particular piece of equipment. For example is an exposure of 400 mA at 70 kVp for 0.2 s allowed? Reference to figure 2.20 shows that the maximum permissible exposure time for 400 mA at 80 kVp is about 0.8 s so the required conditions can be met. Note that for very long exposures the product kVp × mA × time is converging to the same value for all curves and the heat storage capacity of the anode then becomes the limiting factor (see section 2.5.5).

When full wave rectified and three phase supply rating charts are compared at the same kVp, all other features of anode design being kept constant, the curves actually cross (figure 2.21). For very short exposures higher currents can be used with a three phase than with a single phase supply, but the converse holds at longer exposures.

To understand why this is so, consider the voltage and current wave forms for two tubes with the same kVp and mA settings (figure 2.22). Note:

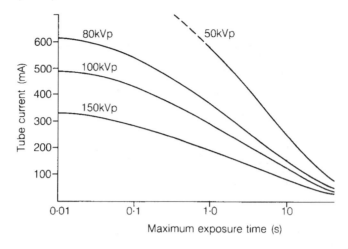

Figure 2.20. *Maximum permissible tube current as a function of exposure time for various tube kilovoltages for a rotating anode. Type PX 306 tube with a 3 inch diameter anode, 15° target angle operating on a single phase with a 60 Hz stator and a 2 mm focal spot—circa 1982. Note: the dotted line indicates that the maximum permissible filament current would probably be exceeded under these conditions. Reproduced by permission of Picker International.*

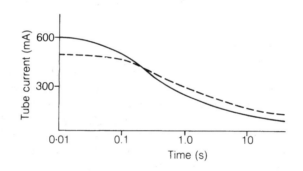

Figure 2.21. *Maximum permissible current as a function of exposure time for 80 kVp single phase full wave rectified (dotted line) and 80 kVp three phase supplies (solid line).*

(1)   The current does not follow the voltage in the full wave rectified tube. As soon as the potential difference is sufficient to attract all the thermionically emitted electrons to the anode, the current remains approximately constant.

(2)   The three phase current remains essentially constant throughout.

(3)   The peak value of the current must be higher for the full wave rectified tube than for the three phase tube, if the average values as shown on the meter are to be equal.

For very short exposures, instantaneous power is important. This is maximum at T_m and since the voltages are then equal, power is proportional to instantaneous current and is higher for the full wave rectified system. Inverting the argument, if power dissipation cannot exceed a predetermined maximum value, the average current limit must be lower for the full wave rectified tube.

For longer exposures, average values of kV and mA are important. Average values of current have been made equal, so power is proportional to average voltage and this is seen to be higher for the three phase supply.

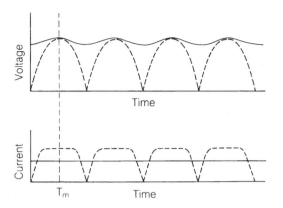

Figure 2.22. *Voltage and current profiles for two tubes with the same kVp and mA settings but with three phase (solid line) or full wave rectified (dotted line) supplies.*

Hence, again inverting the analysis, the current limit must be lower for the three phase tube as predicted by the rating curve graphs.

It is left as an exercise to the reader to explain, by similar reasoning, why the rating curve for a full wave rectified tube will always be above the curve for a half wave rectified tube.

2.5.4. Overcoming short-exposure rating limits

If a desired combination of kVp, mA, time and focal spot size is unattainable owing to rating limits, several things can be done, although all may degrade the image in some way. Increasing the focal spot size or the time of exposure have already been mentioned. The other possibility is to increase the kVp. At first inspection, this appears to give no benefit. Suppose the rating limit has been reached and the kVp is increased by 10%. The current will have to be reduced by 10% otherwise the total power dissipated as heat will increase. The gain from increasing the kVp appears to be negated by a loss due to reduced mA. However, although X-ray output will fall by 10% as a result of reducing mA, it will increase by about 20% as a result of the 10% increase in kVp (see section 2.2.5). Furthermore, X-ray transmission through the patient is better at the higher kVp and, in the diagnostic range, film sensitivity increases with kVp. Thus film blackening, which is ultimately the relevant criterion, is increased by about 40% by the 10% increase in kVp and reduced by only about 10% due to reduction in mA, yielding a net positive gain. Some image degradation may occur as a result of loss of contrast at the higher kVp but patient dose would be reduced (see section 5.3).

The most effective way to overcome short-exposure rating limits, in the longer term, is by improved anode design. Especially in X-ray CT, the combined requirements for quick scanning sequences, dynamic serial studies and spiral CT have necessitated a high tube current being maintained for several seconds. Cine exposures, especially of blood vessels where the blood flow is very rapid, also put severe demands on the anode.

One of the weak features of the conventional design of rotating anode is the ball-bearings that are necessary for the rotational motion. They cannot be run continuously and thus the anode has to be accelerated for each exposure, often causing an unnecessary time delay. A recent technical advance has been the introduction of anodes mounted on spiral groove bearings. A liquid metal alloy with low melting point (to prevent spoiling the vacuum) is used as the lubricant. In addition to allowing the anode to rotate all day once power is applied to the unit, this design permits good thermal contact so a significant amount of heat may be lost by conduction and the load bearing is greater so an anode of greater diameter may be used.

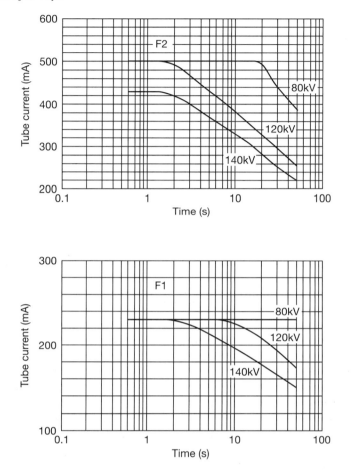

Figure 2.23. *Characteristic rating curves for a Siemens 502 MC tube designed for computed tomography (1995). 8° target angle,* F1 $= 0.6$ mm $\times 0.6$ mm *focal spot,* F2 $= 0.8$ mm $\times 1.1$ mm *focal spot. Reproduced by permission of Siemens.*

The design improvements noted here and elsewhere in this section, coupled with more effective heat transfer from the anode and a greatly increased overall heat storage capacity (see section 2.5.5), have resulted in rating curves such as those shown in figure 2.23. It is instructive to compare these curves for a modern tube with those in figure 2.20 (circa 1982) and figure 2.19 (circa 1966). The maximum tube current is now being maintained for much longer. Note that this pair of curves also shows the effect of focal spot size, not previously illustrated.

2.5.5. *Multiple or prolonged exposures*

If too many exposures are taken in a limited period of time, the tube may overheat for three different reasons:

(1)   the surface of the target can be overheated by repeated exposures before the surface heat has time to dissipate into the body of the anode;

(2)   the entire anode can be overheated by repeating exposures before the heat in the anode has had time to radiate into the surrounding oil and tube housing;

(3)   the tube housing can be overheated by making too many exposures before the tube shield has had time
      to lose its heat to the surrounding air.

The heat capacity of the total system, or of parts of the system, is sometimes expressed in heat units
(HU). By definition, 1.4 HU are generated when 1 J of energy is dissipated.
      The basis of this definition can be understood for a full wave rectified supply:

$$\text{HU} = 1.4 \times \text{energy}$$
$$= 1.4 \times \text{ root mean square (rms) kV} \times \text{ average mA} \times \text{s}.$$

But
$$\text{rms kV} = 0.71 \times \text{kVp}.$$

Hence
$$\text{HU} = \text{kVp} \times \text{mA} \times \text{s}.$$

Thus the HUs generated in an exposure are just the product of voltage × current × time shown on the X-ray
control panel, hence the introduction of the HU was very convenient for single phase generators.
      Unfortunately, this simple logic does not hold for three phase supply. The mean kV is now much higher,
perhaps 0.95 kVp, so:

$$\text{HU} = 1.4 \times 0.95 \text{ kVp} \times \text{mA} \times \text{s} = 1.35 \times \text{kVp} \times \text{mA} \times \text{s}.$$

Hence for three phase supply the product of kVp and mAs as shown on the meters must be multiplied by
1.35 to obtain the heat units generated. With the near universal use of three phase and medium frequency
generators, joules are becoming the preferred unit.
      The rating charts already discussed may be used to check that the surface of the target will not overheat
during repeat exposures. This cannot occur provided that the total heat units of a series of exposures made in
rapid sequence does not exceed the heat units permissible, as deduced from the radiographic rating chart, for
a single exposure of equivalent total exposure duration.
      When the time interval between individual exposures exceeds 20 s there is no danger of focal track
overheating. The number and frequency of exposures is now limited either by the anode or by the tube heat
storage capacity. A typical set of anode thermal characteristic curves is shown in figure 2.24. Two types of
curve are illustrated:

(1)   Input curves showing the heat stored in the anode after a specified, long period of exposure. Also
      shown, dotted, is the line for 470 W input power in the absence of cooling. This line is a tangent to the
      curve at zero time since the anode is initially cold and loses no heat. At constant kVp the initial slope
      is proportional to the current. As the anode temperature increases, the anode starts to lose heat and the
      curve is no longer linear.
(2)   A cooling curve showing the heat stored in the anode after a specified period of cooling. Note that if
      the heat stored in the anode after exposure is only 65×10^3 J, the same cooling curve may be used but
      the point A must be taken as $t = 0$.

Two other characteristics of the anode are important. First, the maximum anode heat storage capacity,
which is 100×10^3 J here, must be known. For low screening currents, the heat stored in the anode is always
well below its heat limit, but for higher input power the maximum heat capacity is reached and screening
must stop.
      The second characteristic is the maximum anode cooling rate. This is the rate at which the anode will
dissipate heat when at its maximum temperature (360 watts) and gives a measure of the maximum current, for

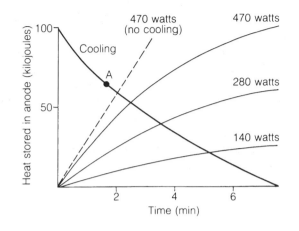

Figure 2.24. *Typical anode thermal characteristic curves, showing the heat stored in the anode as a function of time for different input powers. Reproduced by permission of Picker International.*

a given kVp, at which the tube can operate continuously. Note that under typical modern screening conditions, say 2 mA at 80 kVp, the rate of heat production is only $2 \times 10^{-3} \times 80 \times 10^3 = 160$ W.

During screening, or a combination of short exposures and screening, the maximum anode heat storage capacity must not be exceeded. Exercises in the use of this rating chart are given at the end of this chapter. Note that with the increasing use of microprocessors to control X-ray output, the system will not allow the operator to make an exposure that might exceed a rating limit.

When the total time for a series of exposures exceeds the time covered by the anode thermal characteristic chart, a tube shield cooling chart must be consulted. This is similar to the anode chart except that the cooling time will extend (typically) to 100 min and the maximum tube shield storage capacity in some modern units may be as high as 4×10^6 J.

As a final comment on thermal rating, it is worth noting that a significant amount of power is required to set the anode rotating and this is also dissipated eventually as heat. In a busy accident department taking many short exposures in quick succession, three times as much heat may arise from this source as from the X-ray exposures themselves.

2.5.6. Falling load generators

Although thermal loading considerations indicate that a limit is imposed on the number and rate of multiple exposures that can be taken, in practice this limit is rarely reached with modern X-ray units. In recent years anode design has improved significantly and diagnostic tubes are currently available that are capable of operating at 1 ampere (1000 mA). However, before these improved anodes were available, falling load generators were introduced to enable rapid multiple exposures to be taken by running the tube as close to the maximum rating as possible. This method of operation uses the fact that the rate of heat loss from the anode is greatest when the anode is at its maximum working temperature, so the current through the tube is kept as high as possible without this maximum temperature being exceeded.

The anode temperature is monitored and, if it reaches the maximum allowed, a motor driven rheostat originally set at zero introduces a resistance into the filament circuit thereby reducing the tube current in a step-wise manner. Because the transformer is not ideal, this lowering of tube current causes an increase in the kVp, and this has to be compensated for by increasing the resistance in the primary circuit in the transformer.

In older falling load generators this was performed by introducing fixed resistors, but more modern units use a continuously variable resistor.

By maintaining the current at its maximum possible value, the minimum time will be required for a given exposure. To achieve this the exposure must be set and controlled using a meter calibrated in milliampere seconds or the exposure must be terminated using a photo-timer for automatic exposure control (see section 2.3.6). Note however that the falling load generator will be of little value for short exposure times (say 0.4 s) because there will be insufficient time for the current to fall through many steps. Also there is wear on the tube at high current so lifetime is shortened by falling load operation. Thus a falling load generator might be a possibility for a busy orthopaedic clinic examining spines with heavy milliampsecond loadings and long exposure times. For chest work it would be useless.

2.5.7. *Safety interlocks*

These are provided to ensure that rating limits are not exceeded on short exposures. If a combination of kVp, mA, s and spot size is selected that would cause anode overheating, a 'tube overload' warning light will appear and the tube cannot be energized.

During multiple exposures a photoelectric cell may be used to sample radiant heat from the anode and thereby determine when the temperature of the anode disc has reached a maximum safe value. A visual or audible warning is then triggered. In some modern systems the tube loading is under computer control. Anode temperature is continuously calculated from a knowledge of heat input and cooling characteristics. When the rating limit is reached, generator output is automatically reduced.

2.5.8. *X-ray tube lifetime*

The life of an X-ray tube can be extended by taking steps to avoid thermal stress and other problems associated with heating. For example, the anode is very brittle when cold, and if a high current is used in this condition deep cracks may develop. Thus at the start of operations several exposures at approximately 75 kVp and 400 mA s (200 mA for 2 s) should be made at 1 min intervals. Ideally, if the generator is idle for periods exceeding 30 min, the process should be repeated.

Keeping the 'prepare' time to a minimum will reduce filament evaporation onto the surface of the tube and also bearing wear in the rotating anode. The generator should be switched off when not in use.

The tube should be operated well below its rating limits whenever possible.

2.6. MOBILE X-RAY GENERATORS

There are a number of situations in which it is not practicable to take the patient to the X-ray department and the X-ray set must be taken to the patient. The term mobile X-ray generator applies to machines which can be moved around a hospital but which cannot be dismantled or carried. The latter are strictly called portable X-ray generators and are not considered here. Although mobile units generally take conventional films, there is also a role for mobile image intensifiers both in theatres, for example in orthopaedic practice, and on the wards, for example to monitor the progress of an endoscope as it is inserted into the patient. The manoeuvrability of, say, a C-arm image intensifer allows the radiologist or radiographer to position the arm so that only the region of interest is irradiated, thereby reducing the dose to adjacent organs. Note, however, that a screening procedure should not be adopted if the same information can be obtained from a few short conventional exposures because the latter will almost certainly result in a lower dose to the patient.

Three basic categories of mobile unit are available and these can be distinguished by the type of X-ray generator used, namely single phase full wave rectified, constant potential and capacitor discharge. The fact that such units must be mobile introduces additional constraints and limitations and these will be the main features of interest in this section. For a more detailed review of mobile X-ray generators see the article by Evans *et al* (1985).

2.6.1. *Single phase full wave rectified generators*

These machines operate essentially as ordinary single phase generators powered by the hospital mains with an output wave form similar to that in figure 2.5(*c*). The output and rating limits of a single phase fixed unit described in section 2.5 apply equally to mobile units but the output is dependent in addition on hospital mains supply. Because high currents are drawn, a separate 30 A ring main must be installed for these units. This is expensive and obviously imposes limits on where they can be used. Even with a specially provided circuit, the voltage delivered to the primary of the high voltage transformer, and hence the voltage to the X-ray tube, is dependent on the relative electrical impedances of the input to the transformer and the mains circuit. A mismatch can result in errors of up to 20 kV in the actual kVp produced. Most units now match impedances automatically.

Outside the control of the operator are sudden fluctuations in voltage that can occur in the hospital mains supply following the connection or disconnection of other electrical appliances that draw large currents.

The power output from these machines is limited because of the low available mA (typically 300 mA maximum) and a full range of exposures cannot be obtained from them.

2.6.2. *Constant potential generators*

This category of generator can be further subdivided into two types. The first is truly independent of an electricity supply as it is powered by batteries (which do, however, require charging at a central location). The second type is operated from the normal 13 A hospital mains supply which is used to charge a capacitor.

Battery-powered generators generally use a nickel–cadmium battery which can store a charge equivalent to about 10 000 mA s at normal operating voltages. The direct current voltage of approximately 130 V from this battery must be converted to an alternating voltage before it can be used to supply the generator transformer. This conversion is carried out by an inverter (see section 2.3.4). As with the medium frequency generator described there the AC voltage produced differs significantly from the mains AC voltage in that it is at 500 Hz, some ten times greater than the normal mains frequency. At this high frequency the transformer is more efficient and as a consequence can be much smaller.

Although the transformer output is basically single phase full wave rectified, at 500 Hz the capacitance inherent in the secondary circuit smooths the output to a waveform that is essentially constant. The generator produces a fixed tube current of 100 mA which, as well as simplifying design, allows exposures to be calculated with ease. The battery is depleted and the voltage falls from one exposure to the next so some form of compensation must be applied until the unit is recharged. This compensation can be applied either automatically or manually. Recharging takes place, when necessary, at a low current from any hospital 13 A mains supply.

The second type of constant voltage mobile generator is rather sophisticated. The charge for exposure is stored on a very large capacitor and then this charge is used as a DC supply to an inverter producing a 4.5 kHz supply for the X-ray generator. The large capacitor can be charged using a battery or from the normal hospital 13 A mains supply. During an exposure, the voltage of the X-ray tube is monitored and the output controlled using a microprocessor in such a way that the output is essentially constant.

For both these types of generator the output is continuous, so exposures are shorter than for a single phase mobile unit. Furthermore the output is independent of fluctuations in the hospital mains supply. The output of the battery operated units is low (10 kW) but because of their good stability they are suitable for chest radiography or premature baby units. Mains driven capacitor units have a higher power (23 kW) and thus can be used in any situation.

2.6.3.   *Capacitor discharge units*

In these units the capacitor is used in a quite different way from that in the constant output units. In the latter, charge is fed from the capacitor through an inverter into what is essentially a conventional X-ray tube generator system. In a capacitor discharge unit, the capacitor is connected directly to a grid-controlled X-ray tube. In this type of X-ray tube there is a third electrode shield or grid placed on the cathode assembly. If this electrode is maintained at some 2 kV negative relative to the cathode filament it stops the tube discharging by repelling any electrons emitted from the cathode even when the capacitor is fully charged. This grid control can be turned on and off independently providing instantaneous control of the X-ray tube current and very precisely timed exposures. Note that grid control is useful in any X-ray unit where extremely short exposures (a few milliseconds) are required. It is also useful where rapid repeat exposures are required because the switching mechanism is without inertia.

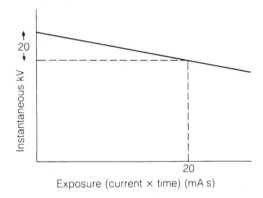

Figure 2.25. *Variation of kV with exposure for a capacitor discharge mobile unit.*

The capacitor can be charged, at low current, from any hospital 13 A mains supply. Charging will continue until the capacitor reaches a preset kilovoltage. Once an exposure has started the mA s delivered must be monitored and the exposure is terminated as required.

The operating kV of a capacitor discharge unit is high at the start of the exposure and relatively low at the end (figure 2.25). This is because the kilovoltage across the tube is reduced as charge is taken off the capacitor. If the capacitor has a value of 1 μF, the reduction in the kilovoltage is 1 kV per 1 mA s of exposure. The output falls accordingly during the exposure and failure to realize this, and the cause, has often led to mistakes being made in the setting of exposure factors with these machines. Consequently they have acquired an unjustified reputation of lacking in output.

Evans *et al* (1985) have proposed that the equivalent kilovoltage of a capacitor discharge unit, i.e. that setting on a constant potential kilovoltage machine which would produce the equivalent radiographic effect, is approximately equal to the starting voltage minus one third of the fall in tube voltage which occurs during the exposure. As the tube voltage drop is numerically equal to the mA s selected, for a 1 μF capacitor

$$\text{equivalent kV} = \text{starting kV} - \text{mA s}/3.$$

If therefore one has a radiographic exposure setting of 85 kV and 30 mA s, the equivalent voltage is 75 kV. If this exposure is insufficient and an under-exposed radiograph is produced, simply increasing the mA s may not increase the blackening on the film sufficiently. For example suppose the exposure is increased to 50 mA s at the same 85 kV. The equivalent kV will now be only 68 kV. Since the equivalent kV is less, some of the increase in mA s will be used to provide soft radiation which does not contribute to the radiograph. The appropriate action is to change the exposure factors to 92 kV and 50 mA s, thereby maintaining the equivalent kV at 75 kV but increasing the exposure to 50 mA s.

If it is desired to reduce the starting kV after the capacitor has been charged or when radiography is finished, the capacitor must be discharged. When the 'discharge' button is depressed, an exposure takes place at a low mA for several seconds until the required charge has been lost. During this exposure the tube produces unwanted X-rays. These are absorbed by a lead shutter across the light beam diaphragm. An automatic interlock ensures that the tube cannot discharge without the lead shutter in place to intercept the beam (figure 2.26). It should be noted, however, that this shutter does not absorb all the X-rays produced, especially when the discharge is taking place from a high kV. Neither patient nor film should be underneath the light beam diaphragm during the discharge operation.

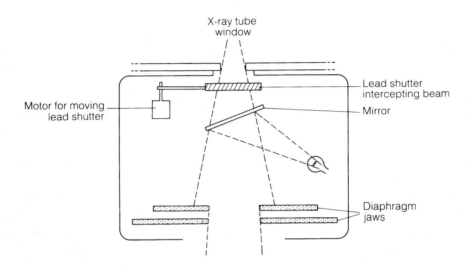

Figure 2.26. *Arrangement of the lead shutter for preventing exposure during capacitor discharge and the light beam diaphragm assembly on a capacitor discharge mobile unit.*

Capacitor discharge units require more operator training to ensure optimum performance than other mobile units, but when used under optimum conditions they have a sufficiently high output to permit acceptably short exposures for most investigations.

2.7.   QUALITY ASSURANCE OF PERFORMANCE FOR STANDARD X-RAY SETS

It is sometimes difficult to identify the boundary between quality assurance and radiological protection in diagnostic radiology. This is because the primary purpose of good quality assurance is to obtain the best diagnostic image required at the first exposure. If this is achieved successfully, not only is the maximum information obtained for the radiation delivered, but also the radiation dose to both the patient and to staff is minimized.

A number of performance checks should be carried out at regular intervals. These will, for example, confirm that:

(1)   the tube kVp is correctly produced;
(2)   the mA s reading is accurate;
(3)   the exposure timer is accurate;
(4)   the radiation output is reproducible during repeat exposures.

All these factors affect the degree of film blackening and the level of contrast.

Tube kV may be measured directly by the service engineer. However, for more frequent checks a non-invasive method is to be preferred. The usual method is based on the fact that different materials show different attenuation properties at different beam energies (see chapter 3) This method, when first suggested by Ardran and Crooks (1968), used a copper step wedge to compensate for the difference in sensitivity between a fast and a slow film screen. With suitable calibration the wedge thickness that gave an exact match in film blackening could be converted into tube kilovoltage. This involves exposure to the special cassette at each kVp with an appropriate mA s.

An early method for checking exposure times was to interpose a rotating metal disc with a row of holes or a radial slit in it between the X-ray tube and the film. Inspection of the pattern of blackening allowed the exposure time to be deduced. Both of the above methods are described more fully in older text books—see for example the first edition of this book.

Modern equipment is now available to simplify both these checks and both kVp and time of exposure are usually measured more directly using a digital timer meter and a digital kV meter which can be attached to a fast responding storage cathode ray oscilloscope (CRO). Several balanced photodetectors are used under filters of different materials and different thicknesses. By using internally programmed calibration curves, a range of kVps may be checked with the same filters. Accuracy to better than 3% should be achievable. This is particularly important for mammography because the absorption coefficient of soft tissue falls rapidly with increasing kV at low energies. Note that the CRO also displays the voltage profile so a fairly detailed analysis of the performance of the X-ray tube generator is possible. For example any delay in reaching maximum kV could affect output at short exposure times. Note that these devices are only accurate at a set tube filtration and readings must be compensated if the tube filtration is different.

Consistency of output may be checked by placing an ion chamber in the direct beam and making several measurements. Both the effect of changes in generator settings (kV, mA s) on the reproducibility of tube output and the repeatability between consecutive exposures at the same setting can be checked. Measurements should be made over a range of clinical settings to confirm a linear relationship between tube output (mGy in air for a fixed exposure time) and the preset mA. It is not normal to make an absolute calibration of tube current.

For automatic exposure control systems a film and suitable attenuating material, e.g. Perspex or a water equivalent slab, should be used to determine the film density achieved when an object of uniform density is exposed under automatic control. A check should be made to ensure that the three ion chambers are matched, i.e. that irrespective of which one is selected to control the exposure, the optical density is similar and repeatable.

There are recommended values for the total beam filtration, 1.5 mm Al for units operating up to 70 kVp and 2.5 mm Al for those operating above 70 kVp, and these should be checked. To do so an ion chamber is placed in the direct beam and readings are obtained with different known thicknesses of aluminium in the beam. The half value layer (HVL) may be obtained by trial and error or graphically. Note that the HVL obtained in this way is *not* the beam filtration (although when expressed in mm of Al the values are sometimes very similar) and the filtration must be obtained from a look-up table (table 2.2).

Table 2.2. *Relationship between the beam filtration and half value layer for a full wave rectified X-ray tube operating at 70 kVp.*

Half value layer (mm Al)	1.0	1.5	2.0	2.5	3.0
Total filtration (mm Al)	0.6	1.0	1.5	2.2	3.0

It is left as an exercise to the reader, after a careful study of chapter 3, to explain why the look-up table will be different at other tube voltages.

Since all operators are urged to use the smallest possible field sizes, it is important to ensure that the optical beam, as defined by the light beam diaphragm, is in register with the X-ray beam. This may be done by placing an unexposed X-ray film on the table and using lead strips or a wire rectangle to define the optical beam. An exposure is made, at a very low mA s because there is no patient attenuation, and the film developed. The exposed area should correspond to the radiograph of the lead strips to better than 1 cm at a focus–film distance of 1 m (figure 2.27). At the same time a check can be made that the axis of the X-ray beam is vertical by arranging two small (2 mm) X-ray opaque spheres vertically one above the other about 20 cm apart in the centre of the field of view. If their images are not superimposed on the developed radiograph, the X-ray beam axis is incorrectly aligned.

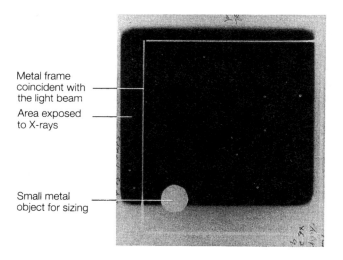

Figure 2.27. *Radiograph showing poor alignment of the X-ray beam and the light beam diaphragm as defined by the metal frame. The small coin is used for orientation and sizing.*

Few centres check focal spot size on general radiographic equipment regularly, perhaps because there is evidence from a range of routine X-ray examinations that quite large changes in spot size are not detectable in the quality of the final image. However, focal spot size is one of the factors affecting tube rating and significant errors in its value could affect the performance of the tube generator. The pin-hole principle illustrated in figure 2.28 may be used to measure the size of the focal spot. The drawing is not to scale but typical dimensions for a 1 mm spot are shown.

By similar triangles

$$\frac{\text{size of image}}{\text{size of focal spot}} = \frac{\text{pin-hole–film distance}}{\text{focus to pin-hole distance}}$$

and for a pin-hole of this size this ratio is usually about 3. Note that the pin-hole must be small—its size affects the size of the image and hence the apparent focal spot size. The 'tunnel' in the pin-hole must be long enough for X-rays passing through the surrounding metal to be appreciably attenuated. This principle is used in the slit camera.

Focal spot size can also be measured and information may be obtained on uniformity of output within the spot, if required, by using a star test pattern. For fuller details see, for example, the book by Curry *et al* (1990). Such information would be important for example if one were attempting to image, say, a 0.4 mm blood vessel at two times magnification because image quality would then be very dependent on both spot

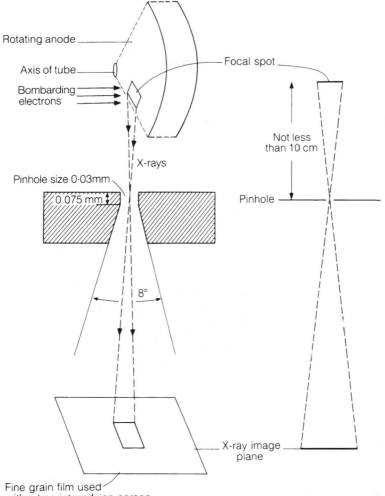

Figure 2.28. *Use of a pin-hole technique to check focal spot size.*

size and shape. Similarly, the high resolution required in mammography (see section 9.6) requires regular checks on the focal spot size.

For further information on quality control, the reader is referred to references given at the end of the chapter, in particular the report by IPEM/CoR/NRPB (1997).

2.8.   SUMMARY

In this chapter the basic principles of X-ray production have been discussed and the most important features may be summarized as follows:

(1)   An X-ray spectrum consists of a continuous component and, if the applied voltage is high enough, characteristic line spectra.

Table 2.3. *Milestones in the development of X-ray tube technology.*

1950s	Rotating X-ray anode
	Control of exposure timer
1960s	High duty X-ray tubes (200 kHU)
	Large anode target disc
	High speed rotation
	Molybdenum/tungsten target
1970s	Rhenium/tungsten target
	Three phase 12 peak generators
	Short exposure times (1 ms)
1980s	Inverter type high voltage generator
	Even higher duty X-ray tubes (500 kHU)
	Microprocessor control
1990s	Anodes mounted on spiral tube bearings
	Variable focal spots
	Enhanced filtration (see section 3.8)

(2)   An important distinction must be made between radiation quantity, which is related to the overall intensity of X-rays produced, and radiation quality which requires a more detailed consideration of the distribution of X-ray intensities with photon energy. The former depends on a number of factors such as tube kilovoltage, time of exposure and atomic number of the target anode. Tube kilovoltage and beam filtration, which will be considered in detail in chapter 3, affect radiation quality. Use of a three phase supply or high frequency generator maintains the tube kilovoltage close to maximum throughout the exposure and both the quantity and quality of X-rays are thereby enhanced.

(3)   There has been steady progress over the past 40 years in X-ray tube technology (see table 2.3) and the high performance of modern X-ray equipment relies on careful design and construction of many components both in the X-ray tube itself and in the associated circuitry. Two features of anode design are particularly important. The first is a consequence of the fact that only about 0.5% of the electron energy is converted into X-rays, whilst the remainder appears as heat which must be removed. The second is the requirement for the X-ray exposure to be as uniform as possible over the irradiated field. This is achieved by careful attention to the anode shape and particularly the angle at which it is presented to the electron flux.

(4)   Notwithstanding careful anode design, generation of heat imposes constraints on X-ray tube performance especially when very short exposures with small focal spot sizes are attempted. The limiting conditions are usually expressed in the form of rating curves. If a rating limit is exceeded, either the duration of exposure (and lens tube current) or the focal spot size must be increased. Occasionally the desired result may be achieved by increasing the tube kilovoltage.

(5)   Careful quality control of the performance of X-ray sets at regular intervals is essential to minimize the need for repeat X-rays, thereby decreasing the overall radiation body burden to the population.

REFERENCES

Ardran G M and Crooks H E 1968 Checking diagnostic X-ray beam quality *Br. J. Radiol.* **41** 193–8
Curry T S, Dowdey J E and Murray R C Jr 1990 *Christensen's Introduction to the Physics of Diagnostic Radiology* 4th edn (Philadelphia, PA: Lea and Febiger)
Evans S A, Harris L, Lawinski C P and Hendra I R F 1985 Mobile X-ray generators: a review *Radiography* **51** 89–107
IPEM/CoR/NRPB 1997 Recommended standards for routine testing of diagnostic X-ray imaging systems *IPEM Report* 77
Meredith W J and Massey J B 1977 *Fundamental Physics of Radiology* 3rd edn (Bristol: Wright)

FURTHER READING

British Institute of Radiology 1988 *Assurance of Quality in the Diagnostic X-ray Department* (London: British Institute of Radiology)

Forster E 1993 Equipment for diagnostic radiography (MTP)

Gifford D 1984 *A Handbook of Physics for Radiologists and Radiographers* (Chichester: Wiley)

Harshbarger-Kelly M E 1985 Devices for measuring peak kilovoltage of diagnostic X-ray equipment, with emphasis on non-invasive electronic meters *Health Care Instrum.* **1** 27–33

Hill D R (ed) 1975 *Principles of Diagnostic X-ray Apparatus* (London: Macmillan)

Institute of Physics and Engineering in Medicine 1996 Measurement of the performance characteristics of diagnostic X-ray systems used in medicine. Part I: X-ray tubes and generators (2nd edn) *IPEMB Report* 32

Institute of Physics and Engineering in Medicine 1997 Measurement of the performance characteristics of diagnostic X-ray systems in medicine. Part IV: Intensifying screens, films, processors and auto exposure control systems (2nd edn) *IPEMB Report* 32

Wilks R 1981 *Principles of Radiological Physics* (Edinburgh: Churchill Livingstone)

EXERCISES

1   Explain why the X-ray beam from a diagnostic set consists of photons with a range of energies rather than a monoenergetic beam.

2   What is meant by 'characteristic radiation'? Describe very briefly three processes in which characteristic radiation is produced.

3   Describe, with the aid of a diagram, the two physical processes that give rise to the production of X-rays from energetic electrons. How would the spectrum change if the target were made thin?

4   Explain why there is both an upper and a lower limit to the energy of the photons emitted by an X-ray tube.

5   What is the source of electrons in an X-ray tube and how is the number of electrons controlled?

6   The cathode of an X-ray tube is generally a small coil of tungsten wire. (a) Why is it a small coil? (b) Why is the material tungsten?

7   Draw a well labelled diagram of the rotating anode X-ray tube as used in diagnosis. Explain the functions of the various parts and the advantages of the materials used.

8   Explain, with a diagram, the action of a timer for a 120 kV diagnostic X-ray set.

9   How would the output of an X-ray tube operating at 80 kVp change if the tungsten anode ($Z = 74$) were replaced by a tin anode ($Z = 50$)?

10   What is the effect on the output on an X-ray set of (a) tube kilovoltage? (b) the material of the anode?

11   It is required to take a radiograph with a very short exposure. Explain carefully why it may be advantageous to increase the tube kilovoltage.

12   What advantages does a rotating anode offer over a stationary anode in an X-ray tube?

13   Discuss the effect of the following on the rating of an X-ray tube: (a) length of exposure, (b) profile of the voltage supply as a function of time, (c) previous use of the tube.

14   Discuss the factors that determine the upper limit of current at which a fixed anode X-ray tube can be used.

15   What do you understand by the thermal rating of an X-ray tube? Explain how suitable anode design may be used to increase the maximum permissible average beam current for (a) short exposures, (b) longer exposures.

16   For a fixed tube kilovoltage and focal spot size, explain why the maximum permissible average current for a three phase supply is sometimes higher and sometimes lower than for a single phase supply.

17   A technique calls for 550 mA, 0.05 s with the kV adjusted in accordance with patient thickness. If the rating chart of figure 2.20 applies, what is the maximum kVp that may be used safely?

18   A technique calls for 400 mAs at 90 kVp. If the possible mA values are 500, 400, 300, 200, 100 and 50 and the rating chart in figure 2.20 applies, what is the shortest possible exposure time?

19   An exposure of 400 mA, 100 kVp, 0.1 s is to be repeated at the rate of six exposures per second for a total of 3 s. Is this technique safe if the rating chart of figure 2.20 applies?

20   A radiographic series consisting of six exposures of 280 mA, 75 kVp and 0.5 s has to be repeated. What is the minimum cooling time that must elapse before repeating the series if the rating chart of figure 2.24 applies?

21   If the series of example 20 is preceded by fluoroscopy at 100 kVp and 3 mA, for how long can fluoroscopy be performed prior to radiography?

22   Suggest reasons why radionuclides do not provide suitable sources of X-rays for medical radiography.

CHAPTER 3

INTERACTION OF X-RAYS AND GAMMA RAYS WITH MATTER

3.1. INTRODUCTION

The radiographic process depends on the fact that when a beam of X-rays passes through matter its intensity is reduced by an amount that is determined by the physical properties, notably thickness, density and atomic number, of the material through which the beam passes. Hence it is variations in these properties from one part of the patient to another that create detail in the final radiographic image. These variations are often quite small, so a full understanding of the way in which they affect X-ray transmission under different circumstances, especially at different photon energies, is essential if image detail is to be optimized. Note that one of the causes of inappropriate X-ray requests is because the clinical symptoms do not suggest there will be any informative changes in thickness, tissue density or atomic structure in the affected region (RCR 1998).

In this chapter an experimental approach to the problem of X-ray beam attenuation in matter will first be presented and then the results will be explained in terms of fundamental processes. Finally some implications of particular importance to radiology will be discussed.

3.2. EXPERIMENTAL APPROACH TO BEAM ATTENUATION

X-rays and gamma rays are indirectly ionizing radiation. When they pass through matter they are absorbed by processes which set electrons in motion and these electrons produce ionization of other atoms or molecules in the medium. The electrons have short, finite ranges (see table 1.2) and their kinetic energy is rapidly dissipated first as ionization and excitation, eventually as heat. With diagnostic X-rays these electrons do not have enough energy to produce 'Bremsstrahlung' (cf X-ray production) especially in low atomic number materials where the process is inefficient.

Conversely, the X-rays and gamma rays themselves do not have finite ranges, whatever their energy. If a fairly well collimated beam passes through different thicknesses of absorbing material, it will be found that equal thicknesses of stopping material reduce the beam intensity to the same fraction of its initial value, but the beam intensity is never reduced to zero. For simplicity a monoenergetic beam of gamma rays will be considered at this stage. The fact that an X-ray beam comprises photons with a range of energies introduces complications that will be considered towards the end of the chapter.

Referring to figure 3.1(a), if a thickness x of material reduces the beam intensity by a fraction α, the beam intensity after crossing a further thickness x will be

$$\alpha \times (\alpha I_0) = \alpha^2 I_0.$$

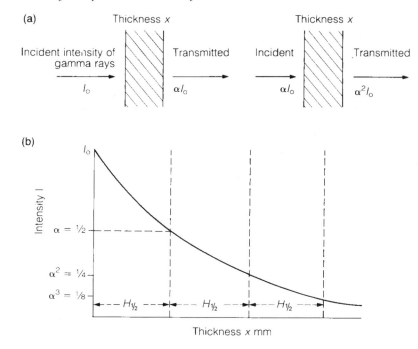

Figure 3.1. *(a) Transmission of a monoenergetic beam of gamma rays through layers of attenuating medium of different thickness. (b) Variation of intensity with thickness of attenuator.*

If, as shown in figure 3.1(*b*), $\alpha = \frac{1}{2}$ it may readily be observed that the variation of intensity with thickness is closely similar to the variation of radioactivity with time discussed in section 1.2. In other words the intensity decreases exponentially with distance, as expressed by the equation $I = I_0 e^{-\mu x}$ where I_0 is the initial intensity and I the intensity after passing through thickness x. μ is a property of the material and is known as the **linear attenuation coefficient**. If x is measured in mm, μ has units mm^{-1}.

A quantity analogous to the half-life of a radioactive material is frequently quoted. This is the **half value layer** or **half value thickness** $H_{1/2}$ and is the thickness of material that will reduce the beam intensity to a half.

The analogy with radioactive decay is shown in table 3.1.

Table 3.1. *Analogy between beam attenuation and radioactive decay.*

Radioactive decay	Attenuation of a monoenergetic gamma ray beam
$A = A_0 e^{-kt}$	$I = I_0 e^{-\mu x}$
$k = \frac{\ln 2}{T_{1/2}} = \frac{0.693}{T_{1/2}}$	$\mu = \frac{\ln 2}{H_{1/2}} = \frac{0.693}{H_{1/2}}$
$A = A_0 e^{-0.693t/T_{1/2}}$	$I = I_0 e^{-0.693x/H_{1/2}}$

Attenuation behaviour may be described in terms of either μ or $H_{1/2}$, since there is a simple relationship between them. If the value of $H_{1/2}$ is known, or is calculated from μ, then the graphical method described in section 1.6 may be used to determine the reduction in beam intensity caused by any thickness of material. Conversely, the method may be used to find the thickness of material required to provide a given reduction in beam intensity. This is important when designing adequate shielding. The smaller the value of μ, the larger the value of $H_{1/2}$ and the more penetrating the radiation. Table 3.2 gives some typical values of μ and $H_{1/2}$ for monoenergetic radiations. The following are the main points to note:

Table 3.2. *Typical values of μ and $H_{1/2}$ (adapted from Johns and Cunningham 1983).*

Energy (keV)	Material	Atomic number	Density (kg m^{-3})	μ (mm^{-1})	$H_{1/2}$ (mm)
30	Water	7.5[a]	10^3	0.036	19
60				0.02	35
200				0.014	50
30	Bone	12.3[a]	1.65 × 10^3	0.16	4.3
60				0.05	13.9
200				0.02	35
30	Lead	82	11.4 × 10^3	33	2 × 10^{-2}
60				5.5	0.13
200				1.1	0.6

[a] These values are effective atomic numbers. For a discussion of the calculation of effective atomic numbers for mixtures and compounds see section 3.5.

(1) In the diagnostic range μ decreases ($H_{\frac{1}{2}}$ increases) with increasing energy, i.e. the radiation becomes more penetrating.
(2) μ increases ($H_{1/2}$ decreases) with increasing density. The radiation is less penetrating because there are more molecules per unit volume available for collisions in the stopping material.
(3) Variation of μ with atomic number is complex although it clearly increases quite sharply with atomic number at very low energies. In table 3.2 some trends are obscured by variations in density.
(4) For water, which for the present purpose has properties very similar to those of soft tissue, $H_{1/2}$ in the diagnostic range is about 30 mm. Thus in passing through the body the intensity of an X-ray beam will be reduced by a factor of two for every 30 mm travelled. If a patient is 18 cm across this represents six half-value thicknesses so the intensity is reduced by 2^6 or 64 times.
(5) At similar energies, $H_{1/2}$ for lead is 0.1 mm or less so quite a thin layer of lead provides perfectly effective shielding for, say, the door of an X-ray room.

It is sometimes convenient to separate the effect of density ρ from other factors. This is achieved by using a **mass attenuation coefficient**, μ/ρ, and then the equation for beam intensity is rewritten

$$I = I_0 e^{-(\mu/\rho)\rho x}.$$

When the equation is written in this form, it may be used to show that the stopping power of a fixed mass of material per unit area is constant, as one would expect since the gamma rays encounter a fixed number of atoms. Consider for example two containers each filled with the same gas and each with the same area A, but of length $5l$ and l (figure 3.2). Let the densities of the two gases be ρ_1 and ρ_2. Furthermore let the mass of gas be the same in each container. Then ρ_2 will be equal to $5\rho_1$, since the gas in container 2 occupies only one-fifth the volume of the gas in container 1.

If the simple expression $I = I_0 e^{-\mu x}$ is used, both μ and x will be different for the two volumes. If $I = I_0 e^{-(\mu/\rho)\rho x}$ is used, then since the product ρx is constant, μ/ρ is also the same for both containers, thus showing that it is determined by the types of molecule and not their number density. Of course both equations

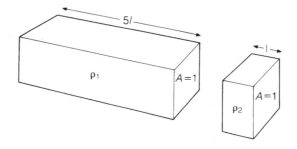

Figure 3.2. *Demonstration that the mass attenuation coefficient of a gas is independent of its density.*

will show that beam attenuation is the same in both volumes. The dimensions of mass attenuation coefficient are $m^2 \ kg^{-1}$.

It is important to emphasize that in radiological imaging the linear attenuation coefficient is the more relevant quantity.

3.3.   INTRODUCTION TO THE INTERACTION PROCESSES

To understand why μ and $H_{1/2}$ vary with photon energy and atomic number in the manner shown in table 3.2, it is necessary to consider in greater detail the nature of the interaction processes between x- and gamma rays and matter. A large number of different processes have been postulated. However, only four have any relevance to diagnostic radiology and need be considered here. As shown in chapter 1, these interactions are essentially collisions between electromagnetic photons and the orbital electrons surrounding the nuclei of matter through which the radiation is passing.

Before considering any interactions in detail, it is useful to discuss some general ideas.

3.3.1.   Bound and free electrons

All electrons are 'bound' in the sense that they are held by positive attractive forces to their respective nuclei, but the binding energy is very variable, being much higher for the K shell electrons than for electrons in other shells. When an interaction with a passing photon occurs, the forces of interaction between the electron and the photon may be smaller than the forces holding the electron to the nucleus, in which case the electron will remain 'bound' to its nucleus and will behave accordingly. Conversely, the forces of interaction may be much greater than the binding forces, in which case the electron behaves as if it were 'free'. Since for the low atomic number elements found in the body, the energy of one photon of diagnostic X-rays is much higher than the binding energy of even K shell electrons, most electrons can behave as if they are free when the interaction is strong enough. However, the interaction is frequently much weaker, a sort of glancing blow by the photon which involves only a fraction of its energy, and thus in many interactions electrons behave as if they were bound. Hence interactions that involve both bound and free electrons will occur under all circumstances and it is frequently the relative contribution of each type of interaction that is important.

This simple picture allows two general statements to be made. First, the higher the energy of the bombarding photons, the greater the probability that the interaction energy will exceed the binding energy. Thus the proportion of interactions involving free electrons can be expected to increase as the quantum energy of the radiation increases. Secondly, the higher the atomic number of the bombarded atom, the more firmly its electrons are held by electrostatic forces. Hence interactions involving bound electrons are more likely when the mean atomic number of the stopping material is high.

3.3.2. Attenuation, scatter and absorption

It is important to distinguish between these three processes and this can also be done on the basis of the simple model of interaction that has already been described (figure 1.8).

When a beam of collimated X-ray photons interacts with matter, some of the X-rays may be **scattered**. This simply means they no longer travel in the same direction as the collimated beam. Some photons have the same energy after the interaction as they had before and lose no energy to the medium. Other photons lose some energy when they are scattered. This energy is transferred to electrons and, as already noted, is dissipated locally. Such a process results in energy **absorption** as well as scattering in the medium. Finally, some photons may undergo interactions in which they are completely destroyed and all their energy is transferred to electrons in the medium. Under the conditions normally obtaining in diagnostic radiology, all this energy is usually dissipated locally and thus the process is one of **total absorption**. Both scatter and absorption result in beam **attenuation**, i.e. a reduction in the intensity of the collimated beam.

If a new term, the **mass absorption coefficient** μ_a/ρ is introduced, it follows that μ_a/ρ is always less than the mass attenuation coefficient μ/ρ although the difference is small at low photon energies. From the viewpoint of good radiology, only absorption is desirable since scatter results in uniform irradiation of the film, which, as will be shown in chapter 5, reduces contrast. Note also that only the absorbed energy contributes to the radiation dose to the patient. This is an undesirable but unavoidable side-effect if good radiographic images are to be obtained.

3.4. THE INTERACTION PROCESSES

Four processes will be considered. Two of these, the photo-electric effect and the Compton effect, are the most important in diagnostic radiology. However, it may be more helpful to discuss the processes in a more logical order, starting with one that is only important at very low photon energies and ending with the one that dominates at high photon energies. Low photon energies are sometimes referred to as 'soft' X-rays, higher photon energies as 'hard' X-rays.

3.4.1. Elastic scattering

When X-rays pass close to an atom, they may cause electrons to take up energy of vibration. The process is one of resonance such that the electron vibrates at a frequency corresponding to that of the X-ray photon. This is an unstable state and the electron quickly re-radiates this energy in all directions and at exactly the same frequency as the incoming photons. The process is one of scatter and attenuation without absorption.

The electrons that vibrate in this way must remain bound to their nuclei, thus the process is favoured when the majority of the electrons behave as bound electrons. This occurs when the binding energy of the electrons is high, i.e. the atomic number of the scattering material is high, and when the quantum energy of the bombarding photons is relatively low. The probability of elastic scattering can be expressed by identifying a mass attenuation coefficient with this particular process, say ε/ρ. Numerically, ε/ρ is expressed as a cross section area. If the effective cross section area for elastic scattering is high, the process is more likely to occur than if the effective cross section area is low. ε/ρ increases with increasing atomic number of the scattering material ($\varepsilon/\rho \propto Z^2$) and decreases as the quantum energy of the radiation increases ($\varepsilon/\rho \propto 1/hf$).

Although a certain amount of elastic scattering occurs at all X-ray energies, it never accounts for more than 10% of the total interaction processes in diagnostic radiology.

3.4.2. Photoelectric effect

At the lower end of the diagnostic range of photon energies, the photoelectric effect is the dominant process. From an imaging view point this is the most important interaction that can occur between X-rays and bound electrons. In this process the photon is completely absorbed, dislodging an electron from its orbit around a nucleus. Part of the photon energy is used to overcome the binding energy of the electron; the remainder is given to the electron as kinetic energy and is dissipated locally (see section 3.2). Although the photoelectric interaction can happen with electrons in any shell, it is most likely to occur with the most tightly bound electron the photon is able to dislodge. The following equation describes the energy changes:

$$hf = W + \tfrac{1}{2}m_e v^2.$$

$$\underset{\text{photon energy}}{hf} = \underset{\substack{\text{binding energy} \\ \text{of electron to} \\ \text{nucleus}}}{W} + \underset{\substack{\text{kinetic energy} \\ \text{of electron}}}{\tfrac{1}{2}m_e v^2.}$$

However, this now leaves the residual atom in a highly excited state since there is a vacancy in one of its orbital electron shells, often the K shell. One possibility is that the vacancy will be filled by an electron of higher energy and characteristic X-ray radiation will be produced in exactly the same way as characteristic radiation is produced as part of the X-ray spectrum from, say, a tungsten anode. This process is summarized in figure 3.3. The number of X-rays emitted, expressed as a fraction of the number of primary vacancies created in the atomic electron shells, is known as the **fluorescence yield**.

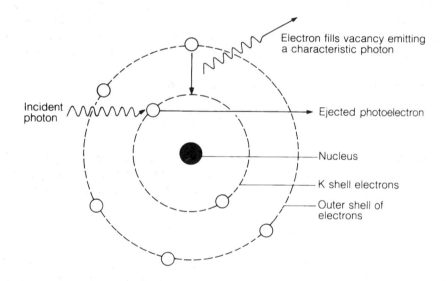

Figure 3.3. *Schematic representation of the photoelectric effect.*

In high atomic number materials the fluorescence yield is high and quite appreciable reradiation of characteristic radiation in a manner similar to X-ray production may occur. This factor is important in the choice of suitable materials for X-ray beam filtration (see section 3.8).

In lower atomic number materials any X-rays that are produced are of low energy (corresponding to low K shell energy) and are absorbed locally. Also the fluorescence yield is low. Production of Auger electrons released from the outer shell of the atom is now more probable. These electrons have energies ranging from a few to several hundred electron volts, so their ranges in tissue are short and the photoelectric interaction process now results in total absorption of the energy of the initial photon. Note that the dense shower of Auger electrons that is emitted deposits its energy in the immediate vicinity of the decay site. The resulting high local energy density can equal or even exceed that along the track of an alpha particle with corresponding radiobiological damage (see section 11.4).

Since the process is again concerned with bound electrons, it is favoured in materials of high mean atomic number and the photoelectric mass attenuation coefficient τ/ρ is proportional to Z^3. The process is also favoured by low photon energies with τ/ρ proportional to $1/(hf)^3$. Notice that, as a result of the Z^3 factor, at the same photon energy lead ($Z = 82$) has a 300 times greater photoelectric coefficient than bone ($Z = 12.3$). This explains the big difference in μ values for these two materials at low photon energies as shown in table 3.2.

The cross section for a photoelectric interaction falls steeply with increasing photon energy although the decrease is not entirely regular because of absorption edges (see section 3.6).

Thus the photoelectric effect is the major interaction process at the low end of the diagnostic X-ray energy range.

3.4.3. Compton effect

The most important effect in radiology involving unbound electrons is inelastic scattering or the Compton effect. This process may be thought of most easily in terms of classical mechanics in which the photon has energy hf and momentum hf/c and makes a billiard ball type collision with a stationary free electron, with both energy and momentum conserved (figure 3.4).

The proportions of energy and momentum transferred to the scattered photon and to the electron are determined by θ and ϕ. The kinetic energy of the electron is rapidly dissipated by ionization and excitation and eventually as heat in the medium and a scattered photon of lower energy than the incident photon emerges from the medium—assuming no further interaction occurs. Thus the process is one of scatter and partial absorption of energy.

The equation used most frequently to describe the Compton process is

$$\lambda' - \lambda = \frac{h}{m_e c}(1 - \cos\phi)$$

where λ' is the wavelength of the scattered photon and λ is the wavelength of the incident photon. This equation shows that the change in wavelength $\Delta\lambda$ when the photon is scattered through an angle ϕ is independent of photon energy. However, it may be shown that the change in energy of the photon ΔE is given by[1]

$$\Delta E = \frac{E^2}{m_e c^2}(1 - \cos\phi). \tag{3.1}$$

Thus the loss of energy by the scattered photon does depend on the incident photon energy. For example when the photon is scattered through $60°$, the proportion of energy taken by the electron varies from about 2% at 20 keV to 9% at 200 keV and 50% at 1 MeV.

Since the process is one of attenuation with partial absorption, the variation in the amount of energy absorbed in the medium, averaged over all scattering angles, with initial photon energy depends on:

(1) the probability of an interaction (figure 3.5(a));

---

[1] If $\lambda = hc/E$, then
$$\lambda + \Delta\lambda = hc/(E - \Delta E)$$
where $\Delta\lambda$ is a small change in wavelength of the photon and ΔE is the corresponding energy change. So
$$\Delta\lambda = \frac{hc}{E - \Delta E} - \frac{hc}{E} \simeq \frac{hc}{E^2}\Delta E$$
or
$$\Delta E = \frac{E^2}{hc}\Delta\lambda = \frac{E^2}{m_e c^2}(1 - \cos\phi).$$

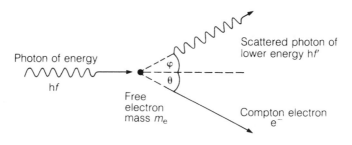

Figure 3.4. *Schematic representation of the Compton effect.*

(2)   the fraction of the energy going to the electron (figure 3.5(*b*));
(3)   the fraction of the energy retained by the photon (figure 3.5(*b*)).

To find out how much energy is absorbed in the medium, the Compton cross section must be multiplied by the percentage of energy transferred to the electron (figure 3.5(*c*)). This shows that there is an optimum X-ray energy for energy absorption by the Compton effect. However, it is well above the diagnostic range.

As shown by equation (3.1), the amount of energy transferred to the electron depends on the scattering angle ϕ. At diagnostic energies, the proportion of energy taken by the electron, i.e. absorbed, is always quite small. For example, even a head-on collision ($\phi = 180°$) only transfers 8% of the photon energy to the electron at 20 keV. Thus at low energies Compton interactions cause primarily scattering and this will have implications when the effects of scattered radiation on film contrast are considered. Although the energy of scattered X-rays is always lower than the energy of the primary beam, in very low energy work, e.g. mammography, the difference can be quite small.

One further consequence of equation (3.1) is that when E is small, quite large values of ϕ are required to produce appreciable changes ΔE. This is important in nuclear medicine where pulse height analysis is used to detect changes in E and hence to discriminate against scattered radiation.

Direction of scatter

After Compton interactions, photons are scattered in all directions and this effect may be displayed by using **polar diagrams** similar to those used in section 2.4 to demonstrate the directions in which X-rays are emitted from the anode (figure 3.6). As the photon energy increases, the scattered photons travel increasingly in the forward direction but this change is quite small in the diagnostic energy range where a significant proportion of X-rays may be back-scattered. Note that, as discussed above, the mean energy of back-scattered photons is lower than the mean energy of forward-scattered photons.

For thicker objects, for example a patient, the situation is further complicated by the fact that both the primary beam and the scattered radiation will be attenuated. Thus, although in figure 3.7 the polar diagrams of figure 3.6 could be applied to each slice in turn, because of body attenuation the X-ray intensity on slice Z may be only 1% of that on slice A. In the example shown in figure 3.7, which is fairly typical, the intensity of radiation scattered back at 150° is 10 times higher than that scattered forward at 30° and is comparable with the intensity in the primary transmitted beam.

Hence a high proportion of the scattered radiation emerging from the patient travels in a backwards direction and this has implications for radiation protection, for example when using an over-couch tube for fluoroscopy.

The effect of scattered radiation on image quality is considered in chapter 5.

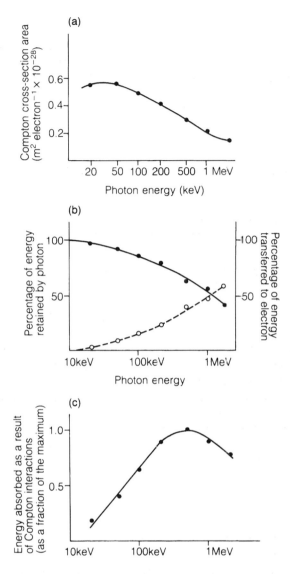

Figure 3.5. *(a) Variation of Compton cross section with photon energy. (b) Percentage of energy transferred to the electron (dotted line) and percentage retained by the photon (solid line) per Compton interaction as a function of photon energy. (c) Product of (a) and (b) to give variation in total Compton energy absorption as a function of photon energy.*

Variation of the Compton coefficient with photon energy and atomic number

For free electrons, the probability of a Compton interaction (σ/ρ) decreases steadily as the quantum energy of the photon increases. However, when the electrons are subject to the forces of other atoms, low energy interactions frequently do not give the electron sufficient energy to break away from these other forces. Thus in practice the Compton mass attenuation coefficient is approximately constant in the diagnostic range and only begins to decrease ($\sigma/\rho \propto 1/hf$) for photon energies above about 100 keV.

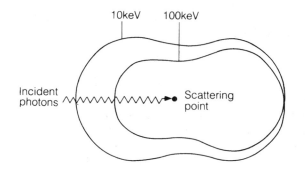

Figure 3.6. *Polar diagrams showing the spatial distribution of scattered X-rays around a free electron at two different energies.*

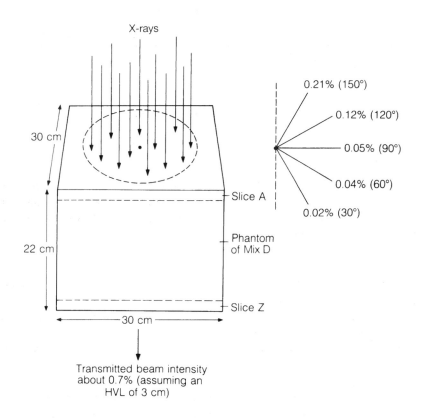

Figure 3.7. *Distribution of scattered radiation around a patient-sized phantom of tissue equivalent material. The phantom measured 30 cm × 30 cm × 22 cm deep and a 400 cm² field was exposed to 100 kVp X-rays. Intensities of scattered radiation at 1 m are expressed as a percentage of the incident surface dose (after Bomford and Burlin 1963, ICRP 1982).*

σ/ρ is almost independent of atomic number. To understand why this should be so, recall the information from chapter 1 on atomic structure. The Compton effect is proportional to the number of electrons in the stopping material. Thus if the Compton coefficient is normalized by dividing by density, σ/ρ should depend on electron density. Now for any material the number of electrons is proportional to the atomic number Z and the density is proportional to the atomic mass A. Hence

$$\sigma/\rho \propto Z/A.$$

Examination of table 3.3 shows that Z/A is almost constant for a wide range of elements of biological importance, decreasing slowly for higher atomic number elements.

Hydrogen is an important exception to the 'rule' for biological elements. Materials that are rich in hydrogen exhibit elevated Compton interaction cross sections and this explains, for example, the small but measurable difference in mass attenuation coefficients in the Compton range of energies between water and air, even though their mean atomic numbers of approximately 7.5 and 7.8 (see section 3.5) respectively are almost identical.

The Compton effect is a major interaction process at diagnostic X-ray energies, particularly at the upper end of the energy range.

Table 3.3. *Values of charge/mass ratio for the atoms of various elements in the periodic table.*

	H	C	N	O	P	Ca	Cu	I	Pb
Z	1	6	7	8	15	20	29	53	82
A	1	12	14	16	31	40	63	127	208
Z/A	1	0.5	0.5	0.5	0.48	0.5	0.46	0.42	0.39

3.4.4. *Pair production*

When a photon with energy in excess of 1.02 MeV passes close to a heavy nucleus, it may be converted into an electron and a positron. This is one of the most convincing demonstrations of the equivalence of mass and energy. The well defined threshold of 1.02 MeV is simply the energy equivalence of the electron and positron masses m_{e^-} and m_{e^+} respectively according to the equation:

$$E = m_{e^-}c^2 + m_{e^+}c^2.$$

Any additional energy possessed by the photon is shared between the two particles as kinetic energy. Each particle can receive any fraction between all and nothing.

The electron dissipates its energy locally and has a range given by table 1.2. The positron dissipates its kinetic energy but when it comes to rest it undergoes the reverse of the formation reaction, annihilating with an electron to produce two 0.51 MeV gamma rays which fly away simultaneously in opposite directions

$$e^+ + e^- \rightarrow 2\gamma\,(0.51\ \text{MeV}).$$

These gamma rays (or **annihilation radiation**) are penetrating radiations which escape from the absorbing material and 1.02 MeV of energy is re-irradiated. Hence the pair production process is one of attenuation with partial absorption.

Pair production is the only one of the four processes considered that shows a steady increase in the chance of an interaction with increasing photon energy above 1.02 MeV (but see section 3.6). Since a large, heavy nucleus is required to remove some of the photon momentum, the process is also favoured by high atomic number materials ($\pi/\rho \propto Z$).

Although pair production has no direct relevance in diagnostic radiology, the subsequent annihilation process is important when positron emitters are used for *in vivo* imaging in positron emission tomography. The characteristic 0.51 MeV gamma rays are detected and, since two are emitted simultaneously, coincidence circuits may be used to discriminate against stray background radiation.

3.5. COMBINING INTERACTION EFFECTS AND THEIR RELATIVE IMPORTANCE

Table 3.4 summarizes the processes that have been considered.

Table 3.4. *A summary of the four main processes by which X-rays and gamma rays interact with matter.*

Process	Normal symbol for process	Type of interaction	Variation with photon energy (hf)	Variation with atomic number (Z)
Elastic	ε/ρ	Bound electrons	$\propto 1/hf$	$\propto Z^2$
Photoelectric	τ/ρ	Bound electrons	$\propto 1/(hf)^3$	$\propto Z^3$
Compton	σ/ρ	Free electrons	Almost constant 10–100 keV; $\propto 1/hf$ above 100 keV	Almost independent of Z
Pair production	π/ρ	Promoted by heavy nuclei	Rapid increase above 1.02 MeV	$\propto Z$

Each of the processes occurs independently of the others. Thus for the photoelectric effect:

$$I = I_0 e^{-(\tau/\rho)\rho x}$$

and for the Compton effect:

$$I = I_0 e^{-(\sigma/\rho)\rho x}$$

etc.

Hence effects can be combined simply by multiplying the exponentials to give

$$I = I_0 e^{-(\tau/\rho)\rho x} e^{-(\sigma/\rho)\rho x} \cdots$$

or

$$I = I_0 e^{-(\tau/\rho + \sigma/\rho + \cdots)\rho x}$$

leading to the simple relationship

$$\mu/\rho = \tau/\rho + \sigma/\rho + \cdots$$

where additional interaction coefficients can be added if they contribute significantly to the value of μ/ρ. Hence the total mass attenuation coefficient is equal to the sum of all the component mass attenuation coefficients obtained by considering each process independently.

In the range of energies of importance in diagnostic radiology, the photoelectric effect and the Compton effect are the only two interactions that need be considered. Since the latter process generates unwanted scattered photons of lower energy but the former does not, and the former process is very dependent on atomic number but the latter is not, it is clearly important to know the relative contributions of each in a given situation. Figures 3.8(a), (b) show photoelectric and Compton cross sections for nitrogen, which has approximately the same atomic number as soft tissue, and for aluminium, with approximately the same atomic number as bone, respectively. Because the photoelectric coefficient is decreasing rapidly with photon energy (note the logarithmic scale) there is a sharp transitional point. By 30 keV the Compton effect is already the more important process in soft tissue.

Understanding the difference in relative importance of these two effects at different energies is quite fundamental to appreciating the origins of radiological image contrast that is based on differences in atomic number in body materials. The photoelectric effect is very dependent on Z; the Compton effect is not. Thus at energies where the photoelectric effect dominates, e.g. mammography, small changes in both mean atomic number and in kV can have a big effect on contrast. At much higher energies, e.g. 120 kV where the Compton

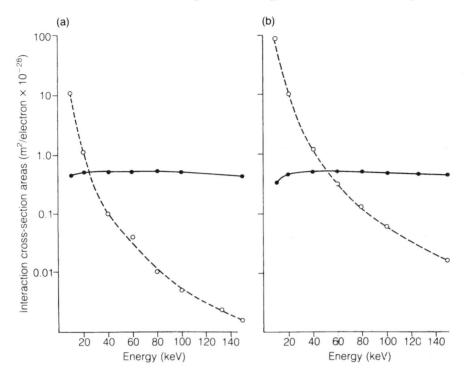

Figure 3.8. *(a) Compton (●) and photoelectric (○) interaction coefficients for nitrogen (Z = 7). (b) Compton (●) and photoelectric (○) interaction coefficients for aluminium (Z = 13).*

effect dominates, image contrast will be relatively insensitive to differences in atomic number and small changes in kV. Note that contrast which is based on differences in density between different structures is not susceptible to this effect.

For the higher atomic number material the photoelectric curve is shifted to the right but the Compton curve remains almost unchanged and the cross-over point is above 50 keV. This trend continues throughout the periodic table. For iodine, which is the major interaction site in a sodium iodide scintillation detector, the cross-over point is about 300 keV and for lead it is about 500 keV. Note that even for lead, pair production does not become comparable with the Compton effect until 2 MeV and for soft tissues not until about 20 MeV.

Note that although calculating the combined effect of different interaction processes to estimate attenuation is fairly straightforward, it is not so easy when trying to work out an **effective atomic number**. This is a useful concept in both radiological imaging and in radiation dosimetry when dealing with a mixture or compound. One wishes to quote a single atomic number for a hypothetical material that would interact in exactly the same way as the mixture or compound. Unfortunately, the effective atomic number of a mix of elements will change as the photon energy increases from the region where the photoelectric effect dominates to the region where the Compton effect dominates. This is because in the photoelectric region the effective atomic number must be heavily weighted in favour of the high Z components. No such weighting is necessary in the Compton region. For further discussion on this point see Johns and Cunningham (1983).

Finally, it is now possible to interpret more fully the data in table 3.2. For any material, the linear attenuation coefficient will decrease with increasing energy, initially because the photoelectric coefficient is decreasing and subsequently because the Compton coefficient is decreasing. At low energy there is a very big difference between the μ values for water and lead, partly because of the density effect but also because the photoelectric effect dominates and this depends on Z^3.

3.6. ABSORPTION EDGES

Whenever the photon energy is slightly greater than the energy required to remove an electron from a particular shell around the nucleus, there is a sharp increase in the photoelectric absorption coefficient. This is known as an absorption edge, and absorption edges associated with K shell electrons have a number of important applications in radiology—the edges associated with the L shell and outer shells are at energies that are too low to be of any practical significance.

As shown in figure 3.9 there will be a substantial difference in the attenuating properties of a material on either side of the absorption edge. There are two reasons for the sudden increase in absorption with photon energy. First the number of electrons available for release from the atom increases. However, in the case of lead the number available only increases from 80 to 82 since the K shell only contains two electrons, and the increase in absorption is proportionately much bigger than this. Thus a more important reason is that a resonance phenomenon occurs whenever the photon energy just exceeds the binding energy of a given shell. Since at 88–90 keV the photon energy is almost exactly equal to that required to remove K shell electrons from lead, a disproportionately large number of K shell interactions will occur and absorption by this process will be high.

Because of absorption edges, there will be limited ranges of photon energies for which a material of low atomic number actually has a higher absorption coefficient than a material of higher atomic number and this has a number of practical applications. For example the presence of K absorption edges has an important influence on the selection of suitable materials for intensifying screens, where a high absorption efficiency by the photoelectric process is required. Although tungsten in a calcium tungstate screen has a higher atomic number than the rare earth elements and therefore has an inherently higher mass absorption coefficient, careful comparison of the appropriate absorption curves (figure 3.10) shows that in the important energy range from 40 to 70 keV where for many investigations there will be a high proportion of photons, absorption by rare earth elements is actually higher than for tungsten.

Other examples of the application of absorption edges are the following.

(1)   In the use of iodine ($Z = 53$, K edge = 33 keV) and barium ($Z = 56$, K edge = 37 keV) as contrast agents. Typical diagnostic X-ray beams contain a high proportion of photons at or just above these energies, thus ensuring high absorption coefficients.
(2)   The use of a selenium plate ($Z = 34$, K edge = 13 keV) for xeroradiography. This K edge value makes selenium a good absorber for the low energy radiation ($\sim$20 keV) used for mammography.
(3)   The presence of absorption edges also has a significant effect on the variation in sensitivity of photographic film with photon energy (see section 6.11).

An important consequence of the absorption edge effect is that a material is relatively transparent to radiation that has a slightly lower energy than the absorption edge, including the material's own characteristic radiation. This factor is important when choosing materials for X-ray beam filtration (see section 3.8). It should also be considered when examining the properties of materials for shielding. For example, although lead is normally used for shielding, if a particular X-ray beam contains a high proportion of photons of energy approaching but just less than the K-edge for lead ($Z = 82$, K edge = 88 keV), some other material, e.g. tin ($Z = 50$, K edge = 29 keV), may be a more effective attenuator on a weight for weight basis in the given situation.

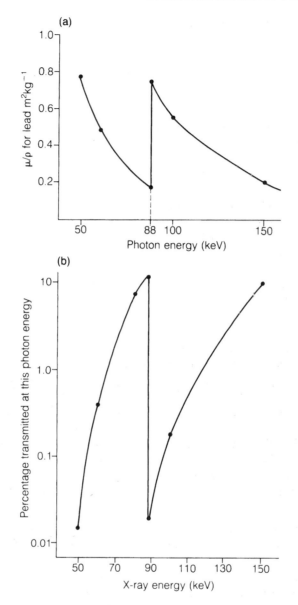

Figure 3.9. *(a) Variation in the mass attenuation coefficient for lead across the K-edge boundary. (b) Corresponding variation in transmission through 1 mm of lead. Note the use of the logarithmic scale; at the absorption edge the transmitted intensity falls by a factor of about 500.*

3.7. BROAD BEAM AND NARROW BEAM ATTENUATION

As already explained, the total attenuation coefficient is simply the sum of all the interaction processes and this should lead to unique values for μ, the linear attenuation coefficient, and $H_{1/2}$, the half-value thickness.

However, if a group of students were each asked to measure $H_{1/2}$ for a beam of radiation in a given attenuating material, they would probably obtain rather different results. This is because the answer would

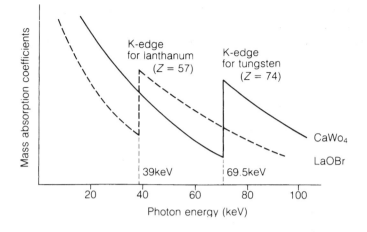

Figure 3.10. *Curves showing the relative absorptions of lanthanum oxybromide (LaOBr) and calcium tungstate (CaWO₄) as a function of X-ray energy in the vicinity of their absorption edges (not to scale).*

be dependent on the exact geometrical arrangement used and whether or not any scattered radiation reached the detector.

Two extremes, narrow beam and broad beam conditions, are illustrated in figures 3.11(*a*), (*b*). For narrow beam geometry it is assumed that the primary beam has been collimated so that the scattered radiation misses the detector. For broad beam geometry, radiation scattered out of the primary beam that would otherwise have

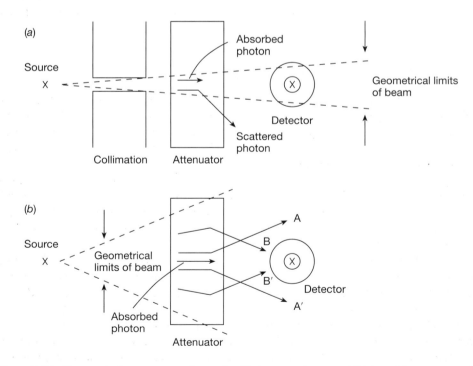

Figure 3.11. *Geometrical arrangement for study of (a) narrow beam and (b) broad beam attenuation.*

reached the detector, labelled A and A′, is not recorded. However, radiation such as B and B′ which would normally have missed the detector is scattered into it and multiple scattering may cause a further increase in detector reading. Hence a broad beam does not appear to be attenuated as much as a narrow beam and, as shown in figure 3.12, the value of $H_{1/2}$ will be different. Absorbed radiation is of course stopped equally by the two geometries. $H_{1/2}$ values are always quoted for narrow beam conditions unless otherwise stated.

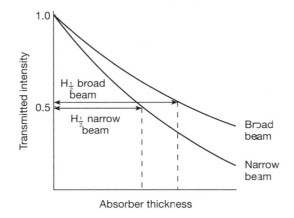

Figure 3.12. *Attenuation curves and values of $H_{1/2}$ corresponding to the geometries shown in figure 3.11.*

An important application of this phenomenon in radiology is the practice of reducing the field size to the smallest value consistent with the required image. The detector is now an X-ray film and if a broader beam than necessary is used extra scattered radiation reaches the film thereby reducing contrast (see section 5.5). Reducing the field size of course also reduces the total radiation energy absorbed in the patient and in interventional procedures radiation scattered to staff.

3.8. FILTRATION AND BEAM HARDENING

Thus far in this chapter, attenuation has been discussed in terms of a beam of photons of a single energy. As a reminder the description 'gamma rays' has frequently been used. Radiation from an X-ray set has a range of energies and these radiations will not all be attenuated equally. Since, in the diagnostic range, the lower energy radiations are the least penetrating, they are removed from the beam more quickly. In other words the attenuating material acts like a filter (figure 3.13).

The choice of a suitable filter is important to the performance of an X-ray set because this provides a mechanism for reducing the intensity of low energy X-ray photons. These photons would be absorbed in the patient, thereby contributing nothing to the image but increasing the dose to the patient. The effect of an ideal low energy filter is shown in figures 3.13(*a*), (*b*), but, in practice, no filter completely removes the low energy radiation or leaves the useful component unaffected.

The position of the K absorption edge must also be considered when choosing a filter material. For example tin, with a K edge of 29 keV, will transmit 25–29 keV photons rather efficiently and these would be undesirable in, for example, a radiographic exposure of the abdomen. Aluminium ($Z = 13$, K edge = 1.6 keV) is the material normally chosen for filters in the diagnostic range. Aluminium is easy to handle, and 'sensible' thicknesses of a few mm are required. Photons of energy less than 1.6 keV, including the characteristic radiation from aluminium, will either be absorbed in the X-ray tube window or in the air gap between filter and patient. The effect of 2.5 mm of aluminium on a 100 kV beam from a tungsten target is shown in figure 3.13(*c*).

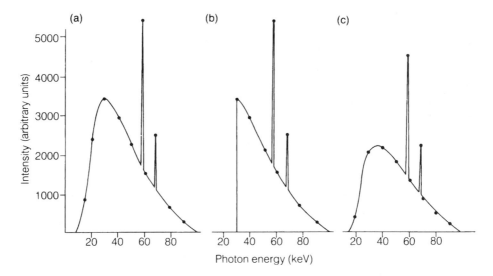

Figure 3.13. *Curves showing the effect of filters on the quality (spectral distribution) of an X-ray beam. (a) Spectrum of the emergent beam generated at 100 kVp with inherent filtration equivalent to 0.5 mm Al. (b) Effect of an ideal filter on this spectrum. (c) Effect of total filtration of 2.5 mm aluminium on the spectrum.*

Finally, the thickness of added filtration will depend on the operating kVp and the inherent filtration. This is the filtration caused by the glass envelope, insulating oil and bakelite window of the X-ray tube itself and is usually equivalent to about 0.5–1 mm of aluminium. Note that the thickness of aluminium that is equivalent to this inherent filtration will vary with kV. The total filtration should be at least 1.5 mm Al for tubes operating up to 70 kVp (e.g. dental units) and at least 2.5 mm Al for tubes capable of operating at higher kVp. Note that if a heavily filtered beam is required at high kV, 0.5 mm copper ($Z = 29$) may be preferred. However, characteristic X-rays at 9 keV will now be produced as a result of the photoelectric effect (section 3.4.2) so aluminium will be required too.

When a beam passes through a filter, it becomes more penetrating or 'harder' and its $H_{1/2}$ increases. If log(intensity) is plotted against the thickness of absorber, the curve will not be a straight line as predicted by $I = I_0 e^{-\mu x}$. It will fall rapidly at first as the 'soft' radiation is removed and then more slowly when only the harder, high energy component remains (figure 3.14(*a*)). Note that the beam never becomes truly monochromatic but after about four half-value layers, i.e. when the intensity has been reduced to about $\frac{1}{16}$ of its original value, the spread of photon energies in the beam is quite small. The value of $H_{1/2}$ then becomes constant within the accuracy of measurement and for practical purposes the beam is monochromatic (figure 3.14(*b*)).

Table 3.5. *Exposure dose to the skin and exposure time for comparable density radiographs of a pelvic phantom (18 cm thick) using a 60 kVp beam with different filtration (adapted from Trout* et al *1952).*

Aluminium filtration (mm)	Skin dose in air (mGy)	Exposure time at 100 mA s
None	20.7	1.4(1)
0.5	16.1	1.6(1)
1.0	11.0	1.6(4)
3.0	4.1	2.1(4)

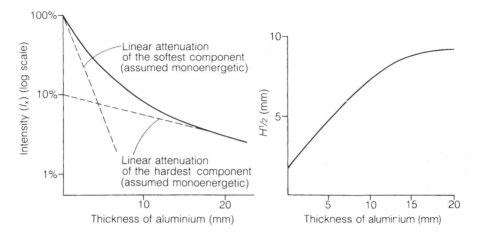

Figure 3.14. *(a) Variation of intensity (I_x) with absorber thickness (x) as an heterogeneous beam of X-rays becomes progressively harder on passing through attenuating material (plotted on a log scale). (b) Corresponding change in the value of $H_{1/2}$.*

The effect of filtration on skin dose is shown in table 3.5. There is clearly a substantial benefit to be gained in terms of patient dose from the use of filters. Note, however, the final column of the table which shows that there is a price to be paid. Although heavy filtration (3 mm Al) reduces the skin dose even more than 1 mm filtration, the X-ray tube output begins to be affected and this is reflected in the increased exposure time. A prolonged exposure may not be acceptable; but if an attempt is made to restore the exposure time to its unfiltered value by increasing the tube current, there may be problems with the tube rating. At 130 kVp, increasing the filtration from 1 mm to 3 mm aluminium would have virtually no effect on exposure time. Also, the technical advances in X-ray tube design discussed in chapter 2 have resulted in powerful X-ray tubes with much greater intensity. The loss of intensity because of filtration then becomes less of a problem. Typically with a 2.5 times increase in tube power an additional 0.5 mm Cu filtration may be used without loss of image quality during screening but with a consequent saving in patient dose. The effect of filtration on patient dose is discussed further in section 6.9.

Thus far the discussion has been concerned with the use of filtration to remove the low energy part of the spectrum. Occasionally, for example when scatter is likely to be a problem, it may be desirable to remove the high energy component of the spectrum. This is difficult because the general trend for all materials is for the linear attenuation coefficient to decrease with increasing keV. However, it is sometimes possible to exploit the absorption edge discussed in section 3.6.

The K shell energy and hence the absorption edge increases with increasing atomic number and for elements in the middle of the periodic table, e.g. gadolinium ($Z = 64$, K edge $= 50$ keV) and erbium ($Z = 68$, K edge $= 57.5$ keV), this absorption may remove a substantial proportion of the higher energies in a conventional spectrum. Figure 3.15 shows the effect of a 0.25 mm thick gadolinium filter on the spectrum shown in figure 3.13(c).

As with low energy filtration, the output of the useful beam is reduced so there is an adverse effect on tube loading. Thus the technique is perhaps best suited to thin body parts where scatter is a problem and a grid is undesirable because of the increased dose to the patient. Paediatric radiology is a good example.

A K edge filter may also sometimes enhance the effect of a contrast agent, for example, iodine. Refer again to figure 3.10 and imagine the lanthanum curve to be replaced by the absorption curve for iodine ($Z = 53$, K edge $= 33$ keV) and the tungsten curve replaced by the curve for erbium. The curves would be

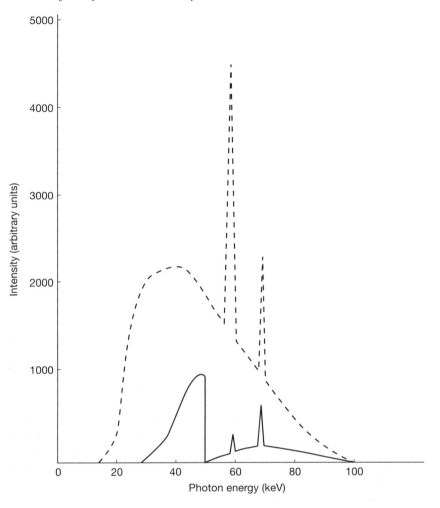

Figure 3.15. *The effect of a 0.25 mm gadolinium filter on a conventionally filtered X-ray spectrum (solid curve). The dotted curve is taken from figure 3.13(c).*

very similar with both edges displaced slightly to the right. Thus in the energy range from 33 to 57.5 keV, the erbium filter would be transmitting X-rays freely, whilst absorbing higher energies, whereas the iodine contrast in the body would absorb heavily in the 33–57.5 keV range.

Because materials are relatively transparent to their own characteristic radiation, the effect of filtration can be rather dramatic when the filter is of the same material as the target anode producing the X-rays. This effect is exploited in mammography and figure 3.16 shows how a 0.05 mm molybdenum filter changes the spectrum from a tube operating at 35 kV constant potential using a molybdenum target. Note that the output is not only near-monochromatic, it contains a high proportion of characteristic radiation. This component of the spectrum does not drift with kVp for example with generator performance, and this helps to keep the soft tissue contrast constant.

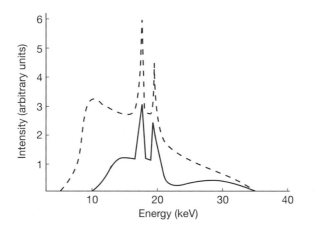

Figure 3.16. *The effect of a 0.05 mm molybdenum filter on the spectrum from a tube operating at 35 kV constant potential using a molybdenum target (dotted line = no filter; solid line = with filter).*

3.9. CONCLUSIONS

In this chapter both experimental and theoretical aspects of the interaction of X-rays and gamma rays with matter have been discussed. In attenuating materials the intensity of such beams decreases exponentially, provided they are monochromatic or near monochromatic, at a rate determined by the density and mean atomic number of the attenuator and the photon energy. In the diagnostic range the trend is for the amount of attenuation to decrease with increasing photon energy.

The two most important interaction processes are the photoelectric effect and the Compton effect. The former is primarily responsible for differences in attenuation (contrast) at low photon energies and its effect is very dependent on atomic number. However, the photoelectric effect decreases rapidly with increasing photon energy and when the Compton effect dominates only differences in density cause any appreciable difference in attenuation.

The Compton effect produces scattered photons of lower mean energy than the primary beam. This scattered radiation is undesirable both because it reduces contrast in the radiograph and also because it constitutes a radiation hazard to staff.

Although the attenuation coefficient generally decreases with increasing photon energies, there are sharp discontinuities, known as absorption edges. The location of an edge along the photon energy axis depends on the atomic number of the absorber, and absorption edges have a number of important consequences, notably in the choice of material for intensifying screens and in determining the variation in sensitivity of photographic film with X-ray energy.

Because the attenuation coefficient decreases with increasing photon energy, low energy photons are preferentially removed from an inhomogeneous or heterochromatic beam—a phenomenon known as beam hardening. This feature is used in filtration to remove soft X-rays that would increase the patient dose.

The effect of the basic interaction processes considered here on contrast and image quality will be discussed in detail in chapter 5.

REFERENCES

Bomford C H and Burlin T E 1963 The angular distribution of radiation scattered from a phantom exposed to 100–300 kVp X-rays *Br. J. Radiol.* **36** 436–9

International Commission on Radiological Protection (ICRP) 1982 Protection against ionizing radiation from external sources used in medicine (ICRP publication 33) *Ann. ICRP.* **9** (1)

Johns H E and Cunningham J R 1983 *The Physics of Radiology* 4th edn (Springfield, IL: Thomas)

Royal College of Radiologists (RCR) Working Party 1998 *Making the best use of a Department of Clinical Radiology* 4th edn (London: RCR)

Sprawls P 1987 *Physical Principles of Medical Imaging* (Aspen)

Trout E D, Kelley J P and Cathey G A 1952 The use of filters to control radiation exposure to the patient in diagnostic radiology *Am. J. Roentgenol.* **67** 946–63

FURTHER READING

Hospital Physicists Association 1979 *Catalogue of Spectral Data for Diagnostic X-rays (Scientific Report Series 30)* (London: Hospital Physicists' Association)

Kohn M L, Gooch A W and Keller W S 1988 Filters for radiation reduction—A comparison *Radiology* **167** 255–7

Meredith W J and Massey J B 1977 *Fundamental Physics of Radiology* 3rd edn (Bristol: Wright)

Nagel H D 1989 Comparison of performance characteristics of conventional and K edge filters in general diagnostic radiology *Phys. Med. Biol.* **34** 1269–87

EXERCISES

1 Explain the terms (*a*) inverse square law, (*b*) linear attenuation coefficient, (*c*) half-value thickness and (*d*) mass absorption coefficient. Indicate the relationships between them, if any.

2 Describe the process of Compton scattering, explaining carefully how both attenuation and absorption of X-rays occur.

3 Describe the variation of the Compton attenuation coefficient and Compton absorption coefficient with scattering angle in the energy range 10–200 keV.

4 How does the process of Compton scattering of X-rays depend on the nature of the scattering material and upon X-ray energy? What is the significance of the process in radiographic imaging?

5 An X-ray beam loses energy by the processes of absorption and/or scattering. Discuss the principles involved at diagnostic X-ray energies and explain how the processes are modified by different types of tissue.

6 Explain why radiographic exposures are usually made with an X-ray tube voltage in the range 50–110 kVp.

7 If the mass attenuation coefficient of aluminium at 60 keV is 0.028 m^2 kg^{-1} and its density is 2.7×10^3 kg m^{-3}, estimate the fraction of a monoenergetic incident beam transmitted by 2 cm of aluminium.

8 A parallel beam of monoenergetic X-rays impinges on a sheet of lead. What is the origin of any lower energy X-rays which emerge from the other side of the sheet travelling in the same direction as the incident beam?

9 What is meant by characteristic radiation? Describe briefly three situations in which characteristic radiation is produced.

10 How would a narrow beam of 100 kV X-rays be changed as it passed through a thin layer of material? What differences would there be if the layer were (*a*) 1 mm lead ($Z = 82$, $\rho = 1.1 \times 10^4$ kg m^{-3}), (*b*) 1 mm aluminium ($Z = 13$, $\rho = 2.7 \times 10^3$ kg m^{-3})?

11 Before the X-ray beam generated by electrons striking a tungsten target is used for radiodiagnosis it has to be modified. How is this done and why?

12 What factors determine whether a particular material is suitable as a filter for diagnostic radiology?

13 A narrow beam of X-rays from a diagnostic set is found experimentally to have a half-value thickness of 2 mm of aluminium. What would happen to (*a*) the half-value of thickness of the beam, (*b*) the exposure rate, if an additional filter of 1 mm aluminium were placed close to the X-ray source?

14 Discuss the advantages and disadvantages of using aluminium as the filter material in X-ray sets at different generating potentials.

15 Compare the output spectra produced by a tungsten target and a copper target operating at 60 kVp. What would be the effect on these spectra of using (*a*) an aluminium filter; (*b*) a lead filter?

	Atomic number	K shell (keV)	L shell (keV)
Al	13	1.6	—
Cu	29	9.0	1.1
W	74	69.5	11.0
Pb	82	88.0	15.0

16 The dose rate in air at a point in a narrow beam of X-rays is 0.3 Gy min^{-1}. Estimate, to the nearest whole number, how many half-value thicknesses of lead are required to reduce the dose rate to 10^{-6} Gy min^{-1}. If $H_{1/2}$ at this energy is 0.2 mm, what is the required thickness of lead?

17 What is the 'lead equivalence' of a material?

18 Explain what you understand by the homogeneity of an X-ray beam and describe briefly how you would measure it.

19   What is meant by an inhomogeneous beam of X-rays and why does it not obey the law of exponential attenuation with increasing filtration?

CHAPTER 4

THE IMAGE RECEPTOR

4.1. INTRODUCTION—BAND STRUCTURE IN SOLIDS

Roentgen discovered X-rays when he noticed that a thin layer of barium platinocyanide on a cardboard screen would fluoresce even when the discharge tube (a primitive X-ray tube) was covered by black paper. Simultaneously he had discovered the first X-ray receptor! Although barium platinocyanide is no longer used, the first stage in the detection process for over 95% of all X-ray imaging devices is now a fluorescent screen or scintillation crystal. For some image receptors, such as film, it is not essential although generally used. In some systems, for example in computerized axial tomography and in nuclear medicine, an essential feature of the scintillation detector is that the first stage of the X-ray detection system produces a flash of light of intensity proportional to the energy of the X-ray photon interacting with the crystal. The new film-less systems use the light generated from the special scintillation screen storing the image when it is scanned by a laser.

To understand adequately the scintillation mechanism, it is necessary to consider briefly the electron levels in a typical solid. The simple model of discrete energy levels introduced in chapter 1 and used to explain features of X-ray spectra in chapter 2 is strictly only true for an isolated atom. In a solid, where the atoms are close together, it applies reasonably well for the innermost K and L shells but for the outer shells the close proximity of other electrons permits each electron to occupy a range or 'band' of energies (see figure 4.1).

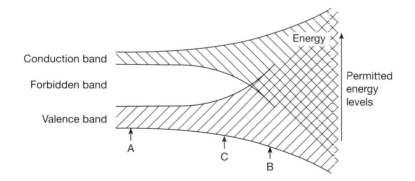

Figure 4.1. *The band structure of energy levels found in solids.*

The highest energy level is the conduction band. Here the electrons have sufficient energy to move freely through the crystalline lattice and in particular to conduct electricity. Next is the forbidden band. In pure materials there are no permitted energy levels here and electrons cannot exist in this region. The next highest energy band is the valence band which contains the valence electrons. Depending on the element, there are further bands of even lower energy. These bands carry their full complement of electrons, with forbidden zones and filled bands alternating like layers in a sandwich, but these are of no interest here.

With this simple model, three types of material may be identified. If the forbidden band between the valence and conduction bands is wide there are no electrons in the conduction band and the material is an insulator (point A in figure 4.1). If the two bands overlap the material is a conductor because electrons can move freely from one atom to the next as the conduction band is continuous between atoms (point B). When the forbidden band is narrow (point C) materials can be made to change from non-conducting to conducting under specific voltage conditions and are termed semiconductors.

However this is the idealized pattern of a pure crystal. In practice real crystals always contain imperfections and impurities which manifest themselves, amongst other ways, as additional energy levels or electron traps in the forbidden energy band. These traps play an important role in all scintillation processes and the manufacture of materials with these properties depends on the production of very pure crystals to which impurities are added under carefully controlled conditions to produce the ideal number and type of traps. The imperfections are called luminescent centres and can be thought of in terms of one localized discrete energy level close to the conduction band, called an electron trap, and one close to the valence band called a hole trap (figure 4.2(*a*)).

With this simple picture of the band structure of a solid, the details of three slightly different luminescent mechanisms are summarized in the next section.

4.2.   FLUORESCENCE, PHOSPHORESCENCE AND THERMOLUMINESCENCE

The essential features of **fluorescence** are as follows:

(1)   Electron traps are normally occupied.
(2)   An X-ray photon interacts by the photoelectric or Compton process to produce a photoelectron which dissipates energy by exciting other electrons to move from the valence band to the conduction band in which they are free to move.
(3)   Holes are thus created in the filled valence band.
(4)   A hole, which has a positive charge numerically equal to that of the electron, moves to a hole trap at a luminescent centre in the forbidden band.
(5)   When both an electron and a hole are trapped at a luminescent centre, the electron may fall into the hole emitting visible light of characteristic frequency f_1 where $E_1 = hf_1$ (see figure 4.2(*b*)).
(6)   The electron trap is refilled by an electron that has been excited up into the conduction band.

In fluorescence, the migration of electrons and holes to the fluorescent centres and the emission of a photon of light happens so quickly that it is essentially instantaneous. Not all transitions of electrons at luminescent centres produce light. The efficiency of the transfer can vary enormously between different materials.

The phenomenon of **phosphorescence** also depends on the presence of traps in the forbidden band but differs in the following respects from fluorescence:

(1)   The electron traps are now normally empty.
(2)   X-ray interactions stimulate transitions from the valance band to the conduction band, as in fluorescence, but for reasons too complex to discuss here, these electrons fall into the traps rather than back into the valence band.
(3)   Furthermore, these electrons must return to the conduction band before they can descend to a lower level.
(4)   Visible light is, therefore, only emitted when the trapped electron acquires sufficient energy to escape from the trap up into the conduction band.

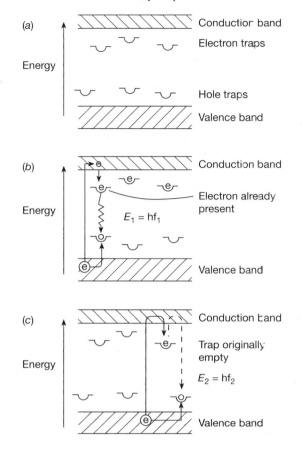

Figure 4.2. *(a) Schematic representation of electron and hole traps in the 'forbidden' energy band in a solid. (b) The change in energy level resulting in light emission in fluorescence. (c) The change in energy level resulting in light emission in phosphorescence.*

(5)   Visible light of a characteristic frequency f_2 (see figure 4.2(*c*)) is now emitted. (As the electron is falling through a larger energy gap than in the fluorescence process f_2 is different from f_1.)

If the electron trap is only a little way below the conduction band, the electron eventually acquires this energy by statistical fluctuations in its own kinetic energy. Thus light is emitted after a time delay (phosphorescence). The time delay that distinguishes phosphorescence from fluorescence is somewhat arbitrary and might range from 10^{-10} s to 10^{-3} s. The two processes can be separately identified by heating the material. Light emission by phosphorescence is facilitated by heating, because the electrons more readily acquire the energy required to escape. Fluorescence is temperature independent. In radiology phosphorescence is sometimes called 'afterglow' and, unless the light is emitted within a very short time when it may contribute to the quantum yield, its presence in a fluorescent screen is detrimental.

If the electron trap is a long way below the conduction band, the chance of the electron acquiring sufficient energy by thermal vibrations at room temperature is negligible and the state is metastable. If the electron is given extra kinetic energy by heating, however, it may be released and then light is emitted. This is essentially the process of **thermoluminescence**. Generally in thermoluminescent material several trapping levels exist and although the light production process is the same for each one it can occur at quite different temperatures resulting in the characteristic glow curve (see section 12.7).

4.3. PHOSPHORS AND FLUORESCENT SCREENS

4.3.1. *Properties of phosphors*

The spectral output required from a phosphor depends on the spectral response of the next stage in the imaging system. It is also desirable for a phosphor to have a high average Z value so that it absorbs a high percentage of the energy from an X-ray beam. However, this is not a sufficient condition, since the overall 'luminescent radiant efficiency' of the phosphor, or light output per unit X-ray beam intensity, will also depend on its efficiency in converting this energy to light output. The chemical properties of a phosphor can also limit how the phosphor is used, e.g. a hygroscopic phosphor must always be used either encapsulated or inside a vacuum. Finally phosphors must also be commercially available in known crystal sizes of uniform sensitivity.

Aside from CT scanning, dealt with in detail in chapter 10, fluorescent screens may be used as the input for four basic imaging systems:

(1)   coupled to film in radiography,
(2)   viewed directly by eye during fluoroscopy (now no longer acceptable),
(3)   in an image intensifier coupled to a photocathode and
(4)   as the image storing medium in a 'film-less' system.

Each of the light receptors used has a different spectral response (figure 4.3) and, as already stated, the spectral output of the phosphor must be matched as closely as possible to the spectral response of the light receptor to which it is coupled.

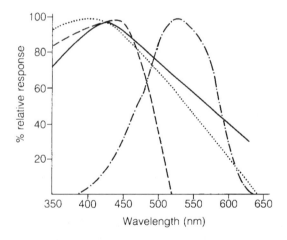

Figure 4.3. *The spectral response of different light receptors. Solid line, S20 photocathode. Dashed line, X-ray film. Dot–dash line, the eye. Dotted line, S11 photocathode.*

Until the mid-1960s, the most widely used phosphors were calcium tungstate in radiography and zinc cadmium sulphide in fluoroscopy and image intensifiers. The spectral outputs of these phosphors are shown in figure 4.4. As a direct result of the American space programme, new phosphors were developed which have since been adopted for medical use. They are mainly crystals of salts of the rare earth elements or crystals of barium salts activated by rare earth elements. A selection of spectral outputs is shown in figure 4.4. Rare earth phosphors containing metallic elements from the lanthanide series emit light at several discrete wavelengths

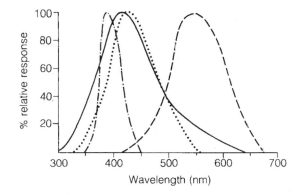

Figure 4.4. *The spectral output of different phosphors. Solid line, CsI:Na. Dashed line, (ZnCd)S:Ag. Dot–dash line, BaFCl:Eu²⁺. Dotted line, CaWO₄.*

between 380 nm and 620 nm with the strongest emissions at about 550 nm. Note that the major exception to the above groups is caesium iodide activated by sodium.

Table 4.1. *Atomic numbers and luminescent radiant efficiencies for some important phosphors. The element after the : sign is the activator to the phosphor salt before the : sign.*

Phosphor	Z of heavy elements	Luminescent radiant efficiency (%)
BaFCl:Eu²⁺	56	13
BaSO₄:Eu²⁺	56	6
CaWO4	74	3.5
CsBr:Tl	35/55	8
CsI:Na	53/65	10
CsI:Tl	53/55	11
Gd₂O₂S:Tb	64	15
La₂O₂S:Tb	57	12
Y₂O₂S:Tb	39	18
(ZnCd)S:Ag	30/48	18
NaI:Tl	53	10

The reason why the rare earth phosphors have replaced calcium tungstate for many purposes can be seen in table 4.1. Although the Z value of the new rare earth phosphors is slightly lower than that of the heavy elements in calcium tungstate phosphor, and thus slightly less energy may, under some conditions, be absorbed from an X-ray beam, the luminescent radiant efficiency for the rare earth phosphors is at least three times higher. The same light output from the phosphor can thus always be achieved with a much lower X-ray dose. In practice, because of absorption edges (section 3.6), at certain photon energies X-ray absorption would also be higher. For some elements in the phosphors (e.g. gadolinium with a K edge at 50.2 keV) this edge is close to the peak in the intensity spectrum of the X-ray beam after it has been transmitted through a patient (see for example figure 5.4).

Figures are also given for comparison for thallium doped sodium iodide NaI:Tl, the primary detector used almost universally in nuclear medicine.

4.3.2. Production of fluorescent screens

Most fluorescent screens are produced by laying down a phosphor crystal/binder suspension onto a paper or metallic substrate and then drying it on the substrate. The size of the crystals affects screen performance with larger crystals producing more light for a given screen thickness. If small crystals are used the effective crystal area is small because there is a high proportion of interstitial space. However, larger crystals produce a screen with a much lower resolution. For a given thickness of screen, with an acceptable resolution, the best packing density (ratio of phosphor to interstitial space) of phosphor crystals that can be achieved is approximately 50%. For high resolution screens the ratio is lower.

A very limited number of phosphors based on halide salts, e.g. CsI:Na, can be vapour deposited in sufficiently thick layers to enable a useful detector to be made. The packing density is almost 100% thus giving a gain of approximately two over crystal deposited screens with a consequently higher X-ray absorption per unit thickness. It is also possible to obtain better resolution and less noise from these phosphors, and this is important in the construction and performance of, for example, an image intensifier (see section 4.11).

4.3.3. Film–phosphor combinations in radiography

For medical applications, radiographic film is almost always used in combination with a fluorescent screen, with only a very few specialized exceptions. However, the properties of the combination are essentially governed by the properties of the film.

Film responds to the light emitted from a fluorescent screen in the same way as it does to X-rays except in one or two minor respects. Thus the many advantages and few disadvantages of using a screen can best be understood by first examining the response of film to X-rays and then considering how the introduction of a screen changes this response.

4.4. X-RAY FILM

4.4.1. Film construction

The basic film construction is shown in figure 4.5. To maintain rigidity and carry the emulsion, which is the sensitive part of the detector, all radiographic films use a transparent base material approximately 0.15 mm thick. The material used is generally 'polyester' and is normally tinted blue although other colours are used for some films. It is completely stable during development. Normally the polyester base has an emulsion bound to both sides (as in figure 4.5). The use of an emulsion on both sides of the base material gives a mechanical advantage during processing as it stops the film curling. However some special films have emulsion on one side only. For example, as described in section 9.6, during mammography a single screen is used on the side of the film remote from the breast. A single emulsion is used next to the screen to eliminate parallax effects. If transparency film is used as in nuclear medicine it is single sided because by that stage in the imaging process only visible light is being recorded. Both sides of the film are coated with a protective layer to prevent mechanical damage.

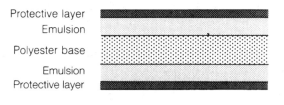

Figure 4.5. *The basic construction of an X-ray film.*

As well as being sensitive to X-rays, the emulsion is sensitive to light and must be kept in a light tight container. It must only be loaded into a cassette using a special daylight loading system or in a dark-room illuminated by a safe light. Short exposures to the wavelength of light in the safe light do not affect the emulsion.

The emulsion is composed of crystals of silver (Ag^+) and bromide (Br^-) ions which are in a cubic lattice and would, in the pure crystal, be electrically neutral. However, the presence of impurities distorts the lattice and produces on the surface of the crystal a spot, called the sensitivity speck, which will attract any free electrons produced within the crystal. When exposed to X-radiation, free electrons are produced by the Compton or photoelectric effect. These electrons, or in the case of visible light from, say, an intensifying screen, light photons, are able to displace further electrons from bromide ions

$$Br^- \quad + \quad hf \quad \rightarrow \quad Br \quad + \quad e^-$$

ion   light   atom   free electron .

Removal of the electron from the bromide ion produces a free bromine atom which is absorbed by the gelatine used to make the emulsion stick to the base.

The electrons move through the crystal and are trapped by the sensitivity speck. An electron at a sensitivity speck then attracts a positively charged silver ion to the speck and neutralizes it to form a silver atom. This occurs many times and the result is an area of the crystal with a number of neutral silver atoms on the surface. This crystal is then said to constitute a latent image. For the crystal to be developable, between 10 and 80 atoms of silver must be produced. During development a reducing alkaline agent is used. Crystals with a latent image in them allow the rest of the silver ions present to be reduced and thus form a dark silver grain speck on the film. Where many X-rays have hit the film, several crystals are affected and thus many silver grains are produced. If the crystals are large it is relatively easy to produce large black specks. If the crystals are small many more X-rays have to hit the film to produce the same amount of blackening. The film is fixed and hardened at the same time using a weakly acidic solution. The crystals which did not contain a latent image are washed off at the fixation stage leaving a light area on the film.

If some time elapses between the production of a latent image and development of the radiograph, some of the crystals revert back to their original state and are no longer developed during processing. The latent image is said to have 'faded'. In practical terms latent image fading is of no significance in radiology as the radiographs are developed immediately following exposure. It can, however, have a small effect when film is used in a personal dosimetry service. For further discussion on this point see section 4.7 on reciprocity.

If the developing agent is too strong it will develop crystals in which no latent image is present. Even in an unexposed film some crystals will be developed to produce a low level of blackening called fog. The 'fog' level can be increased by using inappropriate developing conditions, e.g. too strong a developer or too high a developing temperature.

4.4.2. *Characteristic curve and optical density*

The amount of blackening produced on a film by any form of radiation—visible light or X-rays—is measured by its density. Note the use of density here relates to optical density and must not be confused with mass density (ρ) used elsewhere in the book. The preface 'optical' will not be used elsewhere unless there is some possibility of confusion.

The density of a piece of blackened film is measured by passing visible light through it (figure 4.6(*a*)). Density is defined by the equation

$$D = \log_{10} \frac{I_0}{I}$$

where I_0 is the incident intensity of visible light and I is the transmitted intensity.

Basing the definition on the log of the ratio of incident and transmitted intensities has three important advantages:

(1)   It represents accurately what the eye sees, since the physiological response of the eye is also logarithmic to visible light.
(2)   A very wide range of ratios can be accommodated and the resulting number for the density is small and manageable (see table 4.2).
(3)   The total density of two films superimposed is simply the sum of their individual densities.

Table 4.2. *Relationship between optical density and transmitted intensity.*

Transmitted intensity as a percentage of I_0	$OD = \log_{10} I_0/I$
10%	$\log_{10} 10 = 1.0$
1%	$\log_{10} 100 = 2.0$
0.01%	$\log_{10} 10\,000 = 4.0$

From figure 4.6(*b*),

$$\text{total } D = \log \frac{I_0}{I} = \left(\log \frac{I_0}{I_1} \frac{I_1}{I} \right) = \log \frac{I_0}{I_1} + \log \frac{I_1}{I} = D_1 + D_2.$$

When different amounts of light are transmitted through different parts of the film (figure 4.7), the difference in density between the two parts of the film is called the **contrast**. Hence

$$C = D_1 - D_2 = \log_{10} \frac{I_0}{I_1} - \log_{10} \frac{I_0}{I_2} = \log_{10} \frac{I_2}{I_1}. \tag{4.1}$$

The eye can easily discern differences in density over a range from approximately 0.25 to 2.5, the minimum discernible difference being about 0.02.

If the density produced on a film is plotted against the log of the radiation exposure producing it, the characteristic curve of the film is generated (figure 4.8). The log scale again allows a wide range of exposure to be accommodated. Each type of film has its own characteristic curve although all have the same basic shape. The finite density at zero exposure is due to 'fogging' of the film, i.e. the latent images produced during manufacture, by temperature, humidity and other non-radiation means. This can be kept to a minimum but never completely removed. Note that the apparently horizontal initial portion of the curve arises primarily because one logarithmic quantity (D) has been plotted against another ($\log E$). This has the effect of compressing the lower end of the curve. If the data were replotted on linear axes, there would be a steady increase in film blackening with exposure from zero dose but when using the characteristic curve, a finite dose is required to be given to the film before densities above the fog level are recorded.

The initial curved part of the graph is referred to as the 'toe' of the characteristic curve and this leads into the approximately linear portion of the graph covering the range of densities and doses over which the film is most useful. Eventually, after passing over the shoulder of the curve, the graph is seen to saturate and further exposure produces no further blackening. This decrease in additional blackening is due to the black spots from developed crystals overlapping until eventually the production of more black silver spots has no further effect on the overall density.

At very high exposures (note that the scale is logarithmic), film blackening begins to decrease again, a process known as **solarization**. The mechanism by which this occurs is not fully understood although it may be a result of the release of excess bromine at these high exposures. After the exposure is completed, this

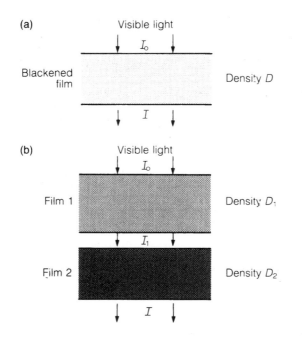

Figure 4.6. *A simple interpretation of optical density: (a) for a simple film; (b) for two films superimposed.*

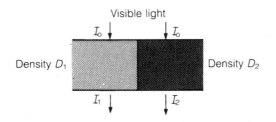

Figure 4.7. *Representation of contrast between two parts of a blackened film as a difference in transmitted light intensities.*

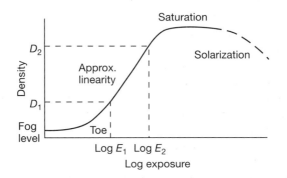

Figure 4.8. *A typical characteristic curve for an X-ray film.*

bromine may recombine with free silver to reconstitute silver bromide which will protect the latent images against development.

Solarization provides a method for obtaining a negative from a negative since the amount of blackening decreases as the intensity of the light transmitted increases. Hence it is used for film copying. The film is specially treated by the manufacturer ('solarized') so that this effect is produced when it is exposed to ultra-violet light. The film to be copied is simply placed over the solarized film and, by adjusting the exposure, it is possible to produce a copy that is lighter or darker than the original if required.

4.4.3.   Film gamma and film speed

The gamma of the film is the maximum slope of the approximately linear portion of the characteristic curve and from figure 4.8 is defined as

$$\gamma = \frac{D_2 - D_1}{\log E_2 - \log E_1}.$$

If no part of the curve is approximately linear, the average gradient may be calculated between defined points on the steepest part of the curve.

The gamma of a film depends on the type of emulsion present, principally the distribution and size of the silver bromide crystals, and secondly on how the film is developed. If the crystals are all the same size a very 'contrasty' film is produced with a large gamma. A wide range of crystal sizes will produce a much lower gamma (figure 4.9). A 'fast' film with large crystals generally also has a wide range of crystal sizes. Finally, increased grain size reduces resolution although unsharpness in the film itself is rarely a limiting factor.

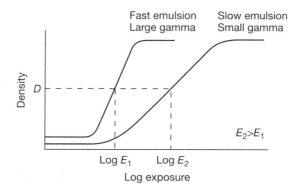

Figure 4.9. *Characteristic curves for films of different gammas and different speeds.*

The correct characteristic curve for a film can only be obtained by using the developing procedure recommended by the film manufacturer, including the concentration of developer, the temperature of the developer, the period of development and even the amount of agitation to be applied to the film. An increase in any of these factors will result in over-development of the film, a decrease will under-develop the film.

Within realistic limits, over-development increases the fog level, the film gamma and the saturation density. Under-development has the opposite effect (figure 4.10).

The amount of radiation to produce a given density is an indication of film speed. The speed is usually taken to be the reciprocal of the exposure that causes unit density above fog so a fast film requires less radiation than a slow film. The speed of the film depends on the size of the crystals making up the emulsion and on the energy of the X-rays striking the film. If the crystals are large then fewer X-ray interactions are required to blacken a film and, because of this, fast films are often called 'grainy' films as the crystals when developed give a 'grainy' picture. This is because the energy deposited by a single X-ray photon is sufficient

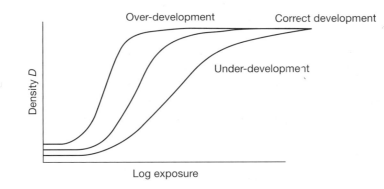

Figure 4.10. *Variation of the characteristic curve for a film, for different development conditions. Note how the gamma and fog level are affected.*

to produce a latent image in a large crystal as well as a small crystal. Fewer large crystals need be developed to obtain a given density. The speed of the film varies with the energy of the X-ray photon, a property that will be considered in detail in section 6.12. In practice, due to the wide range of photon energies present in a diagnostic X-ray beam this variation in sensitivity can be neglected during subjective assessment of X-ray films.

The relative speed of two films is dependent on their characteristic curves and the density at which the speed is compared. In the extreme, if the curves cross then there will be places where one film is faster than the other, one place where they have the same speed and other places where the relative speeds are reversed.

4.4.4. Latitude

Two distinct, but related aspects of latitude are important, film latitude and exposure latitude. Consider first film latitude. The optimum range of densities for viewing, using a standard light box, is between 0.25 and 2.5. Between these two limits the eye can see small changes in contrast quite easily. The latitude of the film refers to the range of exposures that can be given to the film such that the density produced is within these limits. The higher the gamma of the film, the smaller the range of exposures it can tolerate and thus the lower the latitude. For general radiography a film with a reasonably high latitude is used. There is an upper limit however because if the gamma of the film is made too small the contrast produced is too small for reasonable evaluation.

Exposure latitude is related to the object and can be understood by reference to figure 4.8. If a radiograph is produced in which all film densities are on the linear portion of the curve (i.e. the object contains a narrow range of contrasts) the exposure may be altered, shifting these densities up or down the linear portion, without change in contrast. (Radiologists often prefer a darker film to a lighter film although there may in fact be no difference in contrast when the densities are measured and the contrast evaluated.) In other words there is 'latitude' or some freedom of choice over exposure. If, on the other hand, the range of densities on a film covers the whole of the linear range (i.e. the object contains a wide range of contrasts) exposure cannot be altered without either pushing the dark regions into saturation or the light regions into the fog level. There is no 'latitude' on choice of exposure.

Exposure latitude can be restored by choosing a film with a lower gamma (greater film latitude), but only at the expense of loss of contrast.

4.5.   FILM USED WITH A FLUORESCENT SCREEN

Most properties of film described to this point have assumed an exposure to x-radiation. Film also responds to light photons but as the quantum energy of one light photon is only 2–3 eV (see chapter 1) several tens of light photons have to be absorbed to produce one latent image. In contrast, the energy from just one X-ray photon is more than sufficient to produce a latent image.

The advantages of using film in conjunction with fluorescent screens are twofold. First a much greater number of X-ray photons are absorbed by the screen compared to the number absorbed by film alone. The ratio varies between 20 and 40 depending on the screen composition.

Secondly, by first converting the X-ray photon energy into light photons, the full blackening potential is realized. If an X-ray photon is absorbed directly in the film, it will sensitize only one or two silver grains. However, each X-ray photon absorbed in an intensifying screen will release at least 400 photons of light— some screens will release several thousand photons. Thus, although tens of light photons are required to produce a latent image, the final result is that the density on the film for a given exposure is between 30 and 300 times blacker (depending on the type of screen) when a screen is used than when a film alone is exposed.

The increase in blackening when using a fluorescent screen is quantified by the **intensification factor**, defined as:

$$\frac{\text{the exposure required when a screen is not used}}{\text{the exposure required when a screen is used}}$$

for the same film blackening.

A value for the intensification factor is only strictly valid for one density and one kVp. This is because, as shown in figure 4.11, when used in conjunction with a screen, the characteristic curve of the film is altered. Not only is it moved to the left, as one would expect, but the gamma is also increased.

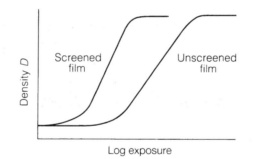

Figure 4.11. *Change in the characteristic curve of film when using a screen.*

When using screens, double sided film has radiological advantages as well as processing advantages since two emulsion layers enable double the contrast to be obtained for a given exposure. This is because the superposition of two densities (one on each side of the film base) produces a density equal to their sum. Although, theoretically, the same contrast could be achieved by doubling the thickness of a single emulsion, this is not possible in practice due to the limited range in emulsion of the light photons from the screens. Note that, because the range of X-ray photons is not limited in this way, the contrast for unscreened films would be almost the same for a one-sided film of double emulsion thickness. The mechanical advantage of double sided film would, of course, still remain.

Reference was made in section 4.4.3 to the speed of the film. In practice it is the speed of the film–screen combination which is important. This is described by a 'speed class'. This system is based on a similar numerical system to the ASA film speed system used in photography. The special feature of ASA film speeds 32, 64, 125 is that the logarithms of these numbers 1.5, 1.8, 2.1 show equal increments. In other words speed classes are equally spaced on the 'log exposure' axis of the characteristic curve. The higher the speed class the more sensitive the film so speed classes for film–screen combinations used in radiography are generally high (table 4.3).

Table 4.3. *The range of speed classes used in film–screen radiography and the corresponding linear increases on a logarithmic scale.*

Speed class	100	200	400	800
Logarithm	2.0	2.3	2.6	2.9

Very high resolution, high contrast combinations are likely to have a speed class in the region of 100–150, whereas very fast combinations (typically 600 or above) will make sacrifices in terms of resolution and possibly contrast. Note that quantum mottle limits the speed of the system that can usefully be used (see section 4.11.2).

4.6. CASSETTES

Radiographic film must be used and contained in a light tight cassette, otherwise it is fogged by ambient light. The front of the cassette is made of a low atomic number material, generally either aluminium or plastic. The back of the cassette is either made of, or lined with, a high atomic number material. This high atomic number material is more likely to absorb totally the X-ray photons passing through both screens and film by the photoelectric effect than to undergo a Compton scatter reaction, which could backscatter photons into the screen.

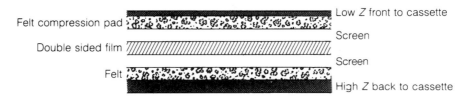

Felt compression pad — Low Z front to cassette / Screen
Double sided film —
Felt — Screen / High Z back to cassette

Figure 4.12. *Cross section through a 'loaded' cassette containing a 'sandwich' of intensifying screens and double sided film.*

The two screens are kept in close contact with the film by the felt pad exerting a constant pressure as shown in figure 4.12. If close contact is not maintained then resolution is lost.

The light emitted from a screen does not increase indefinitely if the screen thickness is increased to absorb more X-ray photons. A point is reached where increasing the thickness produces no more light because internal absorption of light photons in the screen takes place. The absorption of X-ray photons produces an intensity gradient through the screen and significantly fewer leave the screen than enter. The light produced at a given point in the screen is directly proportional to the X-ray intensity at that point.

In the case of the back screen, the reduction in intensity through it is of no significance. This screen must be thick enough to ensure that the maximum amount of light for a given X-ray intensity is produced, but once this is achieved it need be no thicker. The front screen thickness is a compromise between achieving the maximum light output through X-ray photon absorption and not reducing the X-ray photon intensity by too much in the area of effective maximum light production, i.e. in the layers of the screen closest to the film. The compromise thickness for the front screen, giving maximum light production, is in fact somewhat less than optimum for the back screen. Since unsharpness is less from the back screen, attempts to optimize light production can improve image sharpness. For further discussion on the way in which attenuation in the screen can affect image quality see section 4.8.

If the screens are of unequal thickness they must never be reversed but for some modern cassettes there is no difference between the thickness of the front and back screens.

4.7.   RECIPROCITY

The exposure received by a cassette can be considered to depend on two basic parameters, the intensity of the radiation beam striking the cassette and the time of exposure. The intensity at a fixed kVp is proportional to the milliamps (mA) of the exposure, and the exposure is thus proportional to the milliampseconds (mA s). For unscreened film the same mA s will always give the same blackening of the film regardless of the period of exposure. When using screened film, however, it is found that for very short exposures and very long exposures, although the same mA s are given, the blackening of the film is less.

For long exposures this effect is known as latent image fading and the amount of fading depends to a large extent on whether the image has been produced by X-ray interactions or by visible light photons. In general, a single X-ray photon will form a developable grain because it deposits so much energy. Hence image fading does not occur. Conversely, many visible light photons are required to produce a latent image and if their rate of arrival is slow the silver halide lattice may revert to its normal state before sensitization of the speck is completed. This effect is called failure of the reciprocity law.

4.8.   FILM–SCREEN UNSHARPNESS

When screens are used there is some loss of resolution. This is because light produced in the screen travels in all directions (see figure 4.13). Because the screen is of finite thickness, those photons travelling in the direction of the film spread out a little before reaching it. Note that very good contact must be maintained between the screen and film. If they are even slightly separated, the light will be allowed to spread out even further.

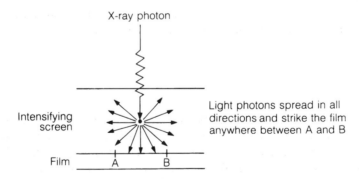

Figure 4.13. *Schematic representation of image unsharpness created by the use of an intensifying screen.*

A major cause of unsharpness in double-sided film is 'cross-over' where light from the upper intensifying screen sensitizes the lower film layer and vice versa. The development of emulsions containing grains that are flat or tubular in shape, with the flat surface facing the X-ray beam, rather than pebble-shaped, has helped to reduce this source of image unsharpness with no loss in sensitivity.

Recent developments using a yttrium tantalate phosphor which emits UV radiation have also reduced both cross-over and screen unsharpness. The UV light is more rapidly absorbed by the screen and film, not diffusing as far from the point of emission as visible light. Thus the UV light is almost completely absorbed by the adjacent film emulsion layer and very little 'cross-over' actually takes place.

4.9. DIGITAL RADIOGRAPHY

Digital radiography is sometimes referred to as film-less radiography because the use of the film as a recording medium is generally dispensed with. One approach is based on phosphor screens which have similar properties and construction to normal screens, but have one additional property which has allowed the development of digital radiography. This property is that, on excitation, only a proportion of the energy is released immediately as light in a luminescent process. The remainder is stored as a latent image and can only be released following stimulation. This latent image can be made to have a lifetime long enough to be useful as an imaging process although eventually it will spontaneously decay to a level which is no longer producing an acceptable image. The latent image is 'read' out by using a laser scanning system to stimulate the phosphor into releasing its stored energy. The wavelength of the stimulating laser is in the red or near infra-red region and the light emitted has a wavelength in the green, blue or even the ultra-violet. This process is known as **photostimulable luminescence**. The amount of light emitted is directly proportional to the number of X-ray photons absorbed hence the screen is completely linear within the normal diagnostic range of exposures.

4.9.1. Read out process

The difference in wavelength between the stimulating light and the output light is essential to the operation of the phosphor because the intensity of the former is of the order of 10^8 larger. In order to 'see' the output, special filters have to be used which only transmit the output wavelength. A laser beam is scanned across the phosphor in a raster pattern (figure 4.14). The stimulated output from each small area (pixel) on the screen is collected by a light pipe which has a photomultiplier attached to it. The photomultiplier amplifies the signal which is then digitized so that each pixel of the phosphor has a light output allocated to it. The image is stored in the computer memory.

To prepare the screen for reuse it has to be exposed to a high intensity flood light for a short period of time which completely removes any remaining energy of excitation from the screen and recreates a uniform response on the screen.

4.9.2. Properties

As digital and standard film–screen systems are essentially both based on the same image forming device, many of the properties are almost the same. The doses required to produce a usable image are similar with the lower limit on dose reduction essentially being the noise produced by quantum mottle. The limit on resolution is also very similar. In a film–screen system the light spreads out isotropically following the conversion of an X-ray photon to light photons. In the storage phosphor system the main contribution to unsharpness comes from the scattering of the stimulating beam as it enters the phosphor thus creating a spread in the luminescence along its path. The diameter of the stimulating laser beam is initially small. The intensity of the stimulating laser light also affects the resolution. A higher intensity beam results in a lower resolution but this is compensated for to some extent by the increase in the amount of stored signal released. The resolution is also limited by the size of the pixels sampled. At the moment an array of approximately 2000 × 2000 is normally used. At this point in time increasing the size produces technical problems which makes them non viable commercially.

For further discussion on receptors for digital imaging see section 9.4.2.

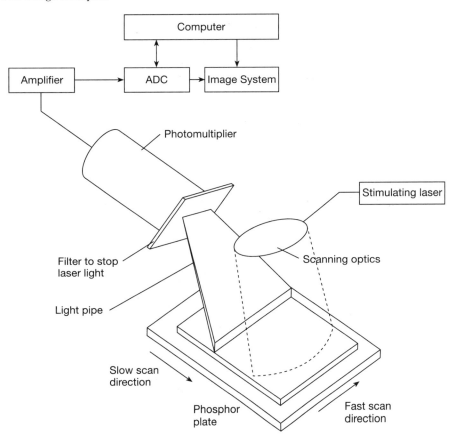

Figure 4.14. *The read-out system for a digital phosphor plate.*

4.10.   EYE–PHOSPHOR COMBINATION IN FLUOROSCOPY

Direct viewing of a fluorescent screen is no longer acceptable because the dose to both the patient and the observer is too high. However, it is instructive to investigate the reason for this. A fluoroscopic screen is shown in figure 4.15. The resolution of the screen varies with crystal size in exactly the same way and for the same reasons as the resolution of a radiographic screen. The wavelength of emitted light is of course independent of the X-ray photon energy. When this technique was used historically it was essential that certain basic preparatory steps were taken. For example, the light output was so low that the room had to be darkened and the eye fully dark-adapted. However, this caused further problems. The spectral output of a cadmium activated zinc sulphide screen closely matches the spectral response of the eye using cone vision at high light intensities (figure 4.16) but the eye is poorly equipped for detecting changes in light levels or resolution at the low intensities of light given out by a fluoroscopic screen. Furthermore, these levels cannot be increased by increasing the X-ray intensity without delivering an even higher dose to the patient. At the levels of light involved, 10^{-3}–10^{-4} millilamberts, the resolution of the eye is no better than about 3 mm at a normal viewing distance of 25 cm and only changes in light levels of approximately 20% can be detected. This is because vision at these levels is by rod vision alone, cone vision having a threshold of brightness perception of 0.1 millilamberts. Also the spectral response of rods peaks at a wavelength of 500 nm, where screen output tends to be poor (figure 4.4). For further discussion see section 8.4.2.

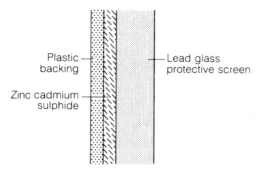

Figure 4.15. *A cross sectional view through a fluoroscopy viewing screen.*

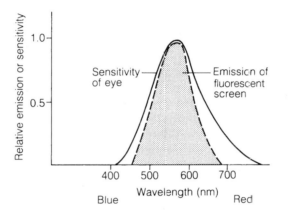

Figure 4.16. *Comparison of the emission spectrum of visible light from a fluorescent screen with the sensitivity of the eye.*

All of the figures quoted above are for a well dark-adapted eye, so the eye had to be at low light levels for 20–40 min before the fluoroscopic screen was viewed. This allowed the visual purple produced by the eye to build up and sensitize the rods. Alternatively, because the rods are insensitive to red light, red goggles could be worn in ambient light, producing the same effect. Exposure to ambient light for even a fraction of a second completely destroyed the build-up of visual purple.

4.11. PHOSPHORS USED WITH IMAGE INTENSIFIERS

4.11.1. *Construction and mode of operation*

The light intensity emitted from a fluorescent screen can only be increased by increasing the exposure rate from the X-ray tube or by increasing the screen thickness. Neither of these methods is acceptable, as the first increases the dose to the patient and the second reduces the resolution. Any increase in signal strength must, therefore, be introduced after the light has been produced. This is achieved by using an image intensifier which increases the light level to such a degree that cone vision rather than rod vision may be used.

The basic image intensifier construction is shown in figure 4.17. The intensifier is partially evacuated and the fluorescent screen is protected by a thin metal housing. This housing also excludes all fluorescent ambient light. The fluorescent screen is laid down on a very thin metal substrate and is now generally CsI.

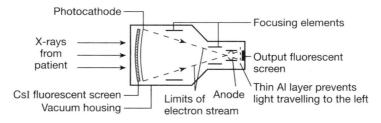

Figure 4.17. *Construction of a simple image intensifier.*

This has two advantages over most other fluorescent materials, as discussed in section 4.3. It can be laid down effectively as a solid layer thus allowing a much greater X-ray absorption per unit thickness to be achieved. As described in section 3.6 the K absorption edges of caesium and iodine are at 36.0 and 33.2 keV respectively. They thus occur at the maximum intensity of the X-ray beam used in most fluoroscopy examinations. The screen is also laid down in needle-like crystals (see figure 4.18). These crystals act as light pipes and any light produced in them is internally reflected along them and does not spread out to cover a large area. CsI can thus be used in thicker layers without any significant loss in resolution although the packing advantage means that a thickness of only 0.1 mm is required.

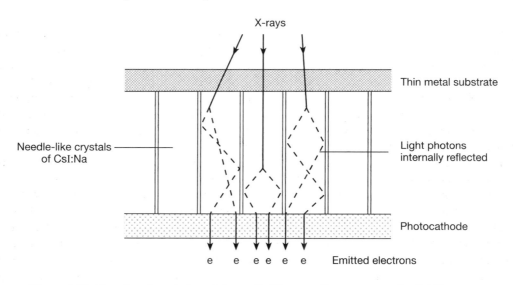

Figure 4.18. *Greatly enlarged view of the needle-like crystalline structure of a CsI:Na screen.*

Intimately attached to the CsI is the photocathode. Just as in radiography and fluorography, the output of the fluorescent screen must be closely matched to the photo-response of the photocathode and CsI is very closely matched to two commonly used photoelectric materials, S11 and S20, as can be seen in figure 4.19. The input phosphor and photocathode are curved to ensure that the electrons emitted have the same distance to travel to the output phosphor. The output of the intensifier is via a second fluorescent screen, often of ZnCdS:Ag, shielded from the internal part of the intensifier by a very thin piece of aluminium. This stops light from the screen entering the image intensifier. This screen is much smaller than the input phosphor. A voltage of approximately 25 kV is maintained between the input and output phosphors.

The mode of action of an image intensifier is as follows. When light from the input fluorescent screen falls on the photocathode it is converted into electrons, and for the present discussion electrons have two

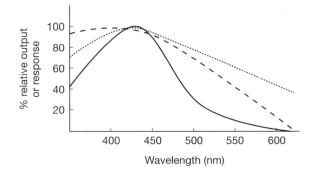

Figure 4.19. *Comparison of the spectral output of a CsI:Na screen (solid line) with the sensitivity response of S11 (dashed line) and S20 (dotted line) photocathodes.*

important advantages over photons:

(1) they can be accelerated, and
(2) they can be focused.

Thus under the influence of the potential difference of 25–30 kV, the electrons are accelerated and acquire kinetic energy as they travel towards the viewing phosphor. This increase in energy is a form of amplification. Secondly, by careful focusing, so as not to introduce distortion, the resulting image can be minified, thereby further increasing its brightness. Even without the increase in energy of the electrons, the brightness of the output screen would be greater by a factor equal to the ratio of the input and output screen areas. The output phosphor is generally made of small crystals of silver-activated zinc–cadmium sulphide which are laid down in a thin layer so as not to affect significantly the resolution of the minified image. The resolution of the image is 3–5 1p mm^{-1} in a new system.

The amplification resulting from electron acceleration is usually about 50. In other words, for every light photon generated at the input fluorescent screen, approximately 50 light photons are produced at the output screen. The increased brightness resulting from reduction in image size depends of course on the relative areas of the input and output screens. If the input screen is approximately 25 cm (10 inches) in diameter and the output screen is 2.5 cm (1 inch), the increase in brightness is $(25/2.5)^2 = 100$ and the overall gain of the image intensifier about 5000. By changing the field on the focusing electrodes it is possible to magnify part of the image. A direct consequence of this is that the intensification factor due to minification is reduced and the exposure of the patient must be increased to compensate. Magnification always leads to a greater patient dose, although careful collimation to the area of interest can restrict the increase.

The use of image intensifiers has not only produced an important improvement in light intensity, but also opened up the way for the introduction of high technology into fluoroscopy. For example, direct viewing of the output of the image intensifier is not very convenient and somewhat restrictive, and although it is possible to increase the size of the image by viewing it through carefully designed optical systems, even with this arrangement there may be a considerable loss of light photons and hence information. In the worst cases this loss in light gathering by the eye can result in a system not much better than conventional fluoroscopy. Thus current practice is generally to view the output directly by a television camera as shown in figure 4.20(*a*). Alternatively, a television camera is used after dividing the output using a partially reflecting mirror as in figure 4.20(*b*). This reflects about 90% of the light towards the recording device letting sufficient through to produce a TV image. This system allows other recording media to be used such as a cine camera or a 100 mm still camera.

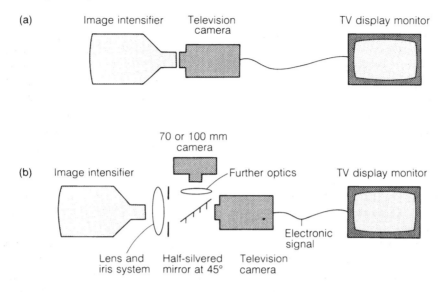

Figure 4.20. *Methods of coupling an image intensifier with a television camera either (a) directly or (b) using a half-silvered mirror to deflect part of the image to a spot or cine camera.*

The use of a television viewing system has several advantages. The first is convenience. The amplification available through the intensifier/television system allows a large image to be viewed under ambient lighting conditions thus eliminating the necessity for dark adaptation. The light output from the television monitor is well above the threshold for cone vision and no limit on resolution is imposed because of the limitation of the human eye. It is also a very efficient system allowing good optical coupling which results in little loss of information after the input stage of the image intensifier. The video signal can be recorded allowing a permanent record of the investigation to be kept. These records are available for immediate play-back but there is, unfortunately, some loss of information and thus a reduction in image quality during the recording/play-back sequence. It is now also possible to use 'frame grabbing' techniques or a last frame hold. The latter allows intermittent screening to be carried out under some circumstances.

The use of 70 mm or 100 mm film or cine film as a substitute for a full sized radiograph has some advantages but also some drawbacks. The quality of the images is high and it is possible to take rapid sequences of films using a motor-driven camera. The radiation dose to the patient per film is lower than for a conventional radiograph but this can be negated if large numbers of films are taken. It is also much cheaper to use 70 mm or 100 mm film than standard radiographs. The major drawback is that the film size is small and requires a magnifying viewing system. Separate developing procedures are also required. To ensure a good image a higher current ($\times 3$ or 4) has to be used.

If the still camera is replaced by a cine camera, it is possible, because of the half-silvered mirror, to take a cine film and view the image on the television screen simultaneously. This allows recording to be restricted to the information that is really required. Although the final image on cine film is better than the image on a video recorder, it does take some time for it to be developed and thus it cannot be viewed immediately following the recording.

4.11.2. Quantum mottle

An amplification of several thousand is available through a modern image intensifier/TV system. However, this does not mean that the dose of radiation delivered to the patient can be reduced indefinitely. Although the brightness of the image can be restored by electronic means, image quality will be lost.

To understand why this is so, it is important to appreciate that image formation by the interaction of X-ray photons with a receptor is a random process. A useful analogy is rain drops coming through a hole in the roof. When a lot of rain has fallen, the shape of the hole in the roof is clearly outlined by the wet patch on the floor. If only a few drops of rain have fallen it is impossible to decide the shape of the hole in the roof.

Similarly, if sufficient X-ray quanta strike a photoreceptor, they produce enough light photons to provide a detailed image but when fewer X-ray quanta are used the random nature of the process produces a mottled effect which reduces image quality. For example a very fast rare earth screen may give an acceptable overall density before the mottled effect is completely eliminated. Thus the information in the image is related to the number of quanta forming the image.

When there are several stages to the image formation process, as in the image intensifier/TV system, overall image quality will be determined at the point where the number of quanta is least, the so-called 'quantum sink'. This will usually be at the point of primary interaction of the X-ray photons with the fluorescent screen. If insufficient X-ray photons are used to form this image, further amplification is analogous to empty magnification in high power microscopy, being unable to restore to the image detail that has already been lost. It follows that neither electron acceleration nor minification in an image intensifier improves the statistical quality of the image if the number of X-ray photons interacting with the fluorescent screen remains the same.

This subject will be considered again under the heading 'quantum noise' in section 8.6.

4.11.3. Factors affecting the image

Two main factors affect the contrast in the image intensifier. The first is the very small amount of light that escapes back into the tube from the output phosphor. Most of this light is stopped by the aluminium backing on the output phosphor but as this is very thin some light can get through to activate the input photocathode, imposing a non-uniform 'fog' of photoelectrons onto the image. The second source of 'fogging' is produced by X-ray photons which are not stopped by the input phosphor and pass through the intensifier tube to excite the output phosphor. As the primary image has now been minified, the image produced by these X-ray photons bears no relationship to the output phosphor image.

The magnification across the intensifier tube is not completely uniform. This has two effects. The first is that the illumination of the screen is not equal, being brighter in the centre than at the periphery (called vignetting). The second is that distortion of the image takes place. The image of a straight wire across the screen will appear curved as it approaches the outside of the intensifier.

4.12. THE VIDICON CAMERA

This is responsible for converting the visual information in the image into electronic form. The two main components (figure 4.21(a)) are a focused electron gun and a specially constructed light sensitive surface. As shown in figure 4.21(b), the light sensitive surface is actually a double layer, the lower of which is the more important. It consists of a photoconductive material, usually antimony trisulphide, but constructed in such a way that very small regions of the photoconductor are insulated one from another by a matrix of mica.

When the camera is directed at visible light, photoelectrons are released from the antimony trisulphide matrix, to be collected by the anode and removed, leaving positive charges trapped there. The amount of charge trapped at any one point is proportional to the light intensity that has fallen on it. Thus the image information has now been encoded in the relative sizes of the positive charges stored at different points in the image plate matrix. These insulated positively charged areas draw a current onto the conductive plate until there is an equal negative charge held there.

The electrons emitted by the cathode are formed into a very narrow beam by the control grid. This beam is attracted towards the fine mesh anode which is at 250 V positive to the cathode. The signal plate has

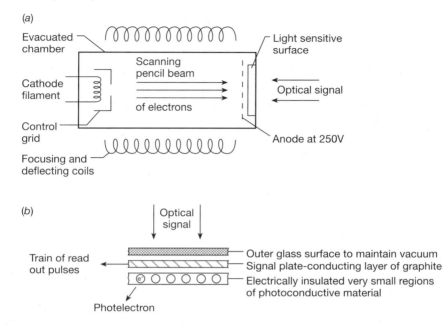

Figure 4.21. *(a) Basic features of the construction of a vidicon camera. (b) The light sensitive surface shown in detail.*

a potential some 225 V less than the anode allowing the photoelectrons to flow from the signal plate to the anode. The electrons from the cathode pass through the mesh and this reversal in the potential field slows them down until they are almost stationary. They have an energy of only 25 eV when they strike the target plate. The field between the anode and the signal plate also straightens out the path of the electron beam so that it is almost at right angles to the signal plate. The electron beam is scanned across the signal plate by the scanning electrodes so that it only interacts with a few of the insulated areas at a time. Electrons flow from the electron beam to neutralize the positive charge on the target. The reduction in the positive charge releases an equivalent charge of electrons from the conducting layer and the flow of electrons from this layer constitutes the video signal from the camera. By strictly controlling where the electron beam is striking the target the location of the area associated with the signal is known and the image data have been coded. As soon as a point on the matrix has been 'read' it is ready to record a new image.

The reverse process of image reconstruction is very similar. The photoconductive material is now replaced by a fluorescent screen and, as the electron beam scans across the screen, its intensity is modulated by the information in the electrical pulses representing the image. The scanning of the electron beams in the camera and the viewing monitor is synchronized.

The following additional points should be noted about a television system:

(1)   Irrespective of the resolving capability of the image intensifier system, the television system will impose its own resolution limit. This is because, whatever the size of the image, the electron beam only executes a fixed number of scan lines (625 in the United Kingdom, equivalent to about 313 line pairs). The effective number of line pairs, for the purpose of determining vertical resolution, is somewhat less than this—probably about 200. Now for a small image, say 5 cm in diameter, this represents four line pairs per mm which is comparable to the resolving capability of an image intensifier (see section 4.11.1). However, any attempt to view a full 230 mm diameter (9 inch) screen would provide only about 0.8 line pairs per mm and would severely limit the resolving capability of the complete system. To display large

fields of view at high resolution it is necessary to use 35 mm cine film, 100 mm spot film or a more modern TV camera with more scan lines.

(2)   Contrast is modified by the use of a TV system. It is reduced by the vidicon camera but increased by the television display monitor, the net result being an overall improvement in contrast (for a discussion of this point see section 5.3.2).

(3)   Rapid changes in brightness seriously affect image quality when using a television system. Thus a photo-cell is incorporated between the image intensifier and the television camera with a feed-back loop to the X-ray generator. If there is a sudden change in brightness as a result of moving to image a different part of the patient, the X-ray output is quickly adjusted to compensate.

For further information on the vidicon camera and TV systems the reader should consult *Christensen's Introduction to the Physics of Diagnostic Radiology* (Curry *et al* 1990).

4.13.   CINEFLUOROGRAPHY

The main image forming components of a cine fluorographic system were shown in figure 4.20(*b*). Not shown in this diagram are the X-ray tube and generator and the electromechanical connection which exists between the cine camera and the X-ray generator.

Cinefluorography may place stringent demands on the X-ray tube. For example, although the cine camera may only need to operate at 10 frames per second to see dynamic movements in the stomach, up to 60 frames per second may be required for coronary angiography. If the X-ray tube emits X-rays continuously during the period the camera operates (probably several seconds), the patient would receive an excessively large dose of radiation and there would be rating problems, especially if a fine focal spot was being used. Both these problems can be reduced by arranging for the X-rays to be pulsed with a fixed relationship between the pulse and the frame movement in the cine camera as shown in figure 4.22. The X-ray pulses are normally between 2 and 5 ms long but the light output from the image intensifier does not fall to zero immediately the X-ray pulse is terminated due to an after-glow in the tube. The after-glow has, however, decreased to zero before the next frame is taken.

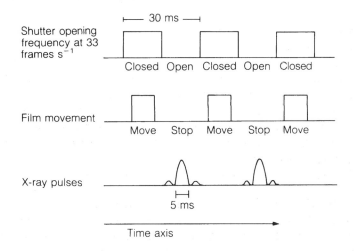

Figure 4.22. *Synchronization of the X-ray pulse with film frame exposure during cinefluorography.*

The dose in air at the front face of the image intensifier required to produce an acceptable image on the film is usually of the order of 0.2 μGy per frame. Modern units operating at 0.1 μGy per frame have been reported to give acceptable results but below 0.15 μGy per frame quantum mottle effects generally start to degrade the image.

The images produced during cinefluorography may suffer from all the artefacts of conventional images (see section 5.10). Depending on the investigation being performed, it is sometimes possible to use a small image intensifier with a correspondingly small field of view. This allows an X-ray tube with an anode angle of only 6° to be used, thus giving a greater output from a smaller effective focal spot. This reduces some of the geometric artefacts. The small field of view also reduces the heel effect.

In addition to the parts of the system responsible for electronic imaging (image intensifier, cine camera), optical imaging processes are also important. Two lens systems are included, the first between the image intensifier output and the half-silvered mirror and the second in front of the cine camera. For a detailed discussion of lens systems the reader is referred elsewhere (e.g. Curry *et al* 1990).

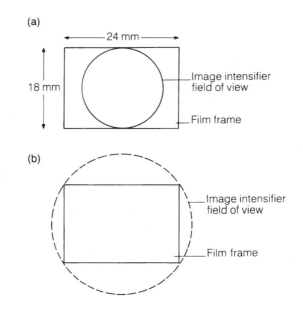

Figure 4.23. *Different framing arrangements with cinefluorography: (a) exact framing, (b) total over-framing.*

In general terms the function of the optical system is to transfer the image, at appropriate magnification, from the image intensifier screen via the mirror on to the film in the camera. Since the image intensifier screen is circular and a frame of film is rectangular, some mismatch is inevitable. Figure 4.23 shows two extreme examples. Figure 4.23(*a*) represents what is known as 'exact framing' with the intensifier screen totally inside the film frame. Figure 4.23(*b*) shows the situation known as 'total over-framing' where the whole of the film frame is within the screen. Clearly total over-framing provides a larger image than exact framing but it is restricted to a smaller field of view. Total over-framing has the further advantage that the area of the patient exposed to radiation can be reduced. The most probable arrangement is intermediate between these two extremes.

The image intensifier uses a caesium iodide input screen (as discussed in section 4.11) and an output phosphor with maximum output in the wavelength range (550 $\pm$ 25) nm. The film used must be sensitive in this wavelength range.

Since cine film has rather a small dynamic range, some form of brightness control is required to compensate for variations in patient thickness. The sensing device, which is analogous to a photo-timer, can either measure the current flowing across the image intensifier or the brightness of the output phosphor. This detector then provides feed-back control of either the tube kilovoltage, the tube current or the effective exposure time per frame (pulse width). Each control mechanism has disadvantages. If the kilovoltage is driven too high contrast is lost, whereas adjustment of the tube current or the pulse width has only a limited brightness range. Variation of pulse width has a much faster response time than either of the other two methods but in practice a combination of controls is normally chosen.

4.14.   SPOT FILMS

If spot films are required, the cine camera is replaced by a single frame film camera using either 70 mm or 100 mm film (see section 4.11.1). As this film is larger than cine film the lens on the front of this camera has a longer focal length than that on the cine camera. The advantages and disadvantages of using spot films for, say, serial angiography can be summarized as follows:

(1)   There is a reduction in procedure time but, because the medical personnel stay close to the patient throughout the investigation, their radiation dose can be greater.
(2)   Production and processing of spot films is easier and the images can be constantly monitored during the study. More experience is required, however, to learn the panning technique.
(3)   Reduced patient exposure occurs because shorter exposure times are used. A consequential secondary effect is that equipment wear is reduced.
(4)   The film is much cheaper and film storage is easier. The smaller film is however not always easy to read.
(5)   The technique cannot be used in some investigations such as peripheral angiography or where the inferior resolution of the image intensifier precludes it.
(6)   The reduction in patient dose may not be as much as anticipated as the radiologist may need longer screening times to compensate for the smaller pictures.

An approximate comparison of patient exposures from the different filming methods is summarized in table 4.4. For more information on dose rates and patient doses see chapter 6.

Table 4.4. *Approximate exposures for various image recording systems.*

Imaging system	Film size	Dose in air per exposure or per frame (μGy)	
		at imager	at the skin surface[a]
Cine	16–35 mm	0.2–0.5[b]	20–50
Spot film	70–100 mm	1–3	100–300
Serial radiography	35 × 35 cm	3–10	300–1000

[a] An attenuation factor of 100 has been assumed between skin dose and film exposure (this is of the right order for the head).
[b] In each instance the smaller number is an approximate lower limit if perceptible image degradation due to quantum mottle (see sections 4.11.2 and 8.6) is to be avoided.

4.15.   QUALITY CONTROL OF RECORDING MEDIA AND IMAGE INTENSIFICATION SYSTEMS

One of the major causes of inferior quality X-ray films is poor processing and strict attention to developing parameters must be paid at all stages of processing. The development temperature must be controlled to better than 0.2 K and the film must be properly agitated to ensure uniform development. All chemicals must be replenished at regular intervals, generally after a given area of film has been processed, and care must be taken to ensure that particulate matter is removed by filtration. Modern processing units monitor the chemicals continuously and automatically replenish them as required. Thorough washing and careful drying are essential if discolourations, streaks and film distortion are to be avoided. The whole process can be controlled by the preparation, at regular intervals, of test film strips obtained using a suitable **sensitometer**. This sensitometer consists of a graded set of filters of known optical density (figure 4.24) which is placed over the film and exposed to a known amount of visible light, in a darkened room of course. After processing, the film can be densitometered and a characteristic curve can be constructed (see figure 4.8). This may be compared with previous curves and the manufacturer's recommendations. Since the construction of complete characteristic curves is a rather time-consuming process, it is normal only to make three or four spot checks routinely. These can then be plotted on a graph and if they stay within ±0.1–0.15 of the base-line density when first set up and adjusted they can be considered acceptable.

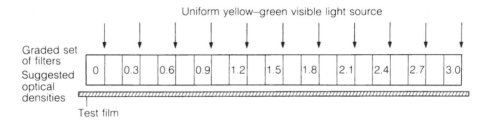

Figure 4.24. *Use of a sensitometer, consisting of a graded set of filters of known optical density, for quality control of film processing.*

If sensitometry is used for quality control of the intensifying screen and film in combination, there are a number of additional problems. For example, it is not possible to use a wide range of timed exposures to X-rays to produce different film densities because of failure of the reciprocity law. Perhaps the best method is to use the inverse square law to vary the X-ray intensity reaching the screen, exposing at different distances and plotting film density versus exposure. Note the need for very reproducible exposure times and a very steady tube output. A slightly less accurate method is to use a step wedge of aluminium or copper filters to provide a range of X-ray intensities. If this method is chosen, fairly heavy filtration must be placed in front of the step wedge to minimize change in beam quality with additional filtration, and lead masks must be used to reduce scatter from the step wedge. As a third possibility, a simulated light sensitometer and a range of neutral density filters may be used, provided that the light spectrum is the same as that from the intensifier screen.

Assessment of the performance of image intensification systems introduces a number of problems in addition to those considered in section 2.7 for standard X-ray sets. Among the more important measurements that should be made are the following.

(i)   *Field size.* A check that the field of view seen on the television monitor is as big as that specified by the manufacturer should be made. The area of the patient exposed to radiation must not exceed that required for effective screening.

(ii) *Image distortion.* The heavy dependence on electronic focusing in the image intensifier and TV system makes distortion much more likely than in a simple radiograph. Measurements on the image of a rectangular grid permit distortion to be checked.

(iii) *Conversion factor.* This is the light output per unit exposure rate, measured in candela per square metre for each μGy s^{-1}. In general this will be markedly lower for older intensifiers, partly due to ageing of the phosphor and partly because better phosphors are being used in the more modern intensifiers. There is little evidence that a poor conversion factor affects image quality directly, but of course to achieve the same light output a bigger radiation dose must be given to the patient. This test is not easily carried out and an alternative indication that a fall in conversion factor is taking place is to measure the input dose rate to the intensifier using a plate ionization chamber and a standard, uniform phantom.

(iv) *Contrast capability and resolution.* A number of test objects, for example the Leeds Test Objects (Hay *et al* 1985), have been devised to facilitate such measurements. One of them which is used to estimate noise in the image consists of a set of discs, each approximately 1 cm in diameter with a range of contrasts from 16% to 0.7% (for a definition of contrast in these terms see section 8.5). They should be imaged at a specified kVp and with a specified amount of filtration (typically 70 kVp and 1 mm Cu to simulate the patient). At a dose rate of about 0.3 μGy s^{-1}, a contrast difference of 2–4% (certainly better than 5%) should be detectable. It is important to specify the input dose rate because, as shown in figure 4.25, the minimum perceptible contrast difference is higher for both sub-optimum and supra-optimum dose rates. When the dose rate is too low this occurs because of quantum mottle effects. When it is too high there is loss of contrast because the video output voltage begins to saturate.

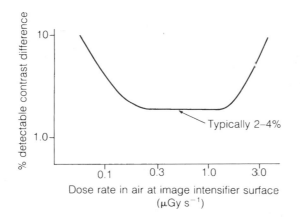

Figure 4.25. *Variation of minimum perceptible contrast with dose rate for an image intensifier screening system.*

The limiting resolution is measured by using a test pattern consisting of line pairs, in groups of three or four, separated by different distances, and viewing the TV monitor for the minimum resolvable separation. Under high contrast conditions (50 kVp), without copper filtration, at least 1.2 line pairs per mm should be resolvable for a 25 cm field of view. Note that the contrast conditions must be controlled carefully since, in common with other imaging systems, the finest resolvable detail varies with contrast level for the image intensifier (see section 8.4.3).

(v) *Automatic brightness control.* It is important to ensure that screening procedures do not result in unacceptably high dose rates to the patient. One way to achieve this is by monitoring the light output from the image intensifier, changing the X-ray tube output (kV and/or mA) by means of a feed-back loop, whenever attenuation in the patient changes. The way in which these are changed can affect the dose received by the patient. Increasing the kVp before the mA will produce a lower patient dose but will reduce contrast. There is also a point of diminishing return where the energy of the peak in the intensity starts to move beyond the K-edge of the input phosphor, and/or contrast medium if being used, and the attenuation coefficient starts to fall rapidly.

The following additional points should be noted:

(a) The light output from an appreciable area of the image intensifier screen, as predetermined by the manufacturer, must be measured.

(b) The brightness of the final image can also be controlled by a technique known as automatic gain control, which uses the video signal to adjust amplification factors in the TV monitor without altering exposure factors or dose rate to the patient. This should not be used instead of automatic brightness control.

(c) Automatic control may mask deterioration in performance somewhere in the system. For example loss of light output from the image intensifier screen could be compensated by increasing kV and/or mA. X-ray output should therefore be checked directly.

(d) Automatic brightness control should be distinguished from automatic exposure control (see section 2.3.6). The latter term is generally used to refer to 'hard copy' control. For example, when using 100 mm film the control might be a photo-cell looking at the image intensifier output.

Most systems should be capable of operating in the range $0.2–1$ μGy s^{-1} but dose rates as high as 5 μGy s^{-1} have been reported for incorrectly adjusted equipment! Note that the skin dose to the patient will be typically 100 times greater than this—about 200 μGy s^{-1} or 30 mGy min^{-1} (see table 4.4).

(vi) *Viewing screen performance.* A calibrated grey-scale step wedge may be used to check that the contrast and brightness settings on the television monitor are correctly adjusted.

For further information on quality control see 'Recommended standards for routine performance testing of diagnostic X-ray imaging systems' (IPEM 1997). A good general rule is that the more sophisticated the unit, the greater the care that must be taken over quality control of the images. In many situations a slow, but steady, deterioration in picture quality may take place over a period of weeks or months and this can be difficult to detect unless quantitative measurements are made on a regular basis.

4.16.   SUMMARY

A primary X-ray photon image cannot be viewed directly by the human eye and in clinical practice nearly all imaging systems convert this X-ray photon image to a light photon image using one of the many phosphors now available. This light photon image should not be viewed directly by the human eye and, indeed, on modern X-ray units this undesirable practice is no longer possible.

In general, the light photon image is either recorded on film or is viewed through an image intensifier/TV viewing system and both of these media have been considered in detail in this chapter. Other forms of recording medium are also used, especially in diagnostic imaging techniques other than plain film radiology. Table 4.5 summarizes the options available. Some of the other methods for recording images are considered in detail elsewhere in the book.

Use of a phosphor allows maximum transfer of image information with minimum radiation dose to the patient. In all viewing systems the spectral output of the phosphor must be closely matched to the next stage of the system. Quantum mottle effects do however impose a lower limit below which further dose reduction produces an image of unacceptable quality. This will occur even if the overall light or density levels are nominally satisfactory.

Each imaging process has its own limits on resolution and contrast and these will be discussed in greater detail in chapters 5 and 8. However, it is important to emphasize here that the inter-relationship between object size, contrast and patient dose is a matter of everyday experience and not a peculiarity of quality control measurements or sophisticated digital techniques. Furthermore, the choice of imaging process cannot always be governed solely by resolution and contrast considerations. The speed at which the image is produced and is made available for display must sometimes be taken into account.

Imaging equipment using optical and electronic imaging systems must be carefully maintained and all imaging systems should be subject to quality control. With many phosphors it is possible for a slow

Table 4.5. *Summary of information on different recording media.*

Recording medium	Use	Special features
Duplitized X-ray film	General purpose radiography	Conventional double sided X-ray film used with a pair of intensifying screens
Single emulsion film (screen type)	Mammography (sec 9.6)	Less sensitive but improved resolution
	High definition extremity film	Anti-halo backing prevents light reflection at film base–air interface
Non-screen film	Dental Kidney surgery Radiation monitoring	Direct exposure film for ultra-high resolution Greatly reduced sensitivity
Single emulsion film	Photo-fluorography	Spot film/rapid sequence camera film—cut and roll Film sensitive to green light—maximum output of image intensifier screen is 500–600 nm
Film for video imaging	Cathode ray tube imaging Radionuclide imaging (ch 7) Digital subtraction imaging (ch 9) CT (ch 10) Ultrasound (ch 13) MRI (ch 14)	Single emulsion negative film generally best for good resolution and a wide density range. Hard copy images from a laser printer are an alternative but generally of poorer quality
Subtraction–duplication film	1. Subtraction masks (sec 9.5) 2. Print film 3. Duplication	1. Gamma must be 1.0 2. Gamma generally greater than 1.0 to increase contrast 3. Gamma of -1 0
Miscellaneous	Miniaturization Microfilm Video tape	Information reproduced in a smaller format may require special fine grain development to retain resolution Usually panchromatic for maximum grey scale range Film speed is not a problem—no patient exposure involved
Special techniques	Xeroradiography (sec 9.7)	

degradation of the final image to occur. This degradation may only be perceptible if images of a test object taken at regular intervals are carefully compared, preferably using quantitative methods.

REFERENCES

Curry T S, Dowdey J E and Murry R C Jr 1990 *Christensen's Introduction to the Physics of Diagnostic Radiology* 4th edn (Philadelphia, PA: Lea and Febiger)

Hay G A, Clark O F, Coleman N J and Cowen A R 1985 A set of X-ray test objects for quality control in television fluoroscopy *Br. J. Radiol.* **58** 335–44

Institute of Physics and Engineering in Medicine 1997 Recommended standards for routine performance testing of diagnostic X-ray imaging systems *IPEM Report* **77**

FURTHER READING

British Journal of Radiology (BJR) 1988 Assurance of quality in the diagnostic X-ray department (London: British Institute of Radiology)

Dressler G, Eriskat H, Schibilla H, Haybittle J L and Secretan L F (ed) 1985 Criteria and methods of quality assurance in medical X-ray imaging *Br. J. Radiol. Suppl.* **18**

Farr R F and Allisy-Roberts P J 1997 *Physics for Medical Imaging* (Phildelphia, PA: Saunders)

Hospital Physicists Association 1979 *Quality Assurance Measurements in Diagnostic Radiology (Conference Report Series 29)* (London: The Hospital Physicists 'Association)

Institute of Physics and Engineering in Medicine 1995 Measurement of the performance characteristics of diagnostic X-ray systems used in medicine. Part VI 'X-ray image intensifier fluorography systems' *IPEMB Report* **32**

Institute of Physics and Engineering in Medicine 1996 Measurement of the performance characteristics of diagnostic X-ray systems used in medicine. Part II 'X-ray image intensifier television systems' *IPEMB Report* **32**

Kodak 1985 *Fundamentals of Radiographic Photography* **1–6** (London: Kodak)

EXERCISES

1 Discuss the use of intensifying screens in the cassettes used for radiography.

2 Explain briefly the effect of increasing the kVp from 50 to 100 on the intensification factor of calcium tungstate screens.

3 Explain how the intensification factors of a set of radiography screens might be compared. Summarize and give reasons for the main precautions that must be taken in the use of such screens.

4 Draw a labelled cross section of an X-ray photographic film. What features make for high sensitivity?

5 Draw on the same axes the characteristic curves for (*a*) a fast film held between a pair of rare earth screens, (*b*) the same film with no screen, and explain the difference between them.

6 Why is it desirable for the gamma of a radiographic film to be much higher than that of a film used in conventional photography and how is this achieved?

7 Explain what is meant by the speed of an X-ray film and discuss the factors on which the speed depends.

8 A radiograph is found to lack contrast. Under what circumstances would increasing the current on the repeat radiograph increase contrast, and why?

9 Make a labelled diagram of the intensifying screen film system used in radiology. Discuss the physical processes that occur from the emergence of X-rays at the anode to the production of the final radiograph.

10 Given that the gamma of an idealized radiographic film–screen combination is 3.5, and the range of acceptable film densities is 2.8, what is the maximum ratio of exposures for which the combination can be used?

11 Discuss the factors which affect the sensitivity and resolution of a screen–film combination used in radiography and their dependence upon each other.

12 How does the difference in diameter of the input and output screens of an X-ray intensifier contribute to the performance of the system?

13 What advantages are associated with a caesium iodide input phosphor in an image intensifier?

14 Discuss the uses made of the brightness amplification available from a modern image intensifier, paying particular attention to any limitations.

15 Compare and contrast the use of fluorescent screens in radiography and fluoroscopy.

16 Discuss the limitations 'quantum mottle' imposes on both image intensifying systems and sensitive film–screen systems. Why are these limits not always reached?

17 Explain how an image intensifier may be used in conjunction with a photoconductive camera to produce an image on a TV screen.

18 Explain what is meant by cinefluorography and give a brief description of the functions of the various parts of such a system.

CHAPTER 5

THE RADIOLOGICAL IMAGE

5.1. INTRODUCTION—THE MEANING OF IMAGE QUALITY

As shown in chapter 3, the fraction of the incident X-ray intensity transmitted by different parts of a patient will vary due to variations in thickness, density and mean atomic number of the body. This pattern of transmitted intensities therefore contains the information required about the body and can be thought of as the primary image.

However, this primary image cannot be seen by the eye and must first be converted to a visual image by interaction with a secondary imaging device. This change can be undertaken in several ways, each of which has its own particular features. The definition of quality for the resultant image in practical terms depends on the information required from it. In some instances it is resolution that is primarily required, in others the ability to see small increments in contrast. More generally the image is a compromise combination of the two, with the dominant one often determined by the personal preference of the radiologist. (This preference can change; the 'contrasty' crisp chest radiographs of several years ago are now rejected in favour of lower contrast radiographs which appear much flatter but are claimed to allow more to be seen.)

The quality of the image can depend as much on the display system as on the way it was produced. A good quality image viewed under poor conditions such as inadequate non-uniform lighting may be useless. The quality actually required in an image may also depend on information provided by other diagnostic techniques or previous radiographs.

This chapter extends the concept of contrast introduced in chapter 4 to the radiological image and then discusses the factors that may influence or degrade the quality of the primary image. Methods available for improving the quality of the information available at this stage are also considered. Other factors affecting image quality in the broader context are discussed in chapter 8.

5.2. THE PRIMARY IMAGE

The primary image produced when X-ray photons pass through a body depends on the linear attenuation coefficient (μ) and the thickness of the tissue they traverse. At diagnostic energies μ is dependent on the photoelectric and Compton effects. For soft tissue, fat and muscle the effective atomic number varies from approximately 6 to 7.5. In these materials μ is thus primarily dependent on the Compton effect, which falls relatively slowly with increasing photon energy (see chapter 3). The photoelectric effect is not completely absent however and at low photon energies forms a significant part of the attenuation process. In mammography low energy X-ray photons are used to detect malignant soft tissue which has a very similar Z value to breast tissue. The difference in attenuation between the two is due to the higher photoelectric attenuation of the higher Z material. For bone with a Z of approximately 14, most of the attenuation is by the photoelectric effect. This falls rapidly with increasing photon energy.

In absolute terms the Compton effect also decreases with increasing energy and the resultant fall in attenuation coefficient with photon energy for tissue and for bone is shown in figure 5.1.

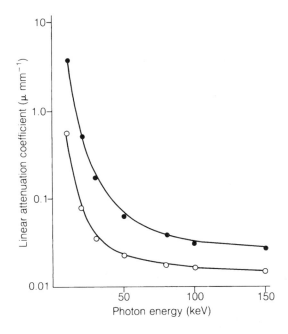

Figure 5.1. *Variation of linear attenuation coefficient with photon energy for (○) muscle and (●) bone in the diagnostic region.*

5.3. CONTRAST

The definition of contrast differs somewhat depending on the way the concept is being applied. For conventional radiology and fluoroscopy, the normal definition is an extension of the definition introduced in chapter 4. In chapter 8 an alternative approach will be considered.

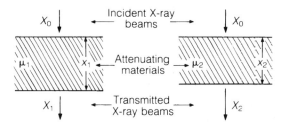

Figure 5.2. *X-ray transmission through materials that differ in both thickness and linear attenuation coefficient.*

Consider the situation shown in figure 5.2. This is clearly very similar to that in figure 4.6. Contrast in the primary image will be due to any difference between X_1 and X_2 and by analogy with equation (4.1) may be defined as:

$$C = \log_{10} \frac{X_2}{X_1}.$$

Converting to Naperian logs:

$$C = 0.43 \ln \frac{X_2}{X_1} = 0.43(\ln X_2 - \ln X_1).$$

Since, from chapter 3

$$X_1 = X_0 \exp(-\mu_1 x_1)$$

and

$$X_2 = X_0 \exp(-\mu_2 x_2)$$

thus

$$C = 0.43(\mu_1 x_1 - \mu_2 x_2).$$

If μ_1 and μ_2 were the same, the difference in contrast would be due to differences in thickness. If $x_1 = x_2$ the contrast is due to differences in linear attenuation coefficient. It is conceivable that the product $\mu_1 x_1$ might be exactly equal to $\mu_2 x_2$ but this is unlikely. Note from figure 5.1 that the difference in μ values decreases on moving to the right, thus contrast between two structures always decreases with increasing kVp.

5.3.1. Contrast on a fluorescent screen

If this primary image is allowed to fall on a fluorescent screen, the light emitted from those parts of the screen exposed to X_1 and X_2, say, L_1 and L_2, will be directly proportional to X_1 and X_2. Hence

$$L_1 = kX_1 \text{ and } L_2 = kX_2.$$

The contrast

$$C(\text{screen}) = \log \frac{L_2}{L_1} = \log \frac{kX_2}{kX_1} = \log \frac{X_2}{X_1}.$$

Hence the contrast on the screen, $C = 0.43(\mu_1 x_1 - \mu_2 x_2)$, is exactly the same as in the primary image. As this is how the eye perceives the image, this is known as the radiation contrast and will be denoted by C_R.

Note that simple amplification, in a fluorescent or intensifying screen, does not alter contrast.

5.3.2. Contrast on a radiograph

If the transmitted intensities X_1 and X_2 are converted into an image on radiographic film, the contrast on the film will be different from that in the primary image because of the imaging characteristics of the film.

As shown in section 4.4 the imaging characteristics of film are described by its characteristic curve. By definition

$$\gamma = \frac{D_2 - D_1}{\log E_2 - \log E_1}.$$

So for ionizing radiation

$$\gamma = \frac{D_2 - D_1}{\log X_2 - \log X_1}. \tag{5.1}$$

Now from section 5.3

$$\log X_2 - \log X_1 = 0.43(\mu_1 x_1 - \mu_2 x_2)$$

and from section 4.4.2

$$D_2 - D_1 = C.$$

Hence substituting in equation (5.1)

$$C = \gamma 0.43(\mu_1 x_1 - \mu_2 x_2).$$

Thus the contrast on film C_F differs from the contrast in the primary image by the factor γ which is usually in the range 3–4. Gamma is often termed the film contrast, thus

radiographic contrast   =   radiation contrast   ×   film contrast
C_F C_R γ

Note that contrast is now modified because the characteristic curve relates two logarithmic quantities. Film can be said to be a '**logarithmic amplifier**'. A TV camera can also act as a logarithmic amplifier.

The output on most digital fluoroscopy systems is changed from a linear/linear system to one using a logarithmic scale where increasing the log of the relative exposure by a set value will increase the grey level by an equal amount within the white and black limits of the defined grey scale.

5.3.3.   *Origins of contrast for real and artificial media*

As discussed in chapter 3, attenuation and hence contrast will be determined by differences in atomic number and density. Typical values for normal tissues are shown in table 5.1.

Table 5.1. *Mean atomic number and density for the major body constituents.*

Material	Mean atomic number Z	Density (kg m^{-3} × 10^3)
Bone	13.8	1.8
Soft tissue } Muscle }	7.4	1.0
Fat	6.0	0.9
Lung	7.4	0.24
Air	7.6	Almost 0

Any agent introduced into the tissues, globally or selectively, in order to modify contrast may be termed a 'contrast enhancing agent'. Contrast may be changed artificially by introducing materials with either a different atomic number or with a different density and enhancement may be positive (more attenuation than other regions) or negative (less attenuation).

The physical principles of positive contrast enhancement have changed little since the earliest days. Iodine ($Z = 53$) is the obvious element to choose for contrast enhancement. Its K shell binding energy is approximately 34 keV so its cross section for a photoelectric interaction with X-ray photons in the diagnostic energy range is high. Barium compounds ($Z = 56$) with a similar K shell energy are used for studies of the stomach and colon.

Negative contrast may be created by the introduction of gas, e.g. CO_2 in the bowel in double contrast studies. Note that modification of atomic number is very kVp dependent whereas modification of density is not.

The number of 'contrast' materials available is limited by the requirements for such materials. They must have a suitable viscosity and persistence and must be miscible or immiscible as the examination requires. Most importantly they must be non-toxic. One contrast material, used for many years in continental Europe, contained thorium which is a naturally radioactive substance. It has been shown in epidemiological studies that patients investigated using this contrast material had an increased chance of contracting cancer or leukaemia.

Iodine based compounds carry a risk for some individuals and most developments in the past 60 years have been directed towards newer agents with lower toxicity (Dawson 1992). These have included achieving the same contrast at reduced osmolarity by increasing the number of iodine atoms per molecule and reducing protein binding capacity by attaching electrophilic side chains.

A recent possibility has been the production of non-ionic dimers which provide an even higher number of iodine atoms per molecule. These rather large molecules impart a high viscosity to the fluid but this can be substantially overcome by warming the fluid to body temperature prior to injection.

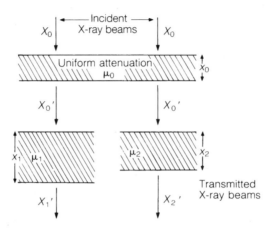

Figure 5.3. *Diagram showing the effect of an overlying layer of uniformly attenuating material on the X-ray transmitted beam of figure 5.2.*

5.4. EFFECTS OF OVERLYING AND UNDERLYING TISSUE

Under scatter-free conditions, it may be demonstrated that a layer of uniformly attenuating material either above or below the region of differential attenuation has no effect on contrast. Consider the situation shown in figure 5.3 where the two regions are shown separated for clarity.

Radiation contrast

$$C_R = \log_{10} \frac{X'_2}{X'_1} = 0.43 \ln \frac{X'_2}{X'_1}.$$

Now

$$X'_1 = X'_0 e^{-\mu_1 x_1}$$

and

$$X'_2 = X'_0 e^{-\mu_2 x_2}.$$

Hence

$$C_R = 0.43 \ln \frac{X'_0 e^{-\mu_2 x_2}}{X'_0 e^{-\mu_1 x_1}}.$$

Thus

$$C_R = 0.43(\mu_1 x_1 - \mu_2 x_2)$$

as before.

The fact that some attenuation has occurred in overlying tissue and that $X'_0 = X_0 e^{-\mu_0 x_0}$ is irrelevant because X'_0 cancels. A similar argument may be applied to uniformly attenuating material below the region of interest.

An alternative way to state this result is that under these idealized conditions logarithmic transformation ensures that equal absorber and/or thickness changes will result in approximately equal contrast changes whether in thick or thin parts of the body.

5.5. REDUCTION OF CONTRAST BY SCATTER

In practice, contrast is reduced by the presence of overlying and underlying material because of scatter. The scattered photons arise from Compton interactions. They are of reduced energy and travel at various angles to the primary beam.

The effect of this scatter, which is almost isotropic, is to produce a uniform increase in blackening across the film. It may be shown, quite simply, that the presence of scattered radiation of uniform intensity invariably reduces the radiation contrast.

If $C_R = \log_{10}(X_2/X_1)$ and a constant X_0 is added to the top and bottom of the equation to represent scatter,

$$C'_R = \log_{10} \frac{X_2 + X_0}{X_1 + X_0}.$$

The value of C'_R will be less than the value of C_R for any positive value of X_0.

The presence of scatter will almost invariably reduce contrast in the final image for the reason given above. The only condition under which scatter might increase contrast would be for photographic film if X_1 and X_2 were so small that they were close to the fog level of the characteristic curve. This is a rather artificial situation.

The amount of scattered radiation can be very large relative to the unscattered transmitted beam. This is especially true when there is a large thickness of tissue between the organ or object being imaged and the film. The ratio of scatter to primary beam in the latter situation can be as high as eight to one but is more generally in the range between two and four to one (see figure 5.4).

5.6. VARIATION IN SCATTER WITH PHOTON ENERGY

If it is necessary to increase the kVp to compensate for loss of intensity due to lack of penetration or to try to reduce the radiation dose to the patient, the amount of scatter reaching the film increases. This is the result of a complex interaction of factors, some of which increase the scatter, others decrease it.

(1)   The amount of scatter actually produced in the patient is reduced because:
      (a) the probability that an individual photon will be scattered decreases as the kVp is increased, although the Compton interaction coefficient only decreases slowly in the diagnostic range;
      (b) a smaller amount of primary radiation is required to produce a given density on the film as film density is proportional to $(kVp)^4$.
(2)   However, the forward scatter leaving the patient will be increased because:
      (a) the fraction of the total scatter produced going in a forward direction increases as the kVp rises;
      (b) the mean energy of the scattered radiation increases and thus less of it is absorbed by the patient.

In practice, as the kVp rises from 50 to 100 kVp, the fall in linear attenuation coefficient of the low energy scattered radiation in tissue is much more rapid than the fall in the scatter producing Compton cross section. Thus factor (2(b)) is more important than factor (1(a)) and this is the prime reason for the increase in scatter reaching the film.

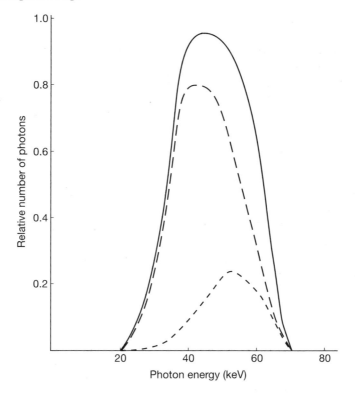

Figure 5.4. *Typical primary, scatter and total spectra when a body-sized object is radiographed at 70 kV:* —————— *total;* — — — *scatter;* – – – – *primary (only the continuous spectrum is shown).*

The increase in scatter is steep between 50 and 100 kVp but there is little further increase at higher kVp and above 140 kVp the amount of scatter reaching the film does start to fall slowly.

5.7.   REDUCTION OF SCATTER

There are several ways in which scatter can be reduced.

5.7.1.   *Careful choice of beam parameters*

A reduction in the size of the beam to the minimum required to cover the area of interest reduces the volume of tissue available to scatter X-ray photons.

A reduction of kVp will not only increase contrast but will also reduce the scatter reaching the film. This reduction is however limited by the patient penetration required and, perhaps more importantly, an increase in patient dose due to the increase in mA s required to compensate for the reduction in kVp. (Whilst an additional 10 kVp allows the mA s to be reduced by approximately a half, a decrease of 10 kVp would require a doubling of the mA s for the same film blackening.)

5.7.2.   Orientation of the patient

The effect of scatter will be particularly bad when there is a large thickness of tissue between the region of interest and the film. When the object is close to the film it prevents both the primary beam and scatter reaching the film. The object stops scatter very effectively since the energy of these photons is lower than those in the primary beam. Thus the region of interest should be as close to the film as possible—see also section 5.10 on geometric effects. In practice other requirements of the radiograph generally dictate the patient orientation.

5.7.3.   Compression of the patient

This is a well known technique that requires some explanation. It is important to appreciate that the process a physicist would call compression, for example a piston compressing a volume of gas, will not reduce scatter. Reference to figure 5.5 will show that the X-ray photons encounter exactly the same number of molecules in passing through the gas on the right as on the left. Hence there will be the same amount of attenuation and the same amount of scatter.

When a patient is 'compressed', soft tissue is actually forced out of the primary beam, hence there is less scattering material present and contrast is improved.

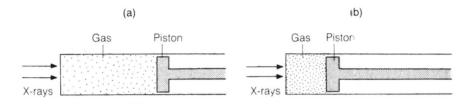

Figure 5.5. *Demonstration that compression in the physical sense will not alter the attenuating properties of a fixed mass of gas: (a) gas occupies a large volume at low density; (b) gas occupies a much smaller volume at a higher density.*

5.7.4.   Use of grids

This is the most effective method for preventing the scatter leaving the patient from reaching the film and is discussed in the next section.

5.7.5.   Air gap technique

If the patient is separated from the film, some obliquely scattered rays miss the film. This technique is discussed more fully in section 9.2.

5.7.6.   Design of intensifying screen and film holder

Since some radiation will pass right through the film, it is important to ensure that no X-ray photons are back-scattered from the film holder. A high atomic number metal backing to the film cassette will ensure that all transmitted photons are totally absorbed by the photoelectric effect at this point. With daylight loading systems and a movement towards a reduction in weight of the cassettes this high Z backing has tended to be reduced or discarded. However storage phosphor systems (see section 4.9) incorporate a lead screen to minimize the effects of backscatter.

The slightly greater sensitivity of intensifying screens to the higher energy primary photons may be only of marginal benefit in reducing the effect of scatter on the film if the position of the K edge of the screen phosphor is well below the peak photon energy in the respective intensity spectrum (see figures 3.10 and 5.4).

5.8.   GRIDS

5.8.1.   Construction

The simplest grid is an array of long parallel lead strips held an equal distance apart by a material with a very low Z value (an X-ray translucent material). Most of the scattered photons, travelling at an angle to the primary beam, will not be able to pass through the grid but will be intercepted and absorbed as shown in figure 5.6. Some of the rays travelling at right angles, or nearly right angles, to the grid are also stopped due to the finite thickness of the grid strips, again shown in figure 5.6. Both the primary beam and the scatter are stopped in this way but the majority of the primary beam passes through the grid along with some scattered radiation. Scatter travelling at an angle of $\theta/2$ or less to the primary beam is able to pass through the grid to point P (figure 5.7).

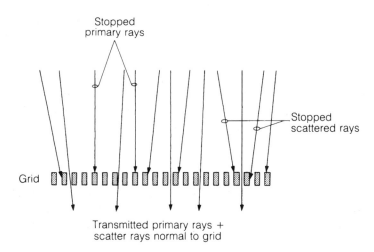

Figure 5.6. *Use of a simple parallel grid to intercept scattered radiation.*

As grids can remove up to 90% of the scatter there is a large increase in the contrast in radiographs when a grid is used. This increase is expressed in the 'contrast improvement factor' K where

$$K = \frac{\text{X-ray contrast with grid}}{\text{X-ray contrast without grid}}.$$

K normally varies between 2 and 3 but can be as high as 4. The higher values of K are normally achieved by increasing the number of grid strips per centimetre. As these are increased, more of the primary beam is removed due to its being stopped by the grid. The proportion stopped is given by

$$\frac{d}{D+d}$$

where d is the thickness of a lead strip and D is the distance between them (figure 5.7). Reduction of the primary beam intensity means that the exposure must be increased to compensate. The use of a grid therefore

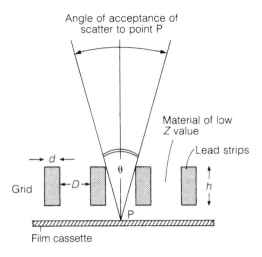

Angle of acceptance of
scatter to point P

Material of low
Z value

Lead strips

Grid

Film cassette

Figure 5.7. *Grid geometry. Number of strips per mm,* $N = 1/(D + d)$; *typically N is about 4 for a good grid. Grid ratio* $r = h/D$; *typically r would be 10 or 12. Fraction of primary beam removed from the beam is* $d/(D+d)$. *Since d might be 0.075 mm and* $(D+d)$ *0.25 mm,* $d/(D+d)$ *will be about 0.3. Tan* $\theta/2 = D/h$.

increases the radiation dose to the patient and thus there is a limit on the number of strips per centimetre that can be used. Additionally the interspace material will also absorb some of the primary beam. Parallel grids remove approximately 30% of the primary radiation but a crossed grid (see section 5.8.2) can remove up to 50% thus requiring the exposure of the patient to be doubled.

5.8.2. Use of grids

As grids are designed to stop photons travelling at angles other than approximately normal to them, it is essential that they are always correctly positioned with respect to the central ray of the primary beam. Otherwise, as shown in figure 5.8, the primary beam will be stopped. The fact that the primary photon beam is not parallel

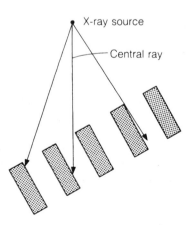

X-ray source

Central ray

Figure 5.8. *Diagram showing that a grid which is not orthogonal to the central X-ray axis may obstruct the primary beam.*

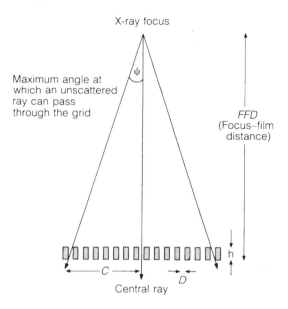

Figure 5.9. *Demonstration that the field of view is limited when using a simple linear grid.*

but originates from a point source limits the size of film that can be exposed due to interception of the primary beam by the grid (figure 5.9). The limit is at ray C where

$$\tan \psi = \frac{C}{\text{FFD}}.$$

Tan ψ can be calculated from the grid characteristics. By similar triangles,

$$\tan \psi = \frac{D}{h}$$

where D is the distance between strips and h is the height of a strip.

Grids are identified by two factors: the grid ratio, which is defined as h/D (from figure 5.7) and the number of strips per centimetre. In practice the two are interdependent. This is because there is an optimum thickness for the strips as they cannot be reduced in thickness to allow more strips per centimetre without reducing their ability to absorb the scattered radiation and thus their efficiency. Decreasing the gap between the strips to increase the number of strips will change the grid ratio unless the height h of the strip is reduced. The grid then becomes too thin to be of any use. The grid ratio varies between 8:1 and 16:1. A ratio of 8:1 is only likely to be used for exposures at kilovoltages of less than approximately 85 kVp. For exposures at higher kVps the choice is between a 10:1 and 12:1 grid. Most departments probably choose the 12:1 grid because although it results in a higher dose to the patient the improvement in contrast for thick sections is thought to be justified. Grid ratios as high as 16:1 should not be used except under exceptional circumstances as the increase in contrast is rarely justified by the increase in dose to the patient.

The grids shown in previous figures are termed linear grids and should be used with the long axis of the grid parallel to the cathode anode axis of the tube so that angled x-radiographs can be taken without the primary beam hitting the lead strips.

The simple linear grid has now been largely replaced by the focused linear grid where the lead strips are progressively angled on moving away from the central axis (figure 5.10). This eliminates the problem

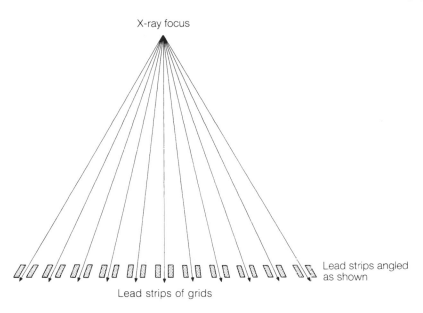

X-ray focus

Lead strips angled
as shown

Lead strips of grids

Figure 5.10. *Construction of a linear focused grid.*

of cut-off at the periphery of the grid but imposes restrictive conditions on the focal film distances (FFDs) that can be used, the centring of the grid under the focal spot and having the correct side of the grid towards the X-ray tube. If any of these is wrong the primary beam is attenuated. If the grid is upside down then a narrow exposed area will be seen on the film with very little blackening on either side of it. Decentring tends to produce generally lighter films which become lighter as the amount of decentring is increased. Using the wrong FFD will not affect the central portion of the radiograph but will progressively increase cut-off at the edge of the film as the distance away from the correct FFD is increased.

Crossed grids with two sets of strips at right angles to each other are also sometimes used. This combination is very effective for removing scatter but absorbs a lot more of the primary beam and requires a much larger increase in the exposure with consequent increase in patient dose.

5.8.3. *Movement of grids*

If a stationary grid is used it imposes on the image a radiograph of the grid as a series of lines (due to the absorption of the primary beam). With modern fine grids this effect is reduced but not removed.

The effect can be overcome by moving the grid during exposure so that the image of the grid is blurred out. Movements on modern units are generally oscillatory, often with the speed of movement in the forward direction different to that on the return. Whatever the detailed design, the movement should be such that it starts before the exposure and continues beyond the end of the exposure. Care must also be taken to ensure that, in single phase machines, the grid movement is not synchronous with the pulses of X-rays from the tube. If this occurs, although the grid has moved between X-ray pulses, the movement may be equal to an exact number of lead strips. The lead strips in the grid are thus effectively in the same position as far as the radiograph is concerned. This is an excellent example of the stroboscopic effect. Medium frequency machines are, of course, not troubled by this effect. One disadvantage of the focused over the simple linear grid is that the decentring of the grid during movement results in a greater absorption of the primary beam.

5.9. RESOLUTION AND UNSHARPNESS

The resolution of a radiological image depends on factors associated with different parts of the imaging system. The most important are geometric unsharpness, patient unsharpness and the resolution of the final imager. For the present discussion, resolution and unsharpness are considered synonymous—a reasonable assumption in most cases. For a more detailed discussion of the relationship between resolution and unsharpness see section 8.8.

An ultimate limit on resolution is provided by the inherent resolution of the image recording system. As discussed in section 4.8 the resolution of film is much better than even the best film–screen combination. The resolution of a fluorescent screen used in fluorography is about 0.25 mm (four line pairs per mm); that of the most up-to-date image intensifier/television systems is also about 0.25 mm. In CT scanning, resolution is limited by the size of the pixels used on the screen.

Except on rare occasions in the case of fluoroscopic screens, the imaging devices used in radiography are not the major cause of loss of resolution. There are various other sources of unsharpness of which the more important are geometric unsharpness and inherent patient unsharpness.

5.9.1. *Geometric unsharpness*

Geometric unsharpness is produced because the focal spot of an X-ray tube has finite size (figure 5.11). Although the focal spot has a dimension, b, on the anode, the apparent size of the focal spot for the central X-ray beam, a, is much reduced due to the slope of the anode. The dimension normal to the plane of the paper is not altered. If, as shown in figure 5.11, a sharp X-ray opaque edge is placed directly under the centre of

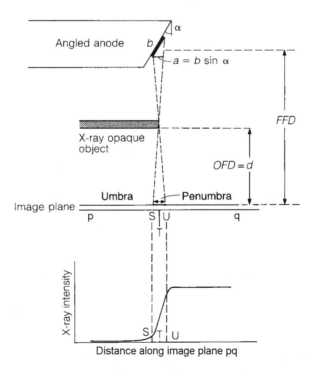

Figure 5.11. *The effect of a finite X-ray focal spot size in forming a penumbral region. FFD = focus–film distance; OFD = object–film distance.*

the focal spot, the image of the edge is not produced directly underneath at T but extends from S to U where S is to the left of T and U to the right of T. By analogy with optics, the shadow of the object to the left of S is termed the **umbra**; the region SU is the **penumbra**. This penumbra, in which on moving from S to U the number of X-ray photons rises to that in the unobstructed beam, is termed the **geometric unsharpness**. The magnitude of SU is given by

$$\text{SU} = b \sin \alpha \frac{d}{(\text{FFD} - d)}. \tag{5.2}$$

If target angle $\alpha = 13°$, $b = 1.2$ mm, FFD = 1 m and $d = 10$ cm then SU = 0.06 mm. Since the focal spot size is strictly limited by rating considerations (chapter 2), a certain amount of geometric unsharpness is unavoidable.

The actual size of the image on the radiograph is only altered significantly when the size of the object to be radiographed approaches or is less than the size of the focal spot.

5.9.2. *Patient unsharpness*

Other sources of unsharpness can arise when the object being radiographed is not idealized, i.e. infinitely thin yet X-ray opaque. When the object is a patient, or part of a patient, it has a finite thickness generally with decreasing X-ray attenuation towards the edges. These features can be considered as part of the geometric unsharpness of the resulting image and are in fact often much larger than the geometric unsharpness described in section 5.9.1. The effect on the number of photons transmitted is shown in figure 5.12. As can be seen there is a gradual change from transmission to absorption, producing an indistinct edge.

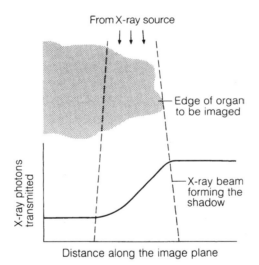

Figure 5.12. *Contribution to image blurring that results from an irregular edge to the organ of interest.*

Another source of unsharpness arises from the fact that during a radiograph many organs within the body can move either through involuntary or voluntary motions. This is shown simply in figure 5.13, where the edge of the organ being radiographed moves from position A to position B during the course of the exposure. Again the result is a gradual transition of film density resulting in an unsharp image of the edge of the organ. The main factors that determine the degree of movement unsharpness are the speed of movement of the region of interest and the time of exposure. Increasing the patient–film distance increases the effect of movement unsharpness.

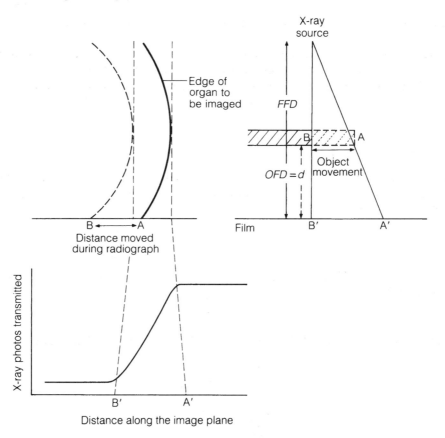

Figure 5.13. *Effect of movement on radiographic blurring. If the object moves with velocity v during the time of exposure t then* AB $= vt$ *and* A$'$B$' =$ AB. FFD$/$(FFD $- d$)*.*

5.9.3.   *Combining unsharpnesses*

It will be apparent from the preceding discussion that in any radiological image there will be several sources of unsharpness and the overall unsharpness will be the combination of all of them.

Unsharpnesses are combined according to a power law with the power index varying between 2 and 3. The power index 2 is most commonly used and should be applied where the unsharpnesses are all of the same order. The power index 3 should be used if one of the unsharpnesses is very much greater than the rest.

Note that, because of the power law relationship, if one contribution is very large it will dominate the expression. If this unsharpness can be reduced at the expense of the others the minimum overall unsharpness will be when all contributions are approximately equal. For example if the geometric unsharpness is U_G, the movement unsharpness U_M and the film–screen unsharpness U_F then the combined unsharpness of the three is

$$U = \sqrt{U_G^2 + U_M^2 + U_F^2}.$$

If $U_G = 0.5$ mm, $U_M = 1.0$ mm, $U_F = 0.8$ mm, then $U = 1.37$ mm.
If $U_G = 0.7$ mm, $U_M = 0.7$ mm, $U_F = 0.8$ mm, then $U = 1.27$ mm.

5.10. GEOMETRIC RELATIONSHIP OF FILM, PATIENT AND X-RAY SOURCE

The interpretation of radiographs eventually becomes second nature to the radiologist who learns to ignore the geometrical effects which can, and do, produce very distorted images with regard to the size and position of organs in the body. Nevertheless, it is important to appreciate that such distortions occur. Most of the effects may be easily understood by assuming that the focus is a point source and that X-rays travel in straight lines away from it.

5.10.1. Magnification without distortion

In the situation shown in figure 5.14 the images of three objects of equal size lying parallel to the film are shown. The images are not the same size as the object. Assume magnification M_1, M_2 and M_3 for objects 1, 2 and 3 respectively given by

$$M_1 = \frac{AB}{ab} \qquad M_2 = \frac{XY}{xy} \qquad M_3 = \frac{GH}{gh}.$$

Consider triangles Fxy and FXY. Angles xFy and XFY are common; xy is parallel to XY. Triangles Fxy and FXY are thus similar. Therefore

$$\frac{XY}{xy} = \frac{FFD}{FFD - d}. \tag{5.3}$$

By the same considerations triangles aFb and AFB are similar and triangles gFh and GFH are similar

$$\frac{AB}{ab} = \frac{FFD}{FFD - d}$$

and

$$\frac{GH}{gh} = \frac{FFD}{FFD - d}.$$

Therefore

$$M_1 = M_2 = M_3$$

i.e. for objects in the same plane parallel to the film, magnification is constant. This is magnification without distortion.

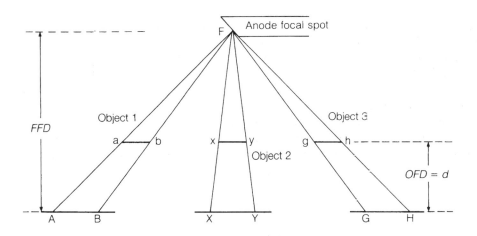

Figure 5.14. *Demonstration of a situation in which magnification without distortion will occur.*

The magnification increases as:

(1) the FFD is decreased,
(2) *d* is increased.

The deliberate use of magnification techniques is discussed in section 9.3.

5.10.2.   *Distortion of shape and/or position*

In general the rather artificial conditions assumed in section 5.10.1 do not apply when real objects are ra-diographed. For example, they are not infinitesimally thin, they are not necessarily orientated normal to the principal axis of the X-ray beam and they are not all at the same distance, measured along the principal axis, from the X-ray source. All of these factors introduce distortions into the resulting image. Figure 5.15(*a*) shows the distortion that results from twisting a thin object out of the horizontal plane, and figure 5.15(*b*)

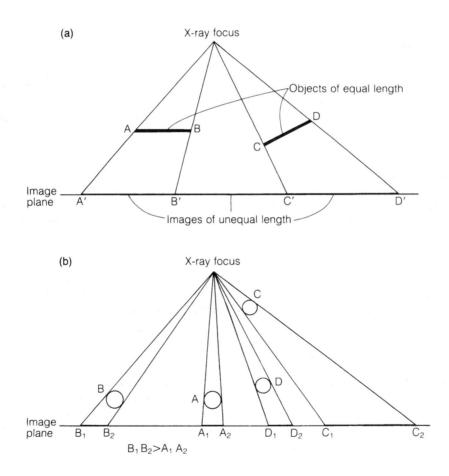

Figure 5.15. *Demonstration of (a) distortion of shape of an object; (b) distortion of shape when objects are of finite thickness (A, B, C) and of relative position when they are at different depths (C, D). Note that the object B has been placed very wide and the object C has been placed very close to the X-ray focus in order to exaggerate the geometrical effects. In particular a patient would not be placed as close to the X-ray focus as C because of the high skin dose (see section 6.13.1).*

shows distortion for objects of finite thickness. Although all the spheres are of the same diameter, the cross sectional area projected parallel to the film plane is now greater if the sphere is off-axis and the image is enlarged more (compare A and B). Note that when the sphere is in a different plane (e.g. C) the distortion may be considerable. Figure 5.15(*b*) also shows that when objects are in different planes, distortion of position will occur. Although C is nearer to the central axis than D, its image actually falls further away from the central axis.

A certain amount of distortion of shape and position is unavoidable and the experienced radiologist learns to take such factors into consideration. The effects will be more marked in magnification radiography.

5.11. REVIEW OF FACTORS AFFECTING THE RADIOLOGICAL IMAGE

It is clear from this and the preceding chapter that a large number of factors can affect a radiological image and these will now be summarized.

5.11.1. *Choice of tube kilovoltage*

A high kV gives a lower contrast and a large film latitude and vice versa. X-ray output from the tube is approximately proportional to $(kVp)^2$ and film blackening proportional to $(kVp)^4$. The higher the kV the lower the patient dose. A higher kV however increases the amount of scatter.

Increasing the kV allows the tube current or the exposure time to be reduced. At 70 kVp an increase of 10 kVp allows the mA s to be approximately halved.

5.11.2. *Exposure time*

In theory, the exposure time should be as short as possible to eliminate movement unsharpness. If movement will not be a problem, exposure time may be increased so that other variables can be optimized.

5.11.3. *Focal spot size*

As the spot size increases so does geometric unsharpness. The minimum size (from a choice of two) should be chosen consistent with the choice of other factors which affect tube rating (kVp, mA, s).

5.11.4. *Quality of anode surface*

As discussed in chapter 2, damage to the anode surface will result in a non-uniform X-ray intensity distribution and an increased effective spot size. Because of the heel effect there is always some variation in X-ray beam intensity in the direction parallel to the line of the anode–cathode. If a careful comparison of the blackening produced by two structures is required they should be orientated at right angles to the electron flux from the cathode to anode.

5.11.5. *Tube current*

In an ideal situation all other variables would be chosen first and then an mA would be selected to give optimum film blackening. If this is not possible, for example because of rating limits, the system becomes highly interactive and the final combination of variables is a compromise to give the best end result.

5.11.6.   *Beam size*

This should be as small as possible, commensurate with the required field of view, to minimize patient dose and scatter. Note that, strictly speaking, collimation reduces the integral dose, i.e. the absorbed dose multiplied by the volume irradiated, rather than the absorbed dose itself.

5.11.7.   *Grids*

These must be used if scatter is significantly reducing contrast, e.g. when irradiating large volumes. Use of grids requires an increased mA s thus increasing patient dose.

5.11.8.   *Focus–film and object–film distance*

A large focus–film distance reduces geometric blurring, magnification and distortion, but the X-ray intensity at the patient is reduced because of the inverse square law. The working distance is thus governed eventually by the tube rating. The object–film distance cannot normally be controlled by the operator but is kept as small as possible by equipment manufacturers. Movement, geometric blurring and magnification can be influenced by patient orientation either anterior/posterior or posterior/anterior.

5.11.9.   *Contrast enhancement*

Modification of either the atomic number of an organ, e.g. by using barium or iodine containing contrast agents, or of its density, e.g. by introducing a gas, alters its contrast relative to the surrounding tissue.

5.11.10.   *Films and screens*

There is generally only a limited choice. A fast film–screen combination will minimize patient dose, geometric and movement unsharpness. Associated screen unsharpness and quantum mottle may be higher than when a slow film–screen combination is used. If extremely fine detail is required, a non-screen film may be used but this requires a much higher mA s and thus gives a higher patient dose.

5.11.11.   *Film processing*

This vital part of image formation must not be overlooked. Quality control of development is extremely important as it can have a profound effect on the radiograph. Bad technique at this stage can completely negate all the careful thought given to selecting correct exposure and position factors. For further information see BIR (1988).

It will be clear from this lengthy list that the quality of a simple, plain radiograph is affected by many factors, some of which are interactive. Each of them must be carefully controlled if the maximum amount of diagnostic information is to be obtained from the image.

REFERENCES

British Institute of Radiology (BIR) 1988 *Assurance of Quality in the Diagnostic X-ray Department* (London: British Institute of Radiology)
Dawson P 1992 X-ray contrast agents. Current status and development prospects *Imaging* **4** 207–16

FURTHER READING

Curry T S, Dowdey J E and Murry R C Jr 1990 *Christensen's Introduction to the Physics of Diagnostic Radiology* 4th edn (Philadelphia, PA: Lea and Febiger)

Gifford D 1984 *A Handbook of Physics for Radiologists and Radiographers* (Chichester: Wiley)

Hay G A 1982 Traditional X-ray imaging *Scientific Basis of Medical Imaging* ed P N T Wells (Edinburgh: Churchill Livingstone) pp 1–53

EXERCISES

1 What is meant by contrast?

2 Why is contrast reduced by scattered radiation?

3 The definition of contrast used in radiography cannot be used in nuclear medicine. Discuss the reasons for this.

4 What are the advantages and disadvantages of having a radiographic film with a high gamma?

5 A solid bone 7 mm diameter lies embedded in soft tissue. Ignoring the effects of scatter, calculate the contrast between the bone (centre) and neighbouring soft tissue.
Film gamma $= 3$
Linear attenuation coefficient of bone $= 0.5$ mm^{-1}
Linear attenuation coefficient of tissue $= 0.04$ mm^{-1}.

6 How can the effect of scattered radiation on contrast be reduced?

7 Why is a low kVp used to take a mammogram?

8 Give a sketch showing how the relative scatter (scattered radiation as a fraction of the unscattered radiation) emerging from a body varies with X-ray tube kV between 30 kVp and 200 kVp and explain the shape of the curve. What measures can be taken to minimize loss of contrast due to scattered radiation?

9 How can X-ray magnification be used to enhance the detail of small anatomical structures? What are its limitations?

10 What are the advantages and disadvantages of an X-ray tube with a very fine focus?

11 List the factors affecting the sharpness of a radiograph. Draw diagrams illustrating these effects.

12 A radiograph is taken of a patient's chest. Discuss the principal factors that influence the resultant image.

13 What factors, affecting the resolution of a radiograph, are out of the control of the radiologist? (assuming the radiographer is performing as required).

14 A radiograph is found to lack contrast. Discuss the steps that might be taken to improve contrast.

CHAPTER 6

RADIATION MEASUREMENT AND DOSES TO PATIENTS

P P Dendy and K E Goldstone

6.1. INTRODUCTION

Lord Kelvin (1824–1907), who is probably best remembered for the absolute thermodynamic scale of temperature, is reported to have stated on one occasion: 'Anything that cannot be expressed in numbers is valueless'. In view of the potentially harmful effect of X-rays it is particularly important that methods should be available to 'express in numbers' the 'strength' or intensity of X-ray beams.

With respect to measurement, three separate features of an X-ray beam must be identified. The first consideration is the flux of photons travelling through air from the anode towards the patient. The ionization produced by this flux is a measure of the **radiation exposure**. If expressed per unit area per second it is the **intensity**. Of more fundamental importance as far as the biological risk is concerned is the **absorbed dose of radiation**. This is a measure of the amount of energy deposited as a result of ionization processes. Finally, it may be important to know about the energy of the individual photons. Because of the mechanism of production, an X-ray beam will contain photons with a wide range of energies. A complete specification of the beam would require determination of the full spectral distribution as shown in figure 2.2. This represents information about the **quality** of the X-ray beam.

Clearly the intensity of an X-ray beam must be measured in terms of observable physical, chemical or biological changes that the beam may cause, so it will be useful to review briefly relevant properties of X-rays. Two of them are sufficiently fundamental to be classified as **primary properties**—that is to say measurements can be made without reference to a standard beam.

(1) *Heating effect*. X-rays are a form of energy which can be measured by direct conversion into heat. Unfortunately, the energy associated with X-ray beams used in diagnostic radiology is so low that the temperature rise can scarcely be measured (see section 11.2).
(2) *Ionization*. X-rays cause ionization by photoelectric, Compton and pair production processes in any material through which they pass. The number of ions produced in a fixed volume under standard conditions of temperature and pressure will be fixed.

A number of other properties of X-rays can be, and often are, used for dosimetry. In all these situations, however, it is necessary for the system to be calibrated by first measuring its response to beams of X-rays of known intensity so these are usually called **secondary properties**.

(3) *Physical effects.* When X-rays interact with certain materials, visible light is emitted. The light may either be emitted immediately following the interaction (**fluorescence**); after a time interval (**phosphorescence**) or, for some materials, only upon heating (**thermoluminescence**).

(4) *Physico-chemical effects.* The action of X-rays on photographic film is well known and widely used.

(5) *Chemical changes.* X-rays have oxidizing properties, so if a chemical such as ferrous sulphate is irradiated, the free ions that are produced oxidize some Fe^{2+} to Fe^{3+}. This change can readily be detected by shining ultra-violet light through the solution. This light is absorbed by Fe^{3+} but not by Fe^{2+}.

(6) *Biochemical changes.* Enzymes rely for their action on the very precise shape associated with their secondary and tertiary structure. This is critically dependent on the exact distribution of electrons, so enzymes are readily inactivated if excess free electrons are introduced by ionizing radiation.

(7) *Biological changes.* X-rays can kill cells and bacteria, so, in theory at least, irradiation of a suspension of bacteria followed by an assay of survival could provide a form of biological dosimeter.

Unless specifically stated otherwise, in the remainder of this chapter references to X-rays apply equally to gamma rays of the same energy.

6.2. IONIZATION IN AIR AS THE PRIMARY RADIATION STANDARD

There are a number of important prerequisites for the property chosen as the basis for radiation measurement.

(1) It must be accurate and unequivocal, i.e. personal, subjective judgement must play no part.

(2) It must be very sensitive to producing a large response for a small amount of radiation energy.

(3) It must be reproducible.

(4) The measurement should be independent of intensity, i.e. an intensity I for time t must give the same answer as an intensity $2I$ for time $t/2$. This is the **law of reciprocity**.

(5) The method must apply equally well to very large and very small doses.

(6) It must be reliable at all radiation energies.

(7) The answer must convert readily into a value for the absorbed energy in biological tissues or 'absorbed dose' since this is the single most important reason for wishing to make radiation measurements.

None of the properties of ionizing radiation satisfies all these requirements perfectly but ionization in air comes closest and has been internationally accepted as the basis for radiation dosimetry. There are two good reasons for choosing the property of ionization.

(1) Ionization is an extremely sensitive process in terms of energy deposition. Only about 34 eV is required to form an ion pair, so if a 100 keV photon is completely absorbed, almost 3000 ion pairs will have been formed when all the secondary ionization has taken place.

(2) As shown in chapter 11, the extreme sensitivity of biological tissues to radiation is directly related to the process of ionization so it has the merit of relevance as well as sensitivity.

There are also good reasons for choosing to make measurements in air:

(1) It is readily available.

(2) Its composition is close to being universally constant.

(3) More important, for medical applications, the mean atomic number of air ($Z = 7.6$) is very close to that of muscle/soft tissue ($Z = 7.4$). Thus, provided ionization and the associated process of energy absorption is expressed per unit mass by using mass absorption coefficients rather than linear absorption coefficients, results in air will be closely similar to those in tissue.

The unit of **radiation exposure** (X) is defined as that amount of radiation which produces in air ions of either sign equal to 1 C (coulomb) kg^{-1}.[1] Expressed in simple mathematical terms:

$$X = \frac{\Delta Q}{\Delta m}$$

where ΔQ is the sum of all the electrical charges on all the ions of one sign produced in air when all the electrons liberated in a volume of air whose mass is Δm are completely stopped in air. The last few words ('are completely stopped in air') are extremely important. They mean that if the electron generated by say a primary photoelectric interaction is sufficiently energetic to form further ionizations (normally it will be), all the associated ionizations must occur within the collection volume and all the electrons contribute to ΔQ.

6.3. THE IONIZATION CHAMBER

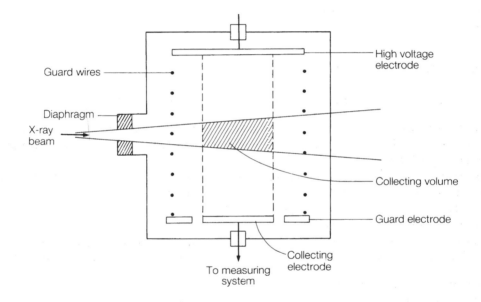

Figure 6.1. *The free air ionization chamber. (From Whyte 1959.)*

Figure 6.1 shows a direct experimental interpretation of the definition of radiation exposure. The diaphragm, constructed of a heavy metal such as tungsten or gold, defines an X-ray beam of accurately known cross section A. This beam passes between a pair of parallel plates in an air-filled enclosure. The upper plate is maintained at a high potential relative to the lower and, in the electric field arising from the potential difference between the plates, all the ions of one sign produced in the region between the dashed lines move to the collecting electrode. This is generally referred to as a free air ionization chamber.

Either the current flow (exposure rate) or the total charge (exposure) may be measured using the simplified electrical circuits shown in figures 6.2(a) and (b).

Since 1 ml of air weighs 1.3×10^{-6} kg at STP, a chamber of capacity 100 ml contains 1.3×10^{-4} kg of air. A typical exposure rate might be 2.5 μC kg^{-1} h^{-1} (a dose rate of approximately 0.1 mGy h^{-1}, see section 6.5) which corresponds to a current flow of $(2.5 \times 10^{-6}/3600) \times 1.3 \times 10^{-4}$ C s^{-1} or about 10^{-13} A.

---

[1] The older, obsolescent unit of radiation exposure is the roentgen (R). One roentgen is that exposure to X-rays which will release one electrostatic unit of charge in one cubic centimetre of air at standard temperature and pressure (STP). Hence 1 R = 2.58×10^{-4} C kg^{-1}.

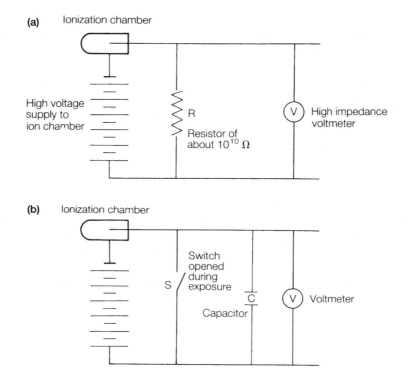

Figure 6.2. *Simplified electrical circuits for measuring (a) current flow (exposure rate), (b) total charge (exposure).*

If $R = 10^{10}$ Ω, since $V = IR$ then $V = 1$ mV which is not too difficult to measure. However, the voltmeter must have an internal resistance of at least 10^{13} Ω so that no current flows through it and this is quite difficult to achieve.

Since the free air ionization chamber is a primary standard for radiation measurement, accuracy better than 1% (i.e. more precision than the figures quoted here) is required. Although it is a simple instrument in principle, great care is required to achieve such precision and a number of corrections have to be applied to the raw data. For example a correction must be made if the air in the chamber is not at STP. For air at pressure P and temperature T, the true reading R_T is related to the observed reading by

$$R_T = R_0 \left(\frac{P_0}{P} \right) \left(\frac{T}{T_0} \right)$$

where P_0, T_0 are STP values.

Rather than trying to memorize this equation, the reader is advised to refer to first principles. Ion pairs are created because X-ray photons interact with air molecules. If the air pressure increases above normal atmospheric, the number of air molecules will increase, the number of interactions will increase, and the reading will be artificially high. Changes in temperature may be considered similarly.

The requirement for precision also creates design difficulties. For example, great care must be taken, using guard rings and guard wires (see figure 6.1), to ensure that the electric field is always precisely normal to the plates. Otherwise, electrons from within the defined volume may miss the collecting plate or, conversely, may reach the collecting plates after being produced outside the defined volume.

Major difficulties arise as the X-ray photon energy increases, especially above about 300 keV, because of the ranges of the secondary electrons (see table 1.2). Recall that all the secondary ionization must occur within the air volume. If the collecting volume is increased, eventually it becomes impossible to maintain field uniformity.

Thus the free air ionization chamber is very sensitive in the sense that one ion pair is created for the deposition of a very small amount of energy. However, it is insensitive when compared to solid detectors that work on the ionization principle because air is a poor stopping material for X-rays. It is also bulky and operates over only a limited range of X-ray energies. However, it is a primary measuring device and all other devices must be calibrated against it.

6.4.   THE GEIGER–MÜLLER COUNTER

If, when using the equipment shown in figure 6.1, the X-ray beam intensity were fixed but the potential difference between the plates were gradually increased from zero, the current flowing from the collector plate would vary as shown in figure 6.3. Initially, all ion pairs recombine and no current is registered. As the potential difference increases (region AB) more and more electrons are drawn to the collector, until, at the first plateau BC, all the ion pairs are being collected. This is the region in which the ionization chamber operates and its potential difference must be in the range BC.

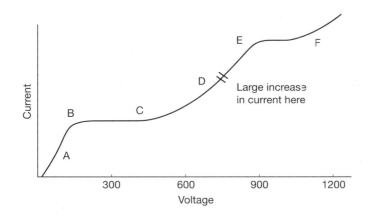

Figure 6.3. *Variation in current appearing across capacitor plates with applied potential difference for a fixed X-ray beam intensity. (AB) loss of ions by recombination. (BC) ionization plateau. (CD) proportional counting. (EF) Geiger–Müller region. (Beyond F) continuous discharge. The voltage axis shows typical values only.*

Beyond C, the current increases again. This is because secondary electrons gain energy from the electric field between the plates and eventually acquire enough energy to cause further ionizations (see figure 6.4). **Proportional counters** operate in the region of CD. They have the advantage of increased sensitivity, the extra energy having been drawn from the electric field, and, as the name implies, the strength of signal is still proportional to the amount of primary and secondary ionization. Hence, proportional counters can be used to measure radiation exposure. However, very precise voltage stabilization is required, since the amplification factor is changing rapidly with small voltage changes, and such a device is unsuitable for precision work with a portable measuring device.

Beyond D the amplification increases rapidly until the so-called Geiger–Müller (GM) plateau is reached at EF. Beyond F there is continuous discharge.

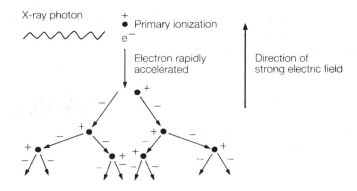

Figure 6.4. *Amplification of ionization by the electric field.*

6.4.1. The Geiger–Müller tube

Essential features of a GM tube are shown in figure 6.5 and the most important details of its design and operation are as follows.

(1)   In figure 6.3, the GM plateau was attained by applying a high voltage, V, between parallel plates. In fact it is the electric field $E = V/d$, where d is the distance between the plates, that accelerates electrons. High fields can be achieved more readily using a wire anode since near the wire E varies as $1/r$, where r is the radius of the wire. Thus the electric field is very high close to a wire anode even for a working voltage of 300–400 V. When working on the GM plateau, EF, count rate changes only slowly with applied voltage so very precise voltage stabilization is not necessary.

(2)   The primary electrons are accelerated to produce an avalanche as in the proportional counter, but in

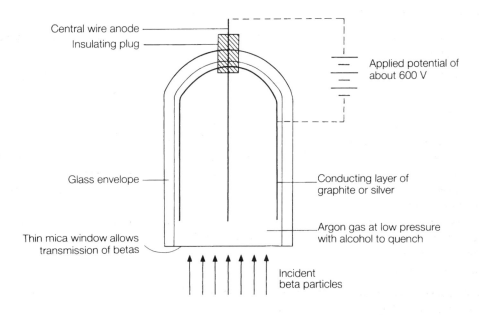

Figure 6.5. *Essential features of an end-window Geiger–Müller tube suitable for detecting beta particles. The thin window is not necessary when detecting X- or gamma-rays.*

the avalanche discharge excited atoms as well as ions are formed. They lose this excitation energy by emitting X-ray and ultra-violet photons which liberate outer electrons from other gas atoms creating further ion pairs by a process of photoionization. As these events may occur some distance from the initial avalanche, the discharge is spread over the whole of the wire. Because of the high electric fields in the GM tube, the positive ions reach the cathode in sufficient numbers and with sufficient energies to eject electrons. These electrons initiate other pulses which recycle in the counter thus producing a continuous discharge.

(3) The continuous discharge must be stopped before another pulse can be detected. This is done by adding a little alcohol or bromine to the counting gas which is either helium or argon at reduced pressure. The alcohol or bromine 'quenches' the discharge because their ionization potentials are substantially less than those of the counting gas. During collisions between the counting gas ions and the quenching gas molecules, the ionization is transferred to the latter. When these reach the cathode, they are neutralized by electrons extracted by field emission from the cathode. The electron energy is used up in dissociating the molecule instead of causing further ionization. The alcohol or bromine also has a small effect in quenching some of the ultra-violet photons.

(4) The discharge is also quenched because a space charge of positive ions develops round the anode, thereby reducing the force on the electrons.

(5) Finally quenching can be achieved by reducing the external anode voltage using an external resistor and this is triggered by the early part of the discharge.

Once discharge has been initiated, and during the time it is being quenched, any further primary ionization will not be recorded as a separate count. The instrument is effectively 'dead' until the externally applied voltage is restored to its full value, typically after about 300 μs. This is known as the **dead time**.

Thus the true count is always higher than the measured count. The difference is minimal at 10 counts per second but at 1000 counts per second the monitor is dead for $1000 \times 300 \times 10^{-6} = 0.3$ s in every second and losses become appreciable.

6.4.2. *Comparison of ionization chambers and Geiger–Müller counters*

Both instruments have important but well defined roles in radiological monitoring so their strengths and weaknesses must be clearly understood.

(i) *Type of radiation.* Both respond to x- and gamma rays and to fast beta particles. By using a thin window the response of the GM tube can be extended to low energy beta particles, but not the very low energy beta particles from H-3 (18 keV max).

(ii) *Sensitivity.* Because of internal amplification, the GM tube is much more sensitive than the ionization chamber and may be used to **detect** low levels of contamination (but see section 6.8).

(iii) *Nature of reading.* The ionization chamber is designed to collect all primary and secondary radiation and hence to give a reading of exposure or exposure rate. With the GM tube there is no proportional relationship between the count rate and the number of primary and secondary ionizations so it is not a radiation **monitor**.

(iv) *Size.* The ionization chamber must be big enough to collect all secondary electron ionizations. Since the GM tube does not have the property of proportionality, there is no point in making it large and it can be much more compact.

(v) *Robustness and simplicity.* Generally favour the GM tube.

In conclusion a GM tube is an excellent **detector** of radiation, but it must not be used to measure radiation exposures for which an appropriate **monitor** is required[1].

---

[1] A 'compensated Geiger' is sometimes used as a radiation monitor. However it is not very suitable for diagnostic radiology because of poor sensitivity at low energies.

6.5.   RELATIONSHIP BETWEEN EXPOSURE AND ABSORBED DOSE

The second aspect of radiation measurement is to obtain the absorbed dose or amount of energy deposited in matter. In respect of the biological damage caused by ionizing radiation, this is more relevant than radiation exposure.

The unit of absorbed dose is the Gray (Gy)[2], where $1 \text{ Gy} = 1 \text{ J kg}^{-1}$.

Whereas radiation exposure, by definition, refers to ionization in air, the dose, or energy absorbed from the radiation, may be expressed in any material. Calculation of the dose in say soft tissue, when the radiation exposure in air is known, may be treated as a two-stage problem as follows:

(1)   conversion of exposure in air to dose in air,
(2)   conversion of dose in air to dose in tissue.

When treated in this way, both parts of the calculation become fairly easy. Note that in future absorbed dose will be simplified to dose unless there is possible confusion.

6.5.1.   *Conversion of exposure in air to dose in air*

A term that is being used increasingly in radiation dosimetry is KERMA. This stands for kinetic energy released per unit mass and must specify the material concerned. Note that KERMA places the emphasis on removal of energy from the beam of indirectly ionizing particles (x- or gamma photons) in order to create secondary electrons. Absorbed dose relates to where those electrons deposit their energy in the medium.

There are two reasons why KERMA in air (K_A) may differ from dose in air (D_A). First, some secondary electron energy may be radiated as bremsstrahlung. Second, the point of energy deposition in the medium is not the same as the point of removal of energy from the beam because of the range of secondary electrons. However, at diagnostic energies bremsstrahlung is negligible and the ranges of secondary electrons are so short that $K_A = D_A$ to a very good approximation.

Now the number of ion pairs generated in each kilogram of air multiplied by the energy required to form one ion pair (W) is equal to the energy removed from the beam. But the first term is the definition of radiation exposure, say E, and the third term is the definition of KERMA in air K_A. Thus

$$E \text{ (C kg}^{-1}) \times W \text{ (J C}^{-1}) = K_A \text{ (J kg}^{-1}).$$

Or, since $K_A = D_A$, expressing dose at the subject

$$D_A \text{ (J kg}^{-1}) = E \text{ (C kg}^{-1}) \times W \text{ (J C}^{-1}).$$

The energy to form one ion pair, W, is close to 34 J C^{-1} (34 electron volts per ion pair) for all types of radiation of interest to radiologists and, coincidentally, over a wide range of materials of biological importance. By definition $1 \text{ Gy} = 1 \text{ J kg}^{-1}$. Thus

$$D_A \text{ (Gy)} = 34E \text{ (C kg}^{-1}).[3]$$

In this book, dose in air will be used in preference to exposure. 'Skin dose' will be used to describe the dose in air at the patient.

---

[2] This has replaced the older unit, rad. One rad was 100 erg g^{-1} and thus $1 \text{ Gy} = 100 \text{ rad}$.
[3] Many older textbooks use the unit roentgen for exposure in air and rad for dose. Since $1 \text{ R} = 2.58 \times 10^{-4} \text{ C kg}^{-1}$, and $1 \text{ Gy} = 100 \text{ rads}$, $D_A \text{ (rad)} = 100 \times 2.58 \times 10^{-4} \times 34E \text{ (roentgen)}$. Hence $D_A \text{ (rad)} = 0.88E \text{ (roentgen)}$.

6.5.2.   *Conversion of dose in air to dose in tissue*

To convert a dose in air to dose in any other material, recall that for a given incident flux of photons, the energy absorbed per unit mass depends only on the mass absorption coefficient of the medium. Hence

$$\frac{D_M}{D_A} = \frac{(\mu_a/\rho)_M}{(\mu_a/\rho)_A}.$$

Note the use of a subscript 'a' to distinguish the mass absorption coefficient from the mass attenuation coefficient.

It follows that only a knowledge of the relative values of μ_a/ρ is required to convert a known dose in air to the corresponding dose in any other material.

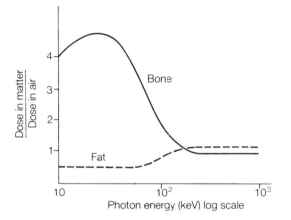

Figure 6.6. *The ratio (dose in matter/dose in air) plotted as a function of radiation energy for bone and fat.*

The ratio D_M/D_A is plotted as a function of different radiation energies for different materials in figure 6.6. It is left as an exercise for the reader to justify the shapes of these curves from a knowledge of:

(1)   the mean atomic number for each material,
(2)   the relative importance of the Compton and photoelectric effects at each photon energy.

6.6.   PRACTICAL RADIATION MONITORS

6.6.1.   *Secondary ionization chambers*

In section 6.3, one of the problems identified for the free air ionization chamber was its large volume. Fortunately, there is a technique which, to an acceptable level of accuracy for laboratory instruments, eliminates this problem.

Imagine a large volume of ethylene gas with dimensions much bigger than the range of secondary electrons. Now compress the gas to solid polyethylene, leaving only a small volume of gas at the centre (figure 6.7). The radiation exposure will be determined by the density of electrons within the gas and if the gas volume is small, the number of secondary electrons either being created in the gas or coming to the end of their range there will be negligible compared to the electron density in the solid.

However, the electron density in the solid is the same as it would be at the centre of the large gas volume. This is because the electron flux across any plane is the product of rate of production per unit thickness and range. The rate of production depends on the number of interactions and is higher in the solid than in the

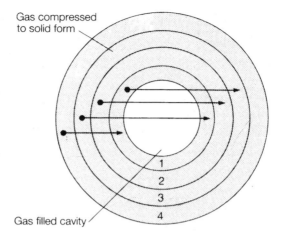

Figure 6.7. *Secondary electron flux in a gas-filled cavity. Consider the solid as a series of layers starting at the edge of the cavity. Layers 1, 2 and 3 contribute to the ionization in the cavity but layer 4 does not.*

gas by ρ_{gas}/ρ_{solid} (the ratio of the densities). But once the electrons are formed they have a fixed energy and lose that energy more rapidly in the solid due to collisions. Thus the ratio *range of electrons in solid/range of electrons in gas* is in the inverse ratio ρ_{gas}/ρ_{solid}.

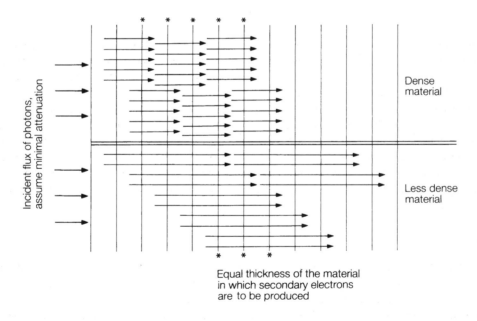

Figure 6.8. *A schematic demonstration that the flux of secondary electrons at equilibrium is independent of the density of the stopping material. The upper material has 2.5 times the density of the lower. It therefore produces 2.5 times more electrons per slice. However, the range of these electrons is reduced by the same factor. When equilibrium is established, shown by a ∗ for each material, ten secondary electrons are crossing each vertical slice (the electron density) in each case. Note that equilibrium is established more quickly in the more dense material.*

Hence the product (rate of production of secondary electrons × range of secondary electrons) is independent of density as shown schematically in figure 6.8.

An alternative way to view this situation is that, provided the atomic composition is the same, the gas in the cavity does not know whether it is surrounded by a big volume of gas or by a much smaller volume of solid resulting from compression of the gas.

The result is precise for polyethylene and ethylene gas because the materials differ only in density. No solid material is exactly like air in terms of its interaction with X-ray photons by photoelectric and Compton processes at all photon energies but good approximations to **air equivalent walls** have been constructed and a correction can be made for the discrepancy.

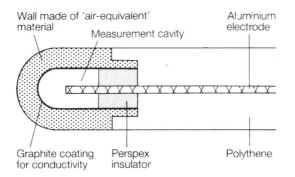

Figure 6.9. *A simple, compact secondary ionization chamber that makes use of the 'air equivalent wall' principle.*

A simple, compact, secondary instrument, suitable for exposure rate measurements around an X-ray set, is shown in figure 6.9. The dimensions now only require that the wall thickness should be greater than the range of secondary electrons in the solid medium. Note that a correction must also be made for attenuation of the primary beam in the wall surrounding the measurement cavity.

An important modern detector that uses the principle of ionization in gas is the high pressure xenon gas chamber. Xenon is chosen because it is an inert gas and its high atomic number ($Z = 54$) ensures a large cross section for photoelectric interactions. It is sometimes mixed with krypton ($Z = 82$). The pressure is increased to about 25 atmospheres to improve sensitivity. Although the latter is still poor by comparison with solid detectors (see sections 6.8 and 6.10), it is possible to pack a large number of xenon gas detectors of very uniform sensitivity into a small space. The use of Xe/Kr high pressure gas detectors in computed tomography is discussed in section 10.3.2.

6.6.2. Dose–area product meters

Dose–area product is defined as the absorbed dose to air averaged over the area of the X-ray beam in a plane perpendicular to the beam axis multiplied by the area of the beam in the same plane. It is usually measured in Gy cm^2 and radiation back-scattered from the patient is excluded.

Provided that the cross sectional area of the beam lies completely within the detector, it may be shown by simple application of the inverse square law (see section 1.12) that the reading will not vary with the distance from the tube focus.

Thus the dose–area product can be measured at any point between the diaphragm housing on the X-ray tube and the patient—but not so close to the patient that there is significant back-scattered radiation.

Dose–area product meters consist of flat, large area parallel plate ionization chambers connected to suitable electrometers which respond to the total charge collected over the whole area of the chamber. The meter is mounted close to the tube focus where the area of the X-ray beam is relatively small and dose rates are high. It is normally mounted on the diaphragm housing where it does not interfere with the examination and is usually transparent so that when fitted to an over-couch X-ray tube the light beam diaphragm device can still be used. The use of dose–area product meters to estimate patient doses is considered in section 6.13.

6.6.3.   *Pocket exposure meters for personnel monitoring*

Although the secondary ionization chamber is much more compact than a free air ionization chamber, it is still too large to be readily portable. However, an individual working in an unknown radiation field clearly may need to have an immediate reading of their accumulated dose or instantaneous dose rate.

An early device for personal monitoring, based on the ionization principle, was the 'fountain pen' dosimeter. This used the principle of the gold leaf electroscope. When the electroscope was charged the leaves diverged because of electrostatic repulsion. If the gas around the leaves became ionized by X-rays, the instrument was discharged and the leaves collapsed.

More widely used nowadays as a practical ionization instrument is the portable radiation monitor or 'bleeper'. As its name implies, it is light enough to carry around in the pocket and gives both an audible warning of radiation dose rate and displays the dose received. The bleep sounds every 15–30 min on background and the bleep rate increases with dose rate becoming continuous in high radiation fields. The instrument is quite sensitive, registering doses as low as 1 μGy X-rays and giving approximately one bleep every 20 s at 10 μGy h^{-1}. Note however:

(1)   although this instrument appears to be an ionization chamber since it gives a direct reading of absorbed dose, it works on a modified Geiger–Müller principle and
(2)   the energy range for this instrument is from 45 keV up to the megavoltage range so it may be unsuitable for use at the lowest diagnostic energies. This poor sensitivity at low energies is a feature of 'compensated Geigers' (see section 6.4).

6.7.   THERMOLUMINESCENT DOSIMETERS (TLDS)

The idea of thermoluminescence was introduced in section 4.2. Electrons excited by the X-ray interactions are trapped in the forbidden energy band but the radiation energy is not emitted spontaneously in the form of visible light. It is stored almost indefinitely and only released when the TLD is heated, generally to about 300–400 °C. TLDs have wide application for personal dosimetry and will be discussed in detail in chapter 12. Here their use as general-purpose dosimeters is considered briefly.

Among their advantages are the following:

(1)   response is linear with dose over a wide range;
(2)   sensitivity is almost energy independent (see section 6.12);
(3)   adequate sensitivity is achieved in a very small volume.

Disadvantages of TLDs are:

(1)   they must be calibrated against standard radiation sources;
(2)   careful annealing is required after read-out to ensure that the TLD material returns to the same condition in respect of the number of available traps otherwise the sensitivity and hence calibration factor of the TLD may change.

6.8. SCINTILLATION DETECTORS AND PHOTOMULTIPLIER TUBES

The fundamental interaction process in a scintillation detector is fluorescence which was discussed in section 4.2. One material widely used for such detectors is sodium iodide to which about 0.1% by weight of thallium has been added—NaI(Tl). The traps generated by thallium in the NaI lattice are about 3 eV above the band of valence electrons so the emitted photon is in the visible range. The detector is carefully designed and manufactured to optimize light yield. Note that whereas the *number* of photons emitted is a function of the energy imparted by the X-ray or gamma ray interaction, the *energy* or wavelength of the photons depends only on the positions of the energy levels in the scintillation crystal.

When a scintillation crystal is used as a monitor, its advantages over the detectors discussed so far are as follows:

(1) Since it is a high density solid, its efficiency, especially for stopping higher energy gamma photons, is greatly increased. A 2.5 cm thick NaI(Tl) crystal is almost 100% efficient in the diagnostic X-ray energy range. Contamination monitors that must be capable of detecting of the order of 30 counts per second from an area of 1000 mm^2 invariably contain scintillation crystals.

(2) It has a rapid response time, in contrast to an ionization chamber which responds only slowly owing to the need to build up charge on the electrodes (see figure 6.1).

(3) Different scintillation crystals can be constructed that are particularly sensitive to low energy X-rays or even to neutrons. Beta particles and alpha particles can be detected using plastic phosphors.

NaI(Tl) detectors are used extensively in nuclear medicine and the properties that render them particularly appropriate for *in vivo* imaging will be discussed in chapter 7. Alternative scintillation detectors are caesium iodide doped with thallium and bismuth germanate. Like NaI (Tl), the latter has a high detection efficiency, and is preferable at high counting rates (e.g. for CT) because it has little 'afterglow'—persistence of the light associated with the scintillation process. Bismuth germanate detectors also exhibit a good dynamic range and long term stability.

The light signal produced by a scintillation crystal is too small to be used until it has been amplified and this is achieved by using either a photomultiplier tube (PMT) or a photodiode.

The main features of the PMT coupled to a scintillation crystal (figure 6.10) are as follows:

(1) An evacuated glass envelope, one end of which has an optically flat surface. Since photon losses must be minimized, the scintillation crystal must either be placed in contact with this surface or if it is impracticable, must be optically coupled using a piece of optically transparent plastic—frequently referred to as a 'light guide'. The magnesium oxide reflecting surface also reduces light losses.

(2) A layer of photoelectric material such as caesium–antimony. The characteristic of such materials is that their work function, i.e. the energy required to release an electron, is very low. Thus electrons are emitted when visible or ultra-violet photons fall on the photocathode, although the efficiency is low with only one electron emitted for every ten incident photons.

(3) An electrode system to provide further amplification. This system consists of a set of plates, each maintained at a potential difference of about 100 V with respect to its neighbours and coated with a metal alloy, say of magnesium–silver, designed to release several electrons for every one incident on it. Each plate is known as a dynode and there may be 12 dynodes in all so the potential difference across the PMT will be in the region of 1200 V. If each dynode releases four electrons for each incident electron and there are 12 dynodes, the amplification in the PM tube is 4^{12} or about 10^7. Furthermore, this figure will be constant to within about 1% provided the voltage can be stabilized to 0.1%.

Note that during the complete detection process in the crystal and PMT, the signal twice takes the form of photons, once as X-ray photons and once as visible light photons, and twice takes the form of electrons.

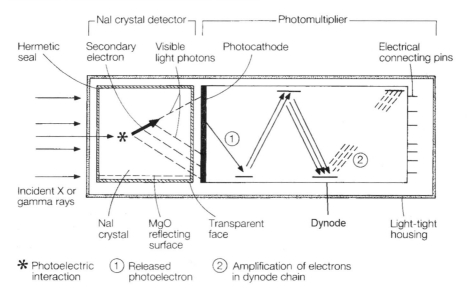

Figure 6.10. *The main features of a PMT coupled to a scintillation crystal for radiation detection.*

Since a PMT is an extremely sensitive light detector, great care must be taken to ensure that no stray light enters the system.

6.9. SPECTRAL DISTRIBUTION OF RADIATION

Thus far nothing has been said about the third factor in a complete specification of a beam of ionizing radiation, namely the energy spectrum of the photons. Under appropriate conditions, a scintillation crystal used in conjunction with a PMT may be used to give such information.

If the crystal is fairly thick, and made of high density material, preferably of high atomic number, most x- or gamma rays that interact with it will be completely stopped within it and each photon will give up all its energy to a single photoelectron. Note that for this discussion it is unimportant whether there is a single photoelectric interaction or a combination of Compton and photoelectric interactions (contrast with the discussion in section 7.2.1).

Since the number of visible light photons is proportional to the photoelectron energy, and the amplification by the PMT is constant, the strength of the final signal will be proportional to the energy of the interacting x- or gamma ray photon. A **pulse height analyser** may now be used to determine the proportion of signals in each predetermined range of strengths and if the pulse height analyser is calibrated against a monoenergetic beam of gamma rays of known energy, the results may be converted into a spectrum of incident photon energies (figure 6.11).

Unfortunately, because of statistical problems, this method of determining the spectral distribution of a beam of x- or gamma rays is not as precise as one would wish. One limitation of the scintillation crystal and PMT combination as a detector of ionizing radiation is that the number of electrons entering the PMT per primary x- or gamma ray photon interaction is rather small. There are two reasons:

(1)   about 30 eV of energy must be dissipated in the crystal for the production of each visible or ultra-violet photon, and

(2)   even assuming no loss of these photons, only about one photoelectron is produced for every 10 photons on the PMT photocathode.

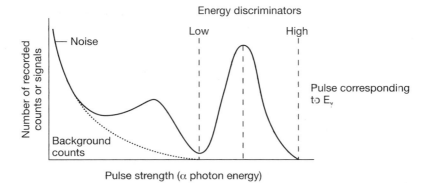

Figure 6.11. *Use of a pulse height analyser to determine the spectrum of photon energies in an X-ray beam. A typical pulse height spectrum obtained after monochromatic gamma rays, E_γ, have passed through scattering material, and the use of energy discriminators to select the peak are shown. The tail of pulses is due to Compton scattering, which are produced fairly uniformly at all energies but selective absorption at low energies creates the maximum. The sharp rise for very low pulses is due to noise.*

Thus to generate one electron at the photocathode requires about 300 eV and a 140 keV photon will produce only 400 electrons at the photocathode. This number is subject to considerable statistical fluctuation ($N^{1/2} = 20$ or 5%). The result is that a monoenergetic beam of gamma rays will produce a range of pulses and will appear to contain a range of energies (figure 6.12). This is a particular problem in the gamma camera and will be considered further in section 7.3.4.

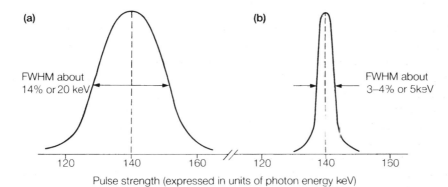

Figure 6.12. *Typical spread in the strength of signals from a monoenergetic beam of gamma rays when using as the primary detector (a) an NaI(Tl) crystal, (b) a solid-state device.*

6.10. SEMICONDUCTOR DETECTORS

In a semiconductor, the forbidden band of energy levels is very narrow and therefore only small quantities of energy, sometimes as little as 2.5–3 eV, are required to release electrons from the filled band into the conduction band. As a result, a 140 keV photon is capable of releasing very large numbers of electrons and the statistical fluctuations are considerably reduced.

In theory, a very pure semiconductor such as silicon or germanium could be used as a detector. If electrodes were attached across a slice of the material, the electrons set free by ionization events would be collected at the anode and electrons from the cathode would drift into the material to neutralize the positive ions. The detector would have a high sensitivity per unit volume because it is solid and, for the reason given above, would have a high energy resolution.

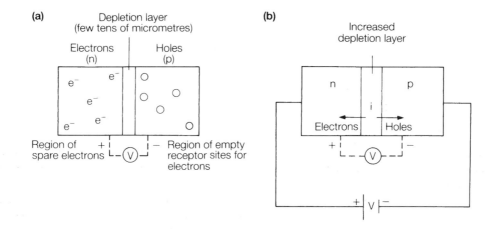

Figure 6.13. *Schematic arrangement of the disposition of mobile electrons and holes in an n–p silicon semiconductor, (a) when no voltage is applied, (b) when the depletion layer is increased by applying a potential difference such that the bulk of the n-type material is at a higher potential than the bulk of the p-type material.*

Until quite recently it has not been possible to manufacture pure enough material to prevent electrons becoming 'lost' in the lattice so more complex structures have been adopted. One possibility is illustrated in figure 6.13. Very small quantities of impurity, say a few parts per million, are added deliberately to the purest obtainable silicon. The material on the left contains an n-type impurity, e.g. antimony or arsenic which has five valence electrons compared to the four in silicon and germanium, and is readily able to make free electrons available to the lattice. The material on the right contains a p-type impurity, e.g. gallium with only three valence electrons. It provides a 'hole' where spare electrons may reside. At the interface between the n-type and p-type materials, a certain amount of diffusion takes place with electrons occupying vacant holes. This process ceases when the electron imbalance has established a potential difference that is sufficient to prevent further flow. A non-conducting depletion zone results as shown in figure 6.13(a).

Such a structure has rectifying properties (see section 2.3.4) because if a potential is applied such that the p material is positive with respect to the n material, the internal potential barrier is reduced and both electrons and holes flow freely. However, if a reverse potential is applied, the electrons and holes are drawn away from the junction, increasing the depletion layer until the internal potential across it is equal and opposite to the applied potential (figure 6.13(b)). No current then flows.

The depletion layer provides an excellent radiation detector, behaving very much like a parallel plate ionization chamber because if any electrons are generated as a result of ionizing interactions, they can migrate to the anode and be registered as a current. The thickness of the depletion layer is determined by the magnitude of V.

In practice only very thin silicon-based detectors, typically about 200 μm, can be constructed to this design. However, volume for volume the silicon detector is about 18 000 times more sensitive than an air-filled ionization chamber. This is partly a density effect, partly because only 2.5–3 eV not 34 eV is required to

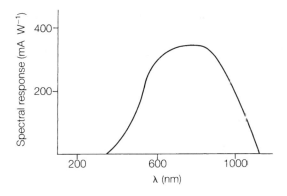

Figure 6.14. *Spectral response of a silicon photodiode.*

release an electron. Thus they may still be used for direct detection of x- or gamma rays. These detectors also respond very well to visible light and near infra-red (figure 6.14) so they can be used as silicon photodiodes in conjunction with a scintillation crystal (see section 6.8).

For better detection of x- and gamma radiation, thicker crystals of germanium with lithium diffused into them are used. They have adequate efficiency but, unlike silicon-based detectors which can operate at room temperature, they must be cooled to liquid nitrogen temperature ($-190\,°C$) during the whole of their working life. Very high purity germanium detectors (only about 10^9–10^{11} electrical impurities per cubic centimetre) which only need to be cooled during operation to reduce noise, are now available and have replaced lithium drifted detectors for most purposes.

Solid-state detectors dispense with the need for relatively bulky PMTs and the requirement for stabilized high voltage supplies. As previously intimated, they have a high ionization yield so the energies of photo-electrons generated by x- or gamma ray absorption can be measured with very high precision (figure 6.12(*b*)).

Silicon diode detectors have an increasingly important role as personal monitors. They have now been developed to the point where they provide both a direct and immediate reading of the dose rate or cumulative dose and are sufficiently reliable to provide a permanent dose record. The direct reading allows the worker to be aware of their radiation environment and to apply the ALARA principle (section 12.2.1). The recorded dose over a period of time satisfies legal requirements.

Typical performance characteristics include a range from 1 μGy to 16 Gy X-rays, a linear response with dose rate from 5 μGy h^{-1} to 3 Gy h^{-1}, calibration to better than 5%, rapid response time (1 μs), and stability over a wide range of temperature (-20 to $+80\,°C$) and humidity. Since the mean atomic number of silicon is very different from that of air, some variation in detector sensitivity with photon energy would be expected, especially in the photoelectric region (see section 6.12). However adequate energy compensation has been applied to give a uniform response ($\pm15\%$) from 20 keV up to megavoltage energies. For further information on electronic personal dosimeters see section 12.7.

6.11. PHOTOGRAPHIC FILM

The properties of photographic film as a recording medium for X-rays were discussed in chapter 4. Here it is sufficient to note that as a dosimeter, apart from its specialized use as a personal monitor (section 12.7), film has a number of disadvantages:

(1)   Because of the shape of its characteristic curve, it is a highly non-linear device. Therefore calibration is required at a large number of radiation exposures.

(2)   Again because of the shape of the characteristic curve, a given film may only be used over a limited range of dose.
(3)   Sensitivity expressed as film blackening per Gy in air is very dependent on energy, especially in the diagnostic range (see section 6.12).

6.12.   VARIATION OF DETECTOR SENSITIVITY WITH PHOTON ENERGY

An important question to consider for any radiation monitor is whether or not its sensitivity will change with photon energy. A good example of a detector which shows marked variation in sensitivity is photographic film (see figure 6.15).

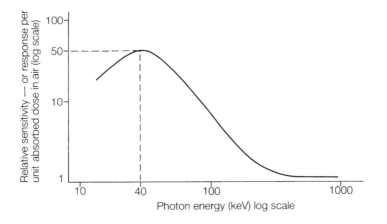

Figure 6.15. *Variation in sensitivity of photographic film as a function of photon energy.*

Since, in general, the response of a detector is proportional to the amount of energy it absorbs, the question can be answered in terms of absorbed energy. However, as shown on the left-hand axis of figure 6.15, 'sensitivity' means detector response per Gy, or per unit absorbed dose in air. The latter depends on the mass absorption coefficient of air. Thus sensitivity is determined by the relative mass absorption coefficients of the detector and air. If both vary with photon energy in a similar manner, detector sensitivity will not change, but if the variation of the two mass absorption coefficients with energy is dissimilar, detector sensitivity will change.

With this information, it is possible to interpret figure 6.15. In the vicinity of 1 MeV, interactions are by Compton processes. The mass absorption coefficient is therefore independent of atomic number and sensitivity is independent of photon energy. However as the photon energy decreases and approaches 100 keV, photoelectric absorption becomes important. The effect is much greater in film, which has a high mean atomic number (Z for silver $= 47$, Z for bromine $= 35$), than it is in air ($Z = 7.6$). Thus the mass absorption coefficient for film $(\mu/\rho)_F$ increases much more rapidly than $(\mu/\rho)_{\text{air}}$ and film sensitivity increases.

Note that below about 40 keV film sensitivity decreases. This is because of absorption edges for silver and bromine at 27 keV and 13 keV respectively. Near an absorption edge, although the incidence of photoelectric interactions is high, conditions are very favourable for the generation of characteristic radiation. Since a material is relatively transparent to its own characteristic radiation, a high proportion of this energy is reradiated and is therefore not available for film blackening (figure 6.16).

Detectors that have a mean atomic number close to that of air or soft tissue, for example lithium fluoride, show little variation in radiation sensitivity with photon energy, and are sometimes said to be 'tissue

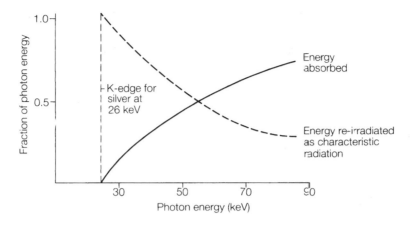

Figure 6.16. *Fraction of photon energy finally absorbed as a function of photon energy in the vicinity of a K absorption edge.*

equivalent'. Note that lithium fluoride does show some variation in sensitivity at very low energies where even small differences in atomic number can be important.

6.13. PATIENT DOSES IN DIAGNOSTIC RADIOLOGY

6.13.1. Principles

Why are doses measured?

It is important to remember that, almost invariably, the reason why doses are measured is to estimated the radiation risk and the reason for dose reduction is to minimize the risk. As discussed elsewhere in the book, especially in section 8.6, reducing the dose of radiation has an adverse effect on image quality. Therefore in the absence of risk to the patients much higher doses would frequently be used.

Where is the dose measured?

There is a very large amount of attenuation as a diagnostic X-ray beam passes through the body (figure 6.17). Thus the exit dose will be typically between 0.1% and 1% of the entrance dose.

For checking the sensitivity of equipment, especially image intensifiers, there is considerable merit in measuring the dose rate entering the imaging system (exit dose) but to assess doses to patients entrance doses must be measured. Also the doses to different organs within the beam will be very dependent on their depth and as shown in figure 6.17 the dose to a critical organ may be substantially different for AP and PA projections.

Clearly the sensitivity of the receptor will have a major effect on the dose to the patient and the relationship between speed class for different film speed combinations and exit dose is shown in table 6.1. As discussed in section 4.5 faster film screens exhibit poorer resolution. Typical values are shown in table 6.1 for comparison. The change in resolution is relatively small compared with the change in dose.

Two practical dose quantities have been recommended (IPSM 1992)—entrance surface dose for individual radiographs and dose–area product for complete examinations.

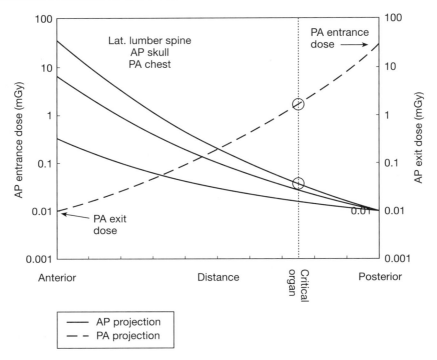

Figure 6.17. *Attenuation of an X-ray beam on passing through the body.*

Table 6.1. *The relationship between speed class and the dose required at the image receptor. The lower limit of visual resolution is shown for comparison (these values were obtained at 80 kV and using a 25 mm aluminium phantom—they will vary slightly with kV and filtration).*

Speed class	Dose at image receptor (μGy)	Lower limit of resolution (line pair mm^{-1})
50	20	4.0
100	10	3.4
200	5	2.8
400	2.5	2.4
800	1.25	2.0

Minimum focus–skin distances

The reasons why minimum focus–skin distances are specified for certain examinations (e.g. 140 cm for chest films, 60 cm for mammography, 20 cm for dental films) are, first to reduce geometric unsharpness (see section 5.9.1) and second to reduce patient dose. The reduction in patient dose is not always appreciated and is an important application of the inverse square law. Consider the situations shown in figures 6.18(*a*) and 6.18(*b*) (in the latter the 'patient' has been placed extremely close to the X-ray focal spot for the purpose of illustration).

The dose at the image receptor, D say, must be the same in both arrangements in order to achieve the same film blackening. Attenuation through the patient will be the same in both cases and can be ignored. However, the effect of the inverse square law is to increase the dose at the front surface of the patient by $(120/100)^2$, i.e. to $1.44D$ on the left but by $(40/20)^2$, i.e. to $4D$ on the right.

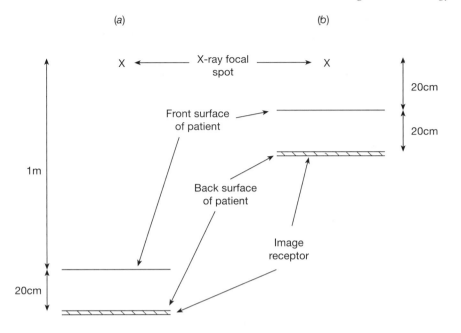

Figure 6.18. *Illustrating how the inverse square law affects the entrance dose to the patient—for explanation see text.*

There is unfortunately an adverse effect of the inverse square law. When longer focal–skin distances are used more X-rays (greater mA s) are required on the left to produce a dose (D) than on the right because the receptor is further from the focal spot.

How is the dose measured?

The entrance surface dose is defined as the absorbed dose in air at the point of intersection of the X-ray beam axis with the entrance surface of the patient, including back-scattered radiation, and a well defined protocol must be followed if results from different hospitals and different countries are to be compared. The Dosimetry Working Party of IPSM (1992) recommended that direct measurements of entrance surface dose should be made using small, independently calibrated, thermoluminescent dosimeters (TLDs). Because TLDs are small and of low atomic number they are unlikely to obscure diagnostic information and they cause no discomfort to the patient. TLDs generally provide the most straightforward and most accurate method of measurement and most major surveys have used them. Their disadvantages are (1) delay between exposure and read-out, (2) possibly slowing patient throughput in a busy department, (3) expensive TLD read-out equipment is required.

A dose estimate may be required when a TLD measurement has not been made. Therefore indirect methods of dose measurement should not be overlooked. To do this the tube output is measured under specified conditions using an ionization chamber in air, generally as part of routine quality assurance (QA) checks. A typical figure would be 100 μGy (mA s)$^{-1}$ for a modern tube operating at 80 kVp and 75 cm FFD. The entrance surface dose can then be calculated from a knowledge of the exposure factors, applying any necessary correction for source–skin distance and a factor to allow for back-scatter from the patient. Indirect methods also allow a large number of dose estimates to be made from a small number of measurements and may be useful at very low doses close to the detection limit of the TLDs. Indirect methods are difficult to apply to automatic control systems where exposure parameters are not known. The direct and indirect approaches are summarized in figure 6.19.

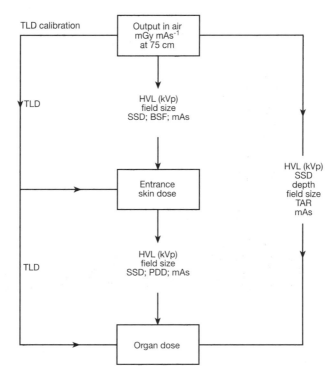

Figure 6.19. *Direct and indirect approaches to estimation of organ dose (reproduced with permission from Harrison 1997). HVL = half-value layer, SSD = source to skin distance, BSF = backscatter factor, PDD = percentage depth dose, TAR = tissue to air ratio.*

Note that the back-scatter factor can be quite high and will typically be in the range 1.2–1.4. It is left as an exercise to the reader to explain why the back-scatter factor will increase with increasing field size and increasing peak applied potential/half-value layer.

6.13.2.   *Entrance doses in plain film radiography*

Most surveys to date have shown a wide range of entrance doses for the same examination. Some of the spread is due to variation in patient thickness which has a significant effect because of rapid exponential attenuation of the X-ray beam in the patient. However, the observed variation cannot be explained by this single factor alone.

Consider for example the results shown in figure 6.20—a lumbar spine examination (L5–S1 projection) taken from a trial conducted by the Commission of the European Communities in 1991. Individual patient doses ranged from 2 mGy to 120 mGy. Equally surprising was that some hospitals had a very narrow range (less than 5 mGy), some a very wide range (up to 80 mGy). This pattern has been repeated, to a greater or lesser degree, for each survey that has been made.

In an attempt to put downward pressure on patient doses the National Radiological Protection Board (NRPB 1990) suggested reference values for entrance doses for standard radiological examinations. These reference values were set at the third-quartile point from an NRPB survey conducted in the mid-1980s and have been adopted by the European Union (table 6.2). All departments with mean dose levels above the reference value should carry out a critical examination of their procedures (see section 6.14). More recent

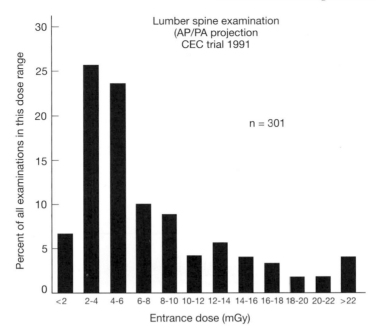

Figure 6.20. *Individual entrance doses registered in the CEC trial (1991) for lumbar spine examinations (Maccia* et al *1993).*

Table 6.2. *National reference doses and third-quartile values from the current UK database.*

Radiograph/examination	National reference dose (mGy)	Third-quartile value in current database (mGy)
Lumbar spine AP	10	7.0
Lumbar spine LAT	30	19.0
Lumbar spine LSJ	40	36.0
Chest PA	0.3	0.19
Chest LAT	1.5	0.66
Abdomen AP	10	7.2
Pelvis AP	10	5.3
Skull AP/PA	5	3.8
Skull LAT	3	1.8
Thoracic spine AP	7	5.1
Thoracic spine LAT	20	16.2

UK measurements (Hart *et al* 1996) show that the mean entrance dose has dropped by between 10% and 50% with an average reduction of around 30% for a range of studies (table 6.2, column 3).

6.13.3. *Entrance doses in fluoroscopy examinations*

As with film–screen combinations, the entrance dose to the patient will depend on the dose rate to the image intensifier. This should be checked at least once a year because deterioration in performance of the image intensifier, especially the resolution, can frequently be overcome by increasing the dose rate.

It follows that in routine use systems with automatic brightness control settings may permit higher input doses to achieve better resolution. The other factor that will affect the input dose rate is the field size. Since the area of the output screen of the image intensifier is fixed, the focusing effect of the intensifier is less when the input area is small so a higher input dose rate is required. These features are summarized in table 6.3.

Table 6.3. *Typical dose rates to the image intensifier of an IGE fluoroscopy unit under automatic brightness control. The maximum dose rate at the skin surface was estimated at* 1 mGy s^{-1}.

Field size (cm)	Dose rate settings (μGy s^{-1})		
	Low	Normal	High
30	0.3	0.5	1.0
23	0.6	1.0	1.8
15	1.0	1.7	3.0

One limitation of the definition of absorbed dose is that it changes very little with change in field size (there is a second-order effect due to a change in the amount of scatter). However, the risk to the patient will clearly be greater if the dose is delivered over a bigger area, so the risk is more closely related to the total energy deposited as ionization than to the absorbed dose. Therefore one of the variables that needs to be known when a surface dose is measured with a TLD is the field size. When screening with image intensifier systems field sizes are non-standard and may vary during a study so risk assessment based on surface TLDs is not appropriate.

Dose–area product, usually expressed in Gy cm^2, is a measure of the total energy fluence incident on the patient and may be related to the total energy absorbed in the patient. Some studies for which dose–area product meters should be used include barium enemas, barium meals, micturating cystograms, cardiac investigations and interventional techniques in neuroradiology and biliary procedures.

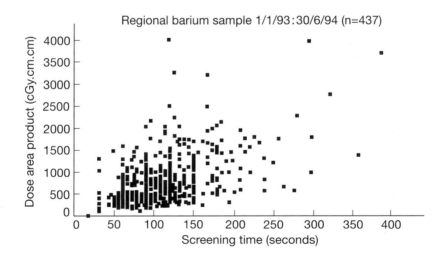

Figure 6.21. *Individual dose–area product meter results for a large sample of barium meal examinations.*

Figure 6.21 shows the results of a survey of barium meal studies carried out on 429 patients in six hospitals in the East Anglian Region of the UK in 1993/94. The spread in dose area product readings is even greater than for plain film studies because investigations vary in complexity and screening time depends on the skill of the operator. National reference doses and current third-quartile values in the NRPB database are shown in table 6.4.

Table 6.4. *National reference dose–area product figures and third-quartile values from the current database.*

Examination	National reference dose (Gy cm^2)	Third-quartile value in current database (Gy cm^2)
Barium enema	60	32
Barium meal	25	17
IVU	40	23

6.13.4. Doses to organs

In order to estimate the radiation risk, the entrance surface dose must be used to calculate the dose to individual organs. This will depend on whether or not the organ is in the primary beam, the depth of the organ and the kV, i.e. penetrating power, of the radiation.

A general consideration of this problem is beyond the scope of the book but one specific situation that requires the calculation of an organ dose is when a patient has been exposed to X-rays during an unsuspected pregnancy. Although the exact position of the foetus depends on gestational age, available data on irradiation during pregnancy are generally not sufficiently refined to make calculations other than for the uterus. Therefore the figures given in table 6.5 are a guideline only (NRPB 1998). Note that the figures must be related to some knowledge of the dose to the mother, otherwise they are meaningless.

Table 6.5. *Typical values for the mean dose to the uterus in radiological examinations.*

Examination	Mean entrance skin dose to mother AP projection (mGy)	Mean total dose to uterus (mGy)
Lumbar spine	6.1	1.7
Abdomen	5.6	1.4
Pelvis	4.4	1.1
Chest, skull, thoracic spine		<0.01
	Mean dose–area product (Gy cm^2)	
Barium enema	26	5.8
Barium meal	13	1.1

The dose to the uterus will be very small when it is not in the primary beam as shown in the table. As a further example, a CT scan of the pelvis could give a dose of 25 mGy to the uterus; a CT scan of the chest would give a dose of only 0.06 mGy.

6.14. PATIENT DOSE REDUCTION IN DIAGNOSTIC RADIOLOGY

In 1990 a joint working party of the Royal College of Radiologists and the National Radiological Board (NRPB 1990) made 21 recommendations for patient dose reduction in diagnostic radiology. Since this book is concerned with the physics of radiology, the emphasis will be on technical factors that affect the dose but other considerations must not be overlooked and will be mentioned briefly.

6.14.1.   Technical factors

(1)   The choice of operating kilovoltage must be optimized and the appropriate amount of filtration must be used. Doses will generally be lower at high kV because of better patient penetration—but note that use of a grid may negate the dose reduction.

(2)   The most sensitive film–screen combination should be chosen that will give the requisite image quality. Note that the sensitivity of the film–screen combination varies with the kVp. Therefore to achieve maximum dose reduction the kV should correspond to the maximum sensitivity.

(3)   A significant cause of retakes, leading to unnecessary patient exposure, is poor film processing. Manufacturer's recommendations for film processing must be followed exactly and a rigorous quality assurance programme is necessary to check that this happens.

(4)   Carbon fibre components—couch, antiscatter grids, cassette fronts—attenuate less radiation than metal components and should be used whenever possible.

(5)   Automatic exposure control devices should be used where practicable and must be checked regularly for reliability.

(6)   If grids are essential, low grid factors should be used. Remove the grid completely whenever possible—consider using an air gap as an alternative.

(7)   Optimize the focus–film distance—see section 6.13.1.

(8)   Use automatic beam collimation whenever possible.

(9)   Use image storage devices with video recorders/spot film photofluorography/pulsed systems to reduce substantially screening times.

Periodic measurements of patient entrance skin dose should be made and the overall quality assurance programme for the department must be such that, when technical problems have been identified and corrected, patient doses are re-audited to demonstrate that the necessary dose reduction has been achieved.

6.14.2.   Non-technical factors

(1)   Probably the most frequent and certainly most avoidable unnecessary exposure to patients is the result of inappropriate radiological examinations. The guidelines in '*Making the Best Use of a Department of Clinical Radiology*' (RCR 1998) contains excellent advice on when X-rays are inappropriate or at least should be deferred for a few weeks.

(2)   In most centres arrangements could be improved considerably to ensure that previous films are available. Hopefully the advent of the digitized department will help to reduce this problem.

(3)   Consideration must be given to using alternative imaging modalities such as ultrasound and MRI thereby avoiding the use of X-rays.

(4)   Employment-related screening programmes should be used sparingly and clinically justified.

(5)   All staff should be given appropriate training, covering both an awareness of the radiological examinations that carry the highest doses and of the procedures that are available to minimize those doses.

6.15.   CONCLUSIONS

A wide range of properties of ionizing radiation is available for radiation measurement. Ionization in air is taken as the reference standard, partly because it can be related directly to fundamental physical processes and partly because it is very sensitive in terms of the number of ion pairs created per unit energy deposition.

The choice of instrument in a given situation will depend on a variety of factors including sensitivity, linearity, dynamic range, response time, performance at high count rates, variation in response with photon energy, uniformity of response between detectors, long term stability, size, operating conditions such as temperature or requirements for stabilized voltage supplies and cost.

It is important to distinguish carefully between radiation measurement with for example an ionization chamber and radiation detection with a Geiger–Müller counter. The latter is very sensitive and may be very suitable for detecting radiation leakage or contamination from spilled radioactivity. It should not normally be used as a radiation measuring device unless specially adapted to do so.

Three aspects of radiation measurement have been identified—exposure, absorbed dose and the quality or spectral distribution of the radiation. If an absolute measure of absorbed dose is required, reference back to the ionization process and cross calibration will be required. However, for many applications, for example in digital radiology and nuclear medicine, there is no need to convert the numerical data into dose routinely so uniformity of response and long term stability are more important. One area in which quantitative dose measurements are required is personal monitoring and this aspect will be discussed in chapter 12.

If information is required on spectral distributions, a detector whose response is proportional to the energy of an individual photon must be used. An NaI(Tl) scintillation detector will normally be chosen and might be used for example to check the homogeneity of an X-ray beam or to identify (and reject) low energy scattered radiation from a monochromatic gamma ray beam. If very precise energy resolution is necessary a cooled solid-state detector may be required.

Routine measurement of doses to patients is becoming increasingly important and there is early evidence to indicate that patient doses are coming down as a result of careful attention to technical detail.

REFERENCES

Harrison R M 1997 Ionizing radiation safety in diagnostic radiology *Imaging* **9** 3–13

Hart D, Hillier M C, Wall B F, Shrimpton P C and Bungay D 1996 *Doses to Patients from Medical Examinations in the UK 1995 Review* NRPB R289 (London: HMSO)

Institute of Physical Sciences in Medicine (IPSM), National Radiological Protection Board (NRPB) and College of Radiographers 1992 *National Protocols for Patient Dose Measurements in Diagnostic Radiology* (Chilton, Didcot, Oxfordshire OX11 0RQ, UK)

Maccia C, Ariche-Cohen M, Severo C and Nadeua X 1993 The 1991 trial on quality criteria for diagnostic radiographic images *Documents for the CEC XII/221/93*

NRPB 1990 Patient dose reduction in diagnostic radiology *Report by the Royal College of Radiologists and the National Radiological Protection Board (Documents of the NRPB 1 No 3)*

—— 1998 Advice on exposure to ionising radiation during pregnancy (Chilton, Didcot, Oxfordshire, OX11 0RQ, UK)

Royal College of Radiologists (RCR) 1998 *Making the Best Use of a Department of Clinical Radiology* 4th edn (London: RCR)

FURTHER READING

Greening J R 1985 *Fundamentals of Radiation Dosimetry (Medical Physics Handbooks 15)* 2nd edn (Bristol: Hilger–Hospital Physicists' Association)

Lovell S 1979 *An Introduction to Radiation Dosimetry* (Cambridge: Cambridge University Press)

McAlister J M 1979 *Radionuclide Techniques in Medicine* (Cambridge: Cambridge University Press)

Tait W H 1980 *Radiation Detection* (London: Butterworths)

Wall B F, Harrison R M and Spiers F W (ed) 1988 Patient dosimetry techniques in diagnostic radiology *IPSM Report* 53

Whyte G N 1959 *Principles of Radiation Dosimetry* (New York: Wiley)

EXERCISES

1   Suggest reasons why ionization in air should be chosen as the basis for radiation measurement.

2   Draw a labelled diagram of a (free air) ionization chamber and explain the principle of its operation.

3   Explain how the following problems are overcome in an ionization chamber: (*a*) recombination of ions, (*b*) definition of the precise volume from which ions are collected.

4   Explain how a (free air) ionization chamber might be used to measure the exposure at a given point in an X-ray beam. How would you expect the exposure to change if thin aluminium filters were inserted in the beam about half way between the source and chamber?

5   Outline briefly the important features of an experimental arrangement for measuring exposure in air and discuss the factors which limit the maximum energy of the radiation that can be measured in this way.

6   Explain from first principles why the reading on a free air ionization chamber will decrease if the temperature increases.

7   Show that under specified conditions the reading on a dose–area product meter will not vary with the distance from the tube focus. What conditions must be satisfied?

8   Explain the importance of the concept of electron equilibrium in radiation dosimetry.

9   Describe a small cavity chamber for measurement of radiation exposure. Discuss the choice of material for the chamber wall and its thickness.

10   Describe the operation of a Geiger–Müller tube and explain what is meant by 'dead time'.

11   Define the gray and show how absorbed dose is related to exposure.

12   Estimate the number of ion pairs created in a cell 10 μm in diameter by a single dose of 10 μGy X-rays.

13   Explain how a scintillation detector works.

14   Show that the energy of a photon in the visible range is about 3 eV.

15   Describe the sequence of events that leads to a pulse of electrons at the anode of a photomultiplier tube if the tube is directed at a NaI scintillation crystal placed in a beam of photons.

16   Explain how a scintillation detector may be used to measure: (*a*) the energy and (*b*) the intensity of a beam of radiation.

17   Explain in as much detail as possible the shape of the curve in figure 6.15.

18   Describe the process of thermoluminescence and explain how a thermoluminescent material may be calibrated for dosimetry.

CHAPTER 7

DIAGNOSTIC IMAGING WITH RADIOACTIVE MATERIALS

7.1. INTRODUCTION

Diagnostic imaging with radioactive materials is the largest single component of nuclear medicine, which has been defined by the World Health Organisation as embracing all applications of radioactive materials in the diagnosis or treatment of patients, and in medical research (with the exception of the use of sealed radiation sources in radiotherapy).

A useful subdivision is in terms of the type of procedure involved. For example some studies are carried out entirely *in vitro*. A specimen, usually blood, is taken from the patient and is incubated with radioactive precursors of the metabolite of interest. Various constituents can then be isolated and their gamma or beta activity counted in a well counter or liquid scintillation counter respectively. Alternatively, the radioactive label may be administered to the patient and measurements are made on samples, e.g. blood, urine, faeces, taken from the patient. For details of the theory of *in vitro* counting see Faires and Boswell (1981).

Other studies are carried out entirely *in vivo* and these may be further subdivided into those which are primarily concerned with counting and those which involve imaging. If it is important to know the precise amount of activity in an organ but an image of the distribution is not required, collimated scintillation detectors directed at the organ of interest may be used (Belcher and Vetter 1971). If the amount of activity is very low, the counter must be designed for maximum sensitivity and will consist of several heavily shielded probe detectors. 'Whole body counting' is used for protection work in nuclear medicine to measure uptakes by staff of small quantities of Tc-99m or I-131. Alternatively the emphasis may be on imaging, with the primary requirement being to obtain information concerning the spatial distribution of activity. This chapter deals with the physical principles involved in obtaining diagnostic quality images after a small quantity of radioactive material has been administered to the patient in a suitable form.

The basic requirements of a good imaging system are:

(1) a device that is able to use the radiation emitted from the body to produce high resolution images, supported by electronics, computing facilities and displays that will permit the resulting image to be presented to the clinician in the manner most suitable for interpretation;

(2) a radionuclide that can be administered to the patient at sufficiently high activity to give an acceptable number of counts in the image without delivering an unacceptably high dose of radiation to the patient and

(3) a radiopharmaceutical, i.e. a radionuclide firmly attached to a pharmaceutical, that shows high specificity for the organ or region of interest in the body.

163

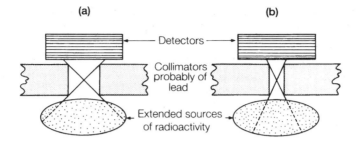

Figure 7.1. *Collimator design showing conflicting requirements of sensitivity and resolution. Arrangement (a) where the detector has a wide acceptance angle will have high sensitivity but poor resolution, whereas arrangement (b) will have much better resolution but greatly reduced sensitivity.*

It is important to recognize that, when detecting *in vivo* radioactivity, sensitivity and spatial resolution are mutually exclusive (see figure 7.1). The arrangement on the left (figure 7.1(*a*)) has high sensitivity because a large amount of radioactivity is in the field of view of the detector, but poor resolution. The arrangement on the right (figure 7.1(*b*)) has better resolution but correspondingly lower sensitivity. Since gamma rays are emitted in all directions, sensitivity can be increased by increasing the detector area and, when this area is maximized in the so-called 'whole body counter', resolution has been effectively reduced to zero. For a recent report on the performance of whole body counters see Fenwick *et al* (1991). In diagnostic imaging spatial resolution is important and sensitivity must be sacrificed. A modern gamma camera (see section 7.2.1) is at least 100 times less sensitive than a typical whole body counter and records no more than 1 in 10^4 of the gamma rays emitted from that part of the patient within the field of view of the camera. Furthermore, any additional loss of counts in the complete system will result in an image of inferior quality unless the imaging time is extended to compensate. Therefore this chapter also considers the factors that limit image quality and the precautions that must be taken to optimize the images obtained using strictly controlled amounts of administered activity and realistic imaging times.

7.2. PRINCIPLES OF IMAGING

The radiations that are most suitable for *in vivo* imaging are medium energy gamma rays in the range 100–200 keV. Lower energy gammas, alpha particles and negative beta particles are stopped in the body resulting in an undesirable patient dose, whilst higher energy gammas are difficult to stop in the detector. Further discussion of this point and its influence on the choice of radionuclide and radiopharmaceutical is deferred until section 7.3 where factors affecting the quality of radionuclide images are considered.

In all commercial equipment currently available, the radiation detector is a scintillation crystal of sodium iodide doped with about 0.1% by weight of thallium—NaI(Tl). The fundamental interaction process in a scintillation detector is fluorescence which was discussed in section 4.2. The sodium iodide has a high density (3.7×10^3 kg m^{-3}) and since iodine has a high atomic number ($Z = 53$) the material has a high stopping efficiency for gamma rays. Furthermore, provided the gamma ray energy is not too high, most of the interactions are by the photoelectric effect (see section 3.4) and result in a light pulse proportional to the gamma ray energy. This is important for discriminating against scatter (see section 7.3.4). The thallium increases the light output from the scintillant, because the traps generated by thallium in the NaI lattice are about 3 eV above the band of valency electrons so the emitted photon is in the visible range and about 10% of the gamma ray energy is converted into light. This yields about 4000 light photons at a wavelength of 410 nm from a 140 keV gamma ray. Note that whereas the number of photons emitted is a function of the energy imparted by the interaction, the energy or wavelength of the photons depends only on the positions of the energy levels in the scintillation crystal. Finally, the light flashes have a short decay time, of the order of 0.2 μs. Thus the crystal has only a short dead time and can be used for quite high counting rates. One disadvantage of the NaI(Tl) detector is that it is hygroscopic and thus must be placed in a hermetically sealed container. Also the large crystals in gamma cameras are easily damaged by thermal or physical shocks.

Alternative scintillation detectors are caesium iodide doped with thallium, bismuth germanate, cadmium tungstate and rare earth ceramic oxides. Desirable properties are a high detection efficiency, little afterglow at high counting rates, a good dynamic range and long-term stability.

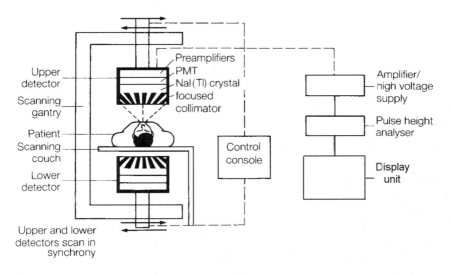

Figure 7.2. *Labelled diagram showing essential features of a scanner. For details of the functions of various parts see text.*

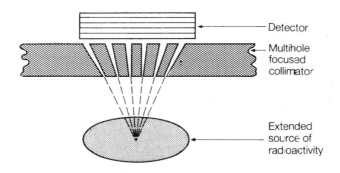

Figure 7.3. *Design of a multihole focused collimator which combines some of the advantages of good resolution with increased sensitivity.*

Two fundamentally different methods have been developed for obtaining information about the spatial distribution of radioactivity in the body. The older and conceptually simpler approach is to 'scan' the patient (figure 7.2). Positional information is obtained by the purely mechanical method of moving two relatively small scintillation crystal detectors positioned one above and one below the patient, in a systematic raster motion. The field of view is defined by a focused collimator (figure 7.3). Typically the holes are tapered so that the collimator is 'focused' at a point 10–15 cm below its lower surface. This design ensures that the system has a resolution comparable with that of a small hole, but sensitivity is increased by a factor N, the number of holes. Note that the collimator is only 'focused' in the sense that all channels are directed to the same point. No refraction of gamma rays takes place as when light is focused.

The light signal produced by a scintillation crystal is too small to be used until it has been amplified and this is almost invariably achieved by using a photomultiplier tube (PMT). The main features of the PMT coupled to a scintillation crystal were discussed in section 6.8—see also figure 6.10.

The technique of **pulse height analysis** is used to select those pulses corresponding to the gamma ray energy of the radionuclide being imaged (see section 6.9). For a radionuclide emitting monoenergetic gamma rays, pulse height analysis should, in principle, discriminate completely between scattered and unscattered rays. Consider for example a 140 keV gamma ray from technetium-99m. When it interacts with an NaI(Tl) crystal, it does so primarily by the photoelectric effect, producing a number of visible photons and hence a final signal that is proportional to the gamma ray energy. Any photon that has been scattered in the patient by the Compton effect will be of lower energy and will produce a smaller pulse that can be identified and rejected (see figure 6.12). Note that there is further discussion on this point in section 7.3.4.

If a pulse is accepted by the pulse height analyser, a signal is passed to the display unit. In the old-fashioned scanner as the detectors moved in a raster motion an image was built up. For routine imaging, scanners have been entirely replaced by gamma cameras which are discussed in detail in the next section.

7.2.1.   The gamma camera

A gamma camera may consist of two units, the collimated detector mounted on a stand to allow it to manoeuvre around the patient, and the console containing pulse processing electronics, counters for controlling the acquisition time, correction circuits and cathode ray tube (CRT) displays. The scintillation crystal is much larger than in a scanner, typically 400 mm across.

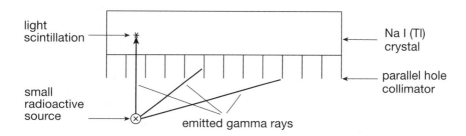

Figure 7.4. *Use of a collimator to encode spatial information. In the absence of the collimator radiation from the source may strike any point in the crystal.*

The principles of operation are as follows. First, collimation is used to establish the spatial relationship between the point of emission of a gamma ray in the patient and the point at which it strikes the crystal (figure 7.4). Note that unlike a grid in conventional radiology, the collimator in radionuclide imaging has no role in discriminating against scatter within the patient. When an incident gamma ray interacts with the crystal (an event), visible light photons are produced. These spread out into the crystal and some go to each PMT. Up to 100 PMTs will be arranged in a close-packed hexagonal array behind the crystal to improve spatial resolution. As shown in figure 7.5, the number of photons reaching each PMT, and hence the strength of the signal, will be determined by the solid angle subtended by the event at that PMT. Hence, by analysing all the PMT signals, it is possible to determine the position of the gamma ray interaction in the crystal. Essential features of the gamma camera may be considered under six headings.

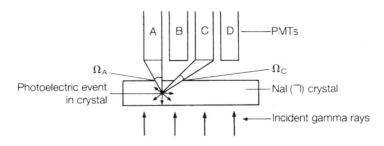

Figure 7.5. *Use of an array of PMTs to obtain spatial information about an event in an NaI (TI) crystal. Light photons spread out in all directions from an interaction and the signal from each PMT is proportional to the solid angle subtended by the PMT at the event. The signal from PMT A is proportional to Ω_A and much greater for the event shown than the signal from PMT C which is proportional to Ω_C.*

The detector system

Components of the detector system are shown in figure 7.6. In the gamma camera, crystal thickness must be a compromise. A very thin crystal reduces sensitivity whereas a very thick crystal degrades resolution (figure 7.7). A camera crystal is typically 6–12 mm thick. As shown in table 7.1 a 12.5 mm crystal stops most of the 140 keV photons from Tc-99m but is less well suited to higher energies. The detector system is protected by lead shielding to stop stray radiation.

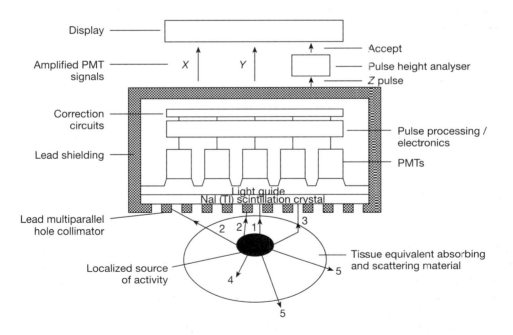

Figure 7.6. *Basic components of a gamma camera detector system. The fates of photons emitted from the source may be classified as follows: 1, useful photon; 2, oblique photon removed by collimator; 3, scattered photon removed by pulse height analyser; 4, absorbed photon contributing to patient dose but giving no information; 5, wasted photons emitted in the wrong direction.*

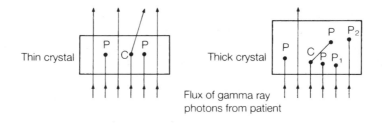

Figure 7.7. *Interactions of gamma rays with thin and thick NaI (TI) crystals. P = photoelectric absorption. C = Compton scattering. With a thin crystal, many photons may pass through undetected, thereby reducing sensitivity. With a thick crystal the image is degraded for two reasons. First, the distribution of light photons to the PMTs for an event at the front of the crystal such as P_1 will be different from the distribution for an event at the rear of the crystal such as P_2. Second, scatter in the crystal degrades image quality since the electronics will position 'the event' somewhere between the two points of interaction in the crystal.*

Table 7.1. *Stopping capability of a 12.5 mm thick NaI (Tl) crystal for photons of different energy.*

Photon energy (keV)	Interactions (%)
80	100
140	89
200	60
350	23
500	15

The collimator

The most common type of collimator, which has parallel holes, is shown in figure 7.8(a). It consists of a thick lead plate in which a series of small holes has been microcast or constructed from stacks of corrugated foil. The axes of the holes are perpendicular to the face of the collimator and parallel to each other.

Performance of the collimator will be determined primarily by its resolution and sensitivity. As shown in figure 7.9 long narrow holes will produce high resolution but low sensitivity so these two variables work against

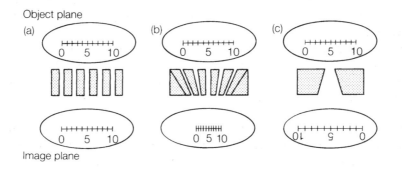

Figure 7.8. *The effect of different collimator designs on image appearance. (a) The parallel hole collimator produces the most faithful reproduction of the object. (b) The diverging collimator produces a minified image but is useful when the required field of view is bigger than the detector area. (c) The pin-hole collimator produces an enlarged inverted image and is useful for very small fields of view.*

each other. A typical compromise would be a resolution of 6 mm and a sensitivity of 0.01–0.02%. As the object is moved away from the collimator face, resolution deteriorates markedly so all imaging should be done with the relevant part of the patient as close as possible to the collimator face. Sensitivity is relatively independent of distance from the collimator face, only decreasing if additional attenuating material is interposed.

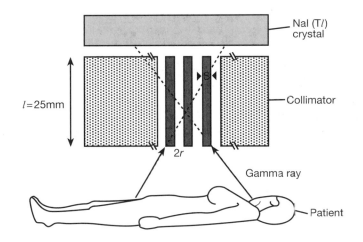

Figure 7.9. *Diagram showing that oblique gamma rays will pass through many lead strips, or septa, before reaching the detector. Typical dimensions for a low energy collimator are $l = 25$ mm, $2r = 3$ mm, $s = 0.2$ mm. The number of holes will be approximately 15 000.*

Figure 7.9 illustrates another problem. Higher energy gamma rays may be able to penetrate the septa and this will cause serious image degradation. Thicker septa are now required and for adequate sensitivity this also means larger holes and correspondingly poorer resolution.

Other collimator designs are used for special purposes. Figure 7.8(*b*) illustrates a diverging collimator which produces a scintillation image that is smaller than the object and this may be useful when the required field of view is larger than the imaging detector. With a modern gamma camera diverging collimators are rarely necessary. A variant in which the holes diverge in one dimension only, a fish-tail collimator, has been used in the scanning gamma camera (see section 7.2.2) but large field of view camera heads, typically 500 mm × 400 mm, can cover the patient adequately. The converse, not illustrated, is a converging collimator which will magnify the image of a small organ. A variation used to image the brain is a cone-beam collimator. This gives improved sensitivity and resolution.

Note that both diverging and converging collimators introduce distortion because the magnification or minification factor depends on the distance from the object plane to the collimator and is therefore different for activity in different planes in the object. There are also variations in resolution and sensitivity across the field of view as the hole geometry varies from being almost parallel at the centre to highly angled near the edge.

To image small objects a pin-hole collimator which functions in a manner analogous to the pin-hole camera may be useful (figure 7.8(*c*)). The pin-hole is a few millimetres in diameter and effectively limits the gamma rays to those passing through a point. The ratio of the size of image to the size of object will depend on the ratio of the distance of the image plane from the hole to the distance of the object plane from the hole. The latter distance must be small if reasonable magnification is to be achieved. The thyroid gland is the organ most frequently imaged in this way.

Note that the pin-hole collimator suffers from the same distortions as converging and diverging collimators and also inverts the image as shown in figure 7.8.

Pulse processing

Pulse arithmetic circuits convert the outputs from the PMTs into three signals, two of which give the spatial co-ordinates of the scintillation, usually denoted by X and Y, and the third the energy of the event Z (see figure 7.6).

Each PMT has two weighting factors applied to its output signal, one producing its contribution to the X co-ordinate, the other to the Y co-ordinate. Several different mathematical expressions have been suggested for the shape of the weighting factors. Those which give the greatest weight to PMTs nearest the event are to be preferred since they will be the largest signals and hence least susceptible to statistical fluctuations due to noise (for fuller discussion see Sharp *et al* 1985). The final X and Y signals are obtained by summing the contributions from all tubes.

The energy signal Z is produced by summing all the unweighted PMT signals. This signal is then subjected to pulse height analysis as described earlier in this section and the XY signal is only allowed to pass to the display system if the Z signal falls within the preselected energy window.

Correction circuits

Image quality has been improved considerably in recent years by using microprocessor technology to minimize some of the defects that are inherent in a gamma camera. Exact methods vary from one manufacturer to another; the examples given below illustrate possible approaches.

Spatial distortion may be corrected by imaging a set of accurately parallel straight lines aligned with either the X or Y axis. The deviation of the measured position of each point on a line from its true position can be measured and stored as a correction matrix which may then be applied to any subsequent clinical image.

Similarly any variation in the energy signal with the position of the scintillation in the crystal can be determined by imaging a flood source and recording the counts in two narrow energy windows situated symmetrically on either side of the photo-peak. If the measured photo-peak coincides exactly with the true photo-peak, the counts in each energy window will be the same. If the measured peak is shifted to one side, this will be reflected in a higher number of counts in the corresponding window. Once again variations in the measured photo-peak from the true photo-peak can be stored as a correction matrix for each part of the crystal.

Finally it is important to monitor and adjust the gains of the PMTs. One way to do this is to use light emitting diodes to flood the crystal with light.

Image display

The method of display widely used in the 1980s was a high quality cathode ray tube (CRT) in which the X and Y signals could be applied to the deflection plates so that the electron beam in the CRT is directed at the point on the screen corresponding to the point of interaction of the gamma ray in the crystal. In this system, if the Z signal shows that there is an acceptable pulse—i.e. it comes from a gamma ray that is likely to be unscattered—a signal is sent to the CRT which briefly increases the intensity of the electron beam resulting in a bright spot on the display. Since these light spots have only a short lifetime, it is necessary to integrate them onto film using a photographic camera. Note that the film used in nuclear medicine is single sided since only light photons are being detected by this stage in the process.

Multiformatting, which permits several views to be displayed on the same piece of film, is sometimes used. This not only saves film but also allows exposures to be made simultaneously at different intensities. Multiformatters also allow several sequential images in a dynamic study to be recorded automatically but this approach is rapidly being superseded by the introduction of digital cameras (see below).

The imaging process discussed above, in which one pulse of light will be allowed to fall on the film at each point corresponding to an event in the sodium iodide crystal results in an **analogue image**. This method of display is normally adequate for static images. However, it has a number of disadvantages:

(1) If incorrect exposure conditions are selected and the film is too dark or too light, the examination will have to be repeated. For a dynamic study (see section 7.4) this means a second injection.
(2) Numerical data cannot be extracted.
(3) The image cannot be manipulated before display—e.g. by altering the contrast or adjusting the background level.

All these problems are overcome in a **digitized image** in which the image space is sub-divided into a matrix of pixels (usually 64×64 or 128×128 or 256×256 in nuclear medicine investigations). The total number of counts in each pixel is then recorded.

Earlier gamma cameras incorporated an analogue-to-digital converter which permitted the final analogue image to be converted into a digital image when necessary. However, the whole process of image formation by a gamma camera is ideally suited to digitization and the latest generation of cameras has dispensed with the analogue facility. Either the X, Y and Z signals are digitized before correction factors are applied or, more recently, analogue to digital converters have been fitted to each PMT thereby digitizing much earlier in the imaging process. Such systems are frequently referred to as *digital cameras*. Block diagrams showing the components of three generations of gamma cameras equipped with reasonably comprehensive data processing facilities are shown in figure 7.10.

Mechanical aspects

The control console enables the operator to set limits on energy selection circuits, control overall intensity on the display and choose either time for collection or total count limit for the image. An anatomical marker is a useful accessory. This is a small closed source (typically about 4 MBc Co-57) which may be held over a particular anatomical point of interest, or used to distinguish left from right where confusion might exist. The gamma rays from the marker source will produce a bright spot on the image. The camera head can be raised, lowered or tilted at any angle and a comfortable examination couch minimizes patient movement.

7.2.2. *Variations on the standard camera*

The mobile camera

The standard gamma camera is not readily moved from one room to another owing to the weight of lead shielding. However, there are a number of clinical conditions where there is a need to take the camera to the patient. Mobile cameras are designed primarily for cardiac work so they have a small field of view (about 30 cm), and since they are only used with radionuclides that emit low energy gamma rays, for example in addition to Tc-99m, currently mainly thallium-201 which emits gamma rays at 80 keV, the NaI(Tl) crystal can be thinner, typically 6–9 mm. Performance is comparable with that of a standard camera.

The scanning camera

This combines the advantages of a camera with some of those of a scanner to obtain a whole body scan, especially a bone scan, in a single image.

Either the detector moves on rails along the length of the patient or the bed moves under the detector and the Y position signals have an offset DC voltage signal applied to them which is a function of detector position. All the data are collected in a single pass of the camera over the patient thereby producing a non-overlapping image, hence facilitating interpretation, in the shortest possible scan time.

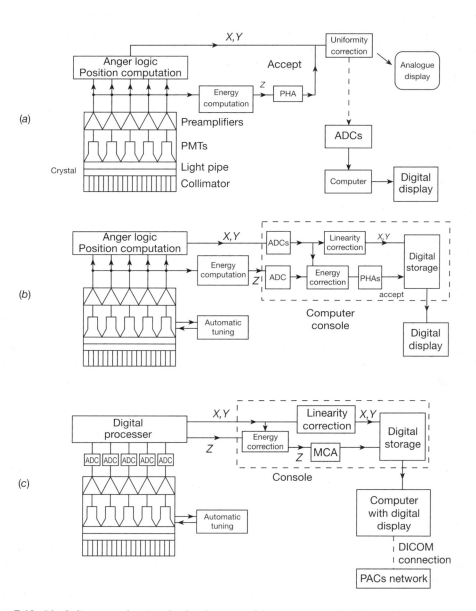

Figure 7.10. *Block diagrams showing the development of data processing facilities with a gamma camera. (a) 1970s. Anger logic used to position events, as described in the text; pulse height analyser (PHA) discriminated against scatter; displays were mainly analogue with the occasional option of analogue-to-digital conversion (ADC). (b) 1980s to mid 1990s. PMTs were tuned individually; stand alone computer consoles were introduced with ADC as standard; two or three PHAs were provided to allow more than one gamma ray energy to be collected; linearity correction was introduced; images stored digitally. (c) Mid-1990s. ADCs fitted to the output from each PMT/preamplifier; signals processed digitally throughout; multichannel analysers allow photons at many gamma ray energies to be collected simultaneously; networking becomes a possibility.*

Although the scanning motion effectively extends the field of view along the axis of the patient, a conventional crystal is not big enough to cover the lateral field of view in a single pass. This problem was first overcome by fitting a fish-tail collimator (see section 7.2.1) and more recently by building cameras with rectangular crystals, specifically for scanning. Such cameras are less versatile since the rectangular field of view is not ideal for other studies.

One final point about scanning cameras is that if an additional separation between detectors is necessary to allow relative movement between the collimator face and the patient, resolution will be degraded. Some systems now incorporate automatic contouring with an infra-red beam or a 'learn mode' to minimize patient–detector separation.

Tomographic camera

By rotating the gamma camera around the patient and collecting data either continuously or at a fixed number of angles, a set of profiles may be collected and reconstructed to form sectional images. The technique of single photon emission computed tomography is discussed further in section 10.4. Slip ring technology can now be used to allow continuous rotation of the camera head.

Dual headed camera

One way to achieve greater sensitivity is to increase the area of crystal available for stopping gamma rays. This is one of the features of dual headed cameras which typically contain two high resolution rectangular detectors capable of acquiring full field of view anterior and posterior whole body scans simultaneously. The increased sensitivity may be used either to permit faster imaging times or to achieve the same counts in the image in the same imaging time with half the administered activity, and hence half the dose to the patient. Triple headed cameras are also available for brain work.

7.3. FACTORS AFFECTING THE QUALITY OF RADIONUCLIDE IMAGES

As with all diagnostic images, the radiologist should always ask of radionuclide images 'Are the pictures of good quality—and if not, why not?'. The numerous factors that affect image quality will now be considered.

7.3.1.   Information in the image and signal-to-noise ratio

It is well known that the quality of an image depends on the number of photons it contains. Figure 7.11 shows three images of a simple phantom frequently used in nuclear medicine with different numbers of counts in each image. Unfortunately, because injected radioactivity spreads to all parts of the body and is retained for several hours (in contrast to X-rays which can be confined to the region of interest and 'switched off' after the study), *in vivo* nuclear medicine investigations are always photon limited by the requirement to minimize radiation dose to the patient. The number of useful gamma rays is further reduced by the heavy collimation that has to be employed. Hence a typical photon density in radionuclide imaging is of the order of 1 mm^{-2} compared with about 10^5 mm^{-2} in radiography and 10^{10} mm^{-2} in conventional photography.

Since the photon count is low, a minimum value for the noise in the image will be the Poisson error $n^{1/2}$ on the measured counts n (there will be other sources of noise). By considering a simple model of a small spherical region of activity in a uniform background of lower activity, it is possible to show (Sharp *et al* 1985) that the signal-to-noise ratio depends on (i) the square root of the total counts, (ii) the lesion to background concentration ratio, (iii) the square root of the sensitivity of the imaging device.

Thus the information in the image can, in theory, always be increased by increasing the time of data collection but the signal-to-noise ratio only increases as the square root of time. Also this time will be limited

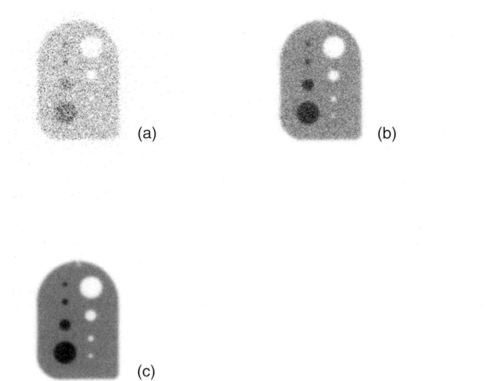

Figure 7.11. *Images of a phantom at different count densities—(a) 20 kilocounts, (b) 200 kilocounts, (c) 2000 kilocounts. The cold areas on the left and hot areas on the right become sharper as the number of counts in the image increases.*

by the length of time the patient can lie still, and the workload on the camera. In some situations physiological factors, e.g. especially movement, will negate the potential gain in image quality from a long data collection time. Thus the primary objective is to obtain the maximum number of counts in the image in a given time with the maximum differential uptake into the organ or lesion of interest, subject to the limitation of an acceptable radiation dose to the patient. In achieving this objective, choice of radionuclide and choice of radiopharmaceutical are the two main factors to be considered.

7.3.2. *Choice of radionuclide*

It is important to use a short half-life radionuclide so that, for a given injected activity, the radiation dose to the patient after the examination is as low as possible. Note, however that the half-life should not be too short compared with the planned duration of the study, and very short half-life materials may create problems of availability.

For the same reason a radionuclide which decays to a non-radioactive or very long half-life daughter should be chosen. If both parent and daughter are radioactive, the ratio of their activities is the inverse ratio of their half-lives. Thus a long half-life daughter is excreted before any significant dose can arise from its decay.

The radionuclide selected should emit no beta particles or, even worse, alpha particles. These would be stopped in the body, adding to the radiation dose, but contributing nothing to the image.

The radionuclide should also have a high 'k' factor (section 11.10). This may seem paradoxical at first since a high 'k' factor implies a large absorbed dose. However, a high 'k' factor also means there are a large number of gamma rays being emitted and hence available to contribute to the image. Some radionuclides decay by more than one mechanism, so decay which produces useful gamma rays is preferable to decay which causes a dose but produces no gamma rays and hence no useful information.

Only gamma rays within a limited energy range are really well suited to *in vivo* imaging. For example they must be sufficiently energetic not to be absorbed in the patient—a lower practical limit is about 80 keV. Conversely, the gamma rays must be stopped in the detector or they will be wasted. The crystal used in a gamma camera becomes inefficient above about 300 keV (table 7.1). For equipment based on the scanning principle, where the scintillation crystal can be made quite thick, gamma ray energies up to 500 keV may be used.

A range of radionuclides is used in diagnostic imaging (table 7.2) but well over 90% of routine investigations are performed with Tc-99m. In addition to its short half-life and near monoenergetic gamma ray at 140 keV, Tc-99m emits no particulate radiations and decays to a long half-life daughter (Tc-99, $T_{1/2} = 2 \times 10^5$ years).

Availability of the 6 h half-life material is not a problem because it is possible to establish a generator system. As explained in section 1.7, equilibrium activity in the decay series is governed by the activity of Mo-99 which has a half-life of 67 h.

$$^{99}_{42}\text{Mo} \xrightarrow{\beta^-\,(67\ \text{h})} {}^{99\text{m}}_{43}\text{Tc} \xrightarrow{\gamma^-\,(6\ \text{h})} {}^{99}_{43}\text{Tc} \xrightarrow{\beta^-\,(2\times10^5\ \text{years})} .$$

An Mo–Tc generator consists of Mo-99 adsorbed onto the upper part of a small chromatographic column filled with high grade alumina (Al_2O_3). When 0.9% saline solution is passed down the column, the Mo-99 remains firmly bound to the alumina but the Tc-99m, which is chemically different, is eluted. Essential features of a generator system are shown in figure 7.12. Since the Tc-99m builds up fairly rapidly (see section 1.7), it is possible to elute the column daily to obtain a ready supply of Tc-99m (figure 7.13). The generator can be replaced weekly, by which time the Mo-99 activity will have decreased significantly.

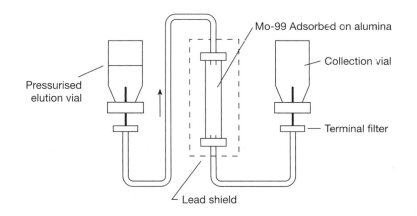

Figure 7.12. *Simplified diagram of a generator system that operates under positive pressure.*

Table 7.2. *Properties of some radionuclides used for in vivo imaging.*

Nuclide	Half-life	Type of emission	Example of use
Carbon-11	20 min[a]	β^+ giving 511 keV γ rays	CO_2 for regional cerebral blood flow[b]
Nitrogen-13	10 min[a]	β^+ giving 511 keV γ rays	Amino acids for myocardial metabolism[b]
Oxygen-15	2 min[a]	β^+ giving 511 keV γ rays	Gaseous studies with labelled O_2, CO_2 and CO, labelled water[b]
Fluorine-18	110 min[a]	β^+ giving 511 keV γ rays	Fluorodeoxyglucose for glucose metabolism[b]
Gallium-67	72 h	92 keV, 182 keV *and* 300 keV γ rays	Soft tissue malignancy and infection
Technetium-99m	6 h[c]	140 keV γ rays	Numerous
Indium-111	2.8 d	173 keV *and* 247 keV γ rays	Labelling blood products
Iodine-123	13 h	160 keV γ rays	Thyroid imaging
Iodine-131	8.0 d	360 keV γ rays *and* β^- particles	Metastases from carcinoma of thyroid
Xenon-133	5.3 d[d]	81 keV γ rays *and* β^- particles	Lung ventilation studies
Thallium-201	73 h	orbital electron capture[e] 80 keV X-rays and Auger electrons	Cardiac infarction and ischaemia

[a] Cyclotron produced positron emitter—see section 10.5.

[b] Not widely used at present, but increasing.

[c] Generator produced. Note short half-life radionuclides that cannot be produced on site are of limited value for *in vivo* imaging.

[d] Since Xe-133 is used in gaseous form, the biological half-life is very short so the β^- particle dose is small.

[e] Tl-201 decays by orbital electron or K shell capture. This is an alternative to positron emission when the nucleus has too many protons and adjusts the balance by capturing an electron from the K shell. The initial capture process may not result in any emission of radiation but characteristic X-rays will be emitted as the vacancy in the K shell is filled. If the atomic number of the element is high enough (e.g. thallium $Z = 81$), this characteristic radiation may be of high enough energy to be useful for imaging.

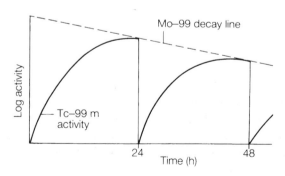

Figure 7.13. *Curve showing the Tc-99m activity in an Mo-99/Tc-99m generator as a function of time, assuming the column is eluted every 24 hours.*

7.3.3.   Choice of radiopharmaceutical

For good counting statistics and a high signal-to-noise ratio, the radionuclide must be firmly bound to an appropriate pharmaceutical and the resulting radiopharmaceutical must achieve a high target:non-target ratio. In addition it must satisfy criteria that are not generally relevant for non-radioactive drugs. It must be easy to produce, inexpensive, readily available for all interested users, have a short effective half-life and be of low toxicity. Very short half-life material may constitute a radiation hazard to the radiopharmacist if it is necessary to start the preparation with a high activity.

Radiopharmaceuticals concentrate in organs of interest by a variety of mechanisms including capillary blockage, phagocytosis, cell sequestration, active transport, compartmental localization, ion exchange and pharmacological localization. The reader is referred to a more specialized text (e.g. Frier 1994) for further details. One disadvantage of Tc-99m is that it is not easily bound to biologically relevant molecules and its chemistry is complex (for a very readable account see the article by Bremer (1984)). Nevertheless in spite of the difficult chemistry, a wide range of pharmaceuticals has been labelled with Tc-99m (Britton 1995) and good target-to-background ratios are sometimes achieved.

However, poor specificity of radiopharmaceuticals for their target organs remains a weak point in nuclear medicine imaging, with most commonly employed radiopharmaceuticals showing very poor selectivity, generally less than 20% in the organ of interest.

Note that the obvious elements to choose for synthesizing specific physiological markers, hydrogen, carbon, nitrogen and oxygen, have no gamma emitting isotopes. Pharmaceuticals containing radioisotopes of some of these elements can be used for positron emission tomography as discussed in section 10.5.

7.3.4.   Performance of the imaging device

Much has been written about the performance of the gamma camera and only the most salient features will be summarized here.

Collimator design

As already explained, resolution and sensitivity are mutually exclusive. The inherently poor sensitivity, which may be as low as 80 cps per MBq for a high resolution, low energy collimator, is a major problem since, as explained in section 7.3.1, the signal-to-noise ratio is proportional to the square root of the sensitivity of the imaging device.

Intrinsic resolution

This is determined primarily by the performance of the scintillation detector crystal. Although a complex problem to treat rigorously, the following simplified explanation contains the essential physics. In principle, by arranging a large number of very small PMTs behind the crystal, one might expect to localize the position of a gamma ray event in the crystal to any required degree of accuracy. However, each 140 keV gamma ray only releases about 4000 light photons and if these are shared between 40 PMTs, the average number, n, reaching each tube is only 100. The process is random, so variations in the signal due to Poisson statistics of $\pm n^{1/2}$ or ± 10 will ensue. Some PMTs will receive more than 100 photons, some will receive a lot less, but the 'error' on the signal from each PMT, which will contribute to the error in positioning the event, will increase rapidly if one attempts to subdivide the original signal too much. Typical intrinsic resolution is about 3–4 mm.

System resolution

The resolution of the complete system can be obtained by imaging a narrow line source of radioactivity—for example, nylon tubing of 1 mm internal diameter filled with an aqueous solution of Tc-99m pertechnetate. A result such as that shown in figure 7.14 would be obtained and the spread of the image can be expressed in terms of the full width at half maximum height (FWHM) calculated as shown. For the arrangement in figure 7.14(c) which is perhaps the most realistic, the FWHM is 9.7 mm which is substantially greater than the intrinsic resolution.

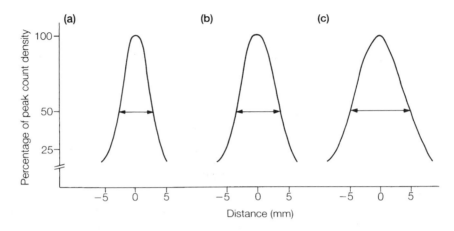

Figure 7.14. *Derivation of system resolution from line spread function measurements and the effect of distance and scattering material on system resolution. The traces are typical images of a 1 mm line source of Tc-99m obtained under different conditions: (a) no scattering material, source on collimator face, FWHM = 5.7 mm; (b) 5 cm tissue equivalent scattering material, FWHM = 7.1 mm; (c) 10 cm tissue equivalent scattering material, FWHM = 9.7 mm.*

Spatial linearity and non-uniformity

The outputs from the PMT array must be converted into the X and Y signals that give the spatial co-ordinates of the scintillation. Any error in this process, caused perhaps by a change in the amplification factor in one PMT, will result in counts being misplaced in the ensuing image. This will result in distortion (or non-linearity) if a narrow line source of radioactivity is imaged, or in non-uniformity for a uniform extended source.

Correction circuits for non-linearity were discussed in section 7.2.1. Non-uniformity in the image of a uniform flood source is a useful overall measure of the performance of the camera.

Several methods of expressing non-uniformity have been suggested. For example, integral non-uniformity is a measure of the difference between the maximum counts (C_{max}) and minimum counts (C_{min}) per pixel in an image

$$\text{Integral non-uniformity} = \frac{(C_{max} - C_{min})}{(C_{max} + C_{min})} \times 100\%.$$

Measurements are usually made over the whole usable field of view and over the centre of the field of view which has the linear dimensions of the whole field of view scaled down by 0.75. The standard deviation (SD) of the pixel counts is also calculated and the coefficient of variation is $100(SD/C_{mean})$ where C_{mean} is the mean of the two values. Alternatively the differential non-uniformity U_D is based on the maximum rate of change of count density,

$$U_D = (\Delta C / C_{\mathrm{mean}}) \times 100\%$$

where ΔC is the maximum difference in counts between any two adjacent pixel elements. From the viewpoint of accurate diagnosis, camera non-uniformities must be minimized or they may be wrongly interpreted as real variations in the image count density. As for linear distortion, it is now possible to collect and store a uniformity correction matrix that can be applied to each image. For a well adjusted modern camera, integral non-uniformity over the centre of the field of view should be less than 2%.

Effect of scattered radiation

Although pulse height analysis ought, in principle, to reject all scattered radiation, discrimination is far from perfect. One limitation of the scintillation crystal and PMT combination is that the number of electrons entering the PMT per primary x- or gamma ray photon interaction is rather small. There are two reasons:

(1)   about 30 eV of energy must be dissipated in the crystal for the production of each visible or ultra-violet photon, and
(2)   even assuming no loss of these photons, only about one photoelectron is produced for every 10 photons on the PMT photocathode.

Thus to generate one electron at the photocathode requires about 300 eV and a 140 keV photon will produce only about 400 electrons at the photocathode. This number is subject to considerable statistical fluctuation ($N^{1/2} = 20$ or 5%). The result is that a monoenergetic beam of gamma rays will produce a range of pulses and will appear to contain a range of energies.

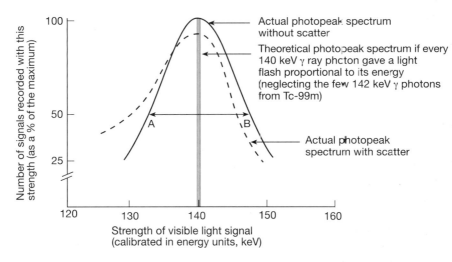

Figure 7.15. *Graph demonstrating the energy resolution of the NaI(Tl) crystal in a gamma camera. The FWHM (AB) is about 14 keV or 10% of the peak energy.*

The result, as shown in figure 7.15, is that even in the absence of scatter monoenergetic gamma rays produce light signals representing a range of energies. This spread, expressed as the ratio of the full width at half maximum height (FWHM) of the photopeak spectrum to the photopeak energy, is a measure of the energy resolution of the system and is about 10% for a gamma camera at 140 keV. The spectrum is then further degraded by scatter in the patient (dotted curve).

Unscattered photons contribute information about the image so a wide energy window (typically about 20%) must be used. Unfortunately a wide energy window permits some gamma photons that have been Compton scattered through quite large angles, and may have lost as much as 20 keV, to be accepted by the pulse height analyser. The problem is greater for low energy gamma photons, for two reasons:

(1)   fewer light photons are produced, so statistical variations are greater,
(2)   the energy lost during a Compton interaction, for fixed scattering angle, is smaller.

Note that semiconductor detectors (see figure 6.12) produce a narrower spread and much better energy discrimination. However, it has not yet been possible to build a gamma camera with a semiconductor detector at a commercially competitive price.

Modern gamma cameras allow more than one energy window to be set, thus accepting several photopeaks. This can be useful when working with a radionuclide which emits gamma rays at more than one energy, e.g. Ga-67 or In-111, or when attempting to image two radionuclides simultaneously.

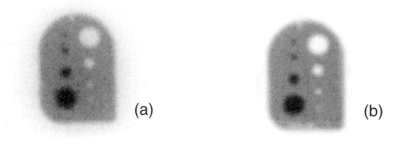

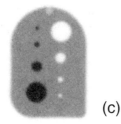

Figure 7.16. *The effect of distance and scatter on image quality: (a) test object in contact with collimator face; (b) 10 cm air separation; (c) 10 cm separation and Perspex scatter material. In all images 2000 k counts were collected (cf figure 7.11(c)).*

As shown in figure 7.16 scattered radiation causes deterioration in the image of a test object, especially when the scattering material also increases the distance to the collimator face. The patient is the major source of scattering material and there is an obvious difference in image quality for say a bone scan of a very thin person when compared with that of an obese person. Careful setting up of the pulse height analyser is essential for good image quality.

High count rates

The reasons for loss of counts at high count rates and implications for quantitative work will be discussed in section 7.4.2. Some degradation of image quality also occurs. The main reason is thought to be that the system fails to separate in time two scattered photons whose summed energy falls within the pulse height analyser window. Thus an event is recorded at the weighted mean position of the two scattered photons but in a position unrelated to the activity in the patient.

7.3.5.   Data display

The quality of the final image is also influenced by the performance of the recording medium. Methods of displaying data will be discussed briefly.

Persistence monitor

When setting up a study, an image of the distribution of activity may greatly assist patient positioning. This may be achieved using a video monitor on which each flash of light representing a collected gamma ray persists long enough for a transient image to be formed. This image is ideal for positioning but is unsuitable for diagnosis.

Hard copy

Hard copy images are obtained with film-based video formatters. As discussed in section 7.3.1, increasing the number of gamma rays will improve image quality. However, the imaging process does suffer from a form of reciprocity failure and a point may be reached where increasing the count rate results in a decrease in film density even though the same number of counts is being collected.

The brightness of the visible light spot must neither saturate the film, nor result in an under-exposed image. If the spot is too bright it may have the effect of blurring the image. Also it is important to check that the optical system does not cause aberrations, particularly at wide aperture settings.

Note that images are increasingly being produced in digital format (see later). Gamma cameras can then be connected to networks in radiology departments and film-less systems are being developed with primary diagnosis directly from the screen and hard copy generated from networked printers.

Analogue versus digitized images

Most of the criticism of digitized images has arisen because a 128×128 matrix over a typical gamma camera face corresponds to a pixel size of about 3 mm and many observers find the underlying matrix pattern visually intrusive when viewing images. However, both theoretical calculations and experimental observations indicate that the inherent resolving capability of the camera does not warrant finer pixellation (see section 8.3). Also the effect can be largely removed by using interpolation techniques to display the data on a finer matrix size than that on which they have been collected. Note also that for some purposes, e.g. outlining regions of interest, fine pixellation is not necessary and simply uses up valuable data storage space.

Digitized images overcome all the difficulties of analogue images discussed in section 7.2.1. They also permit graphical data to be produced and sophisticated forms of image processing are possible. For further information on some of these techniques see Sharp *et al* (1985). If the images collected in a dynamic study are to be analysed quantitatively, they must of course be digitized. This aspect will be discussed in more detail in the next section.

Greyscale versus colour images

When an image has been digitized, a range of count densities may be assigned to either a shade of grey or a spectral colour. Much has been written about the relative merits of greyscale and colour images and this controversial subject cannot be discussed fully here. The following simple philosophy suggests an approach to each type of display. The sharp visual transition from one colour to another may alert the eye to the possibility of an abnormal amount of uptake of radioactivity and colour images can be useful for this purpose. However, by the same token, this colour change may represent an increase or decrease of only one or two counts per pixel and may not be significant statistically. It thus follows that greyscale is generally preferable for unprocessed images such as bone scans but colour can be useful when looking at processed images, especially in functional imaging (see section 7.4.1).

7.4. DYNAMIC INVESTIGATIONS

The potential for performing dynamic studies, in which changes in distribution of the radiopharmaceutical are monitored throughout the investigation, was recognized at an early stage. However, two developments were essential before dynamic imaging became feasible on a routine basis. The first was an imaging device with a reasonably large field of view, sufficiently sensitive to give statistically reliable counts in short time intervals. The gamma camera satisfies these requirements although for dynamic studies it is not uncommon to choose a collimator design that increases sensitivity at the expense of some loss of resolution. The dominance of gamma cameras over scanners in nuclear medicine departments is very largely due to their capability for dynamic studies.

The second development was the availability of reasonably priced data handling hardware and software powerful enough to handle the large number of data collected. Hence dynamic imaging has only been widely available in general hospitals since the late 1970s.

Important features of dynamic imaging will now be considered under two general headings, but with specific reference to some frequent dynamic investigations.

7.4.1.   *Data analysis*

Consider as an example the study of kidney function—nowadays usually performed by administering intra-venously about 75 MBq of Tc-99m labelled MAG3 (mercapto acetyl triglycine) which is actively secreted by the renal tubules. Historically, fixed detectors of small area or 'probes' were used for such studies but the results were critically dependent on probe positioning. Using a gamma camera, a set of images can be obtained and regions of interest for study can be selected retrospectively. Thus methods of data display have been developed that will show both the spatial distribution of radiopharmaceutical and temporal changes in the distribution.

Cine mode

A useful starting point is to examine individual image frames looking for aspects that require further detailed study, for example in renograms to look for evidence of patient movement. This can be done by running a 'cine-film' of the frames using a continuous loop so that the display automatically returns to the start of the study and continues until interrupted. Data are usually collected in about sixty 20 s frames but it is sometimes useful to expand the initial time frame by collecting 1 s frames for the first minute.

Time–activity curves

The system will plot activity as a function of time for regions of interest selected by the operator. Examples of such regions of interest and the resulting curves, again taken from renography, are shown in figure 7.17. Some other features of this apparently simple procedure must be mentioned. To achieve a better image (improved counting statistics) on which the regions of interest can be drawn, it may be necessary to add several sequential frames. Data smoothing will also help to keep noise to a minimum. Also it is important to subtract background counts arising from activity in overlying and underlying tissue. This is usually done by defining a region of interest representative of blood background.

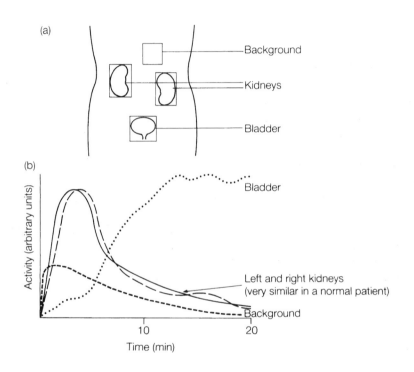

Figure 7.17. *(a) Schematic drawings of regions of interest around kidneys and bladder for a renogram study. A background region is also shown. (b) Typical activity–time curves for such regions of interest.*

Deconvolution

Although radioactivity is injected intravenously as a bolus, after mixing with blood and passing through the heart and lungs, it arrives at the kidneys over a period of time. Also some activity may recycle. Thus the measured activity–time curve is a combination (convolution) of a variable amount of activity and the rate of handling by the organ. The requirement is to measure the mean transit time for the organ and deconvolution is a mathematical technique that offers the possibility for removing arrival time effects and presenting the result as for a single bolus of activity. In nuclear medicine the presence of noise limits the power of deconvolution methods.

Functional imaging

Some dynamic studies now produce a large number of images, and although it may be important to examine these images in cine mode, methods of data compression will be required before a particular quantitative feature can be visualized, perhaps on a single image.

One approach to this problem is to choose a feature thought to be of physiological interest, for example the time of maximum on a renogram curve, and then to examine the data sets, pixel by pixel mapping the time of maximum. The resulting image is a 'functional image' since it displays a feature of physiological rather than anatomical interest.

The most extensive use of functional imaging currently is probably in the analysis of multiple gated acquisition studies of the cardiac blood pool (MUGA). The patient's own blood cells are labelled with Tc-99m and the gamma camera collects data in frames which are gated physiologically to the signal from an electrocardiograph attached to the patient. For a patient with a regular heart rate, 16–24 equally spaced frames between each R wave signal can be collected. The data may be collected in 'list' mode. That is to say the (x, y, t) co-ordinates of every scintillation accepted by the electronics are registered for subsequent sorting into the correct pixel and time frame. The number of counts collected during one cardiac cycle would be far too small for the statistics to be reliable so collection continues for several hundred cycles until there are about 200 000 counts per frame. Protocols vary between centres. One simple, standard procedure collects anterior and left anterior oblique (LAO) views only, with attention focused on the left ventricle, best seen in the LAO view.

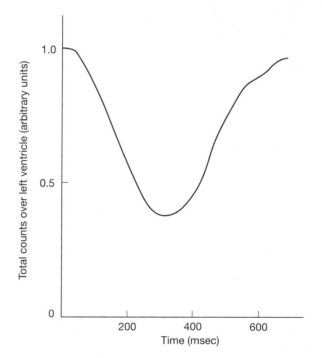

Figure 7.18. *Time–activity curve over the left ventricle in a normal patient.*

One physiological feature of importance is the change in volume of the compartments of the heart through the cycle, especially of the ventricles. This property may be derived from a plot of the time–activity curve (TAC) which is an average cardiac cycle formed from the hundreds of cycles acquired. As shown in figure 7.18 to a first approximation this may be assumed to be sinusoidal and the variation in counts $C(t)$ in a region of interest with time t is given by

$$C(t) = a + b\sin(2\pi f t + \phi)$$

where a is a baseline constant, b is the amplitude of the motion, f is the reciprocal of the number of frames per cardiac cycle and ϕ is the phase of the motion.

The most important quantity derived from the left ventricle TAC is the ejection fraction

$$\text{LVEF} = (\text{ED} - \text{ES})/(\text{ED} - \text{bgnd})\%$$

where ED = region of interest (roi) counts at end diastole, ES = roi counts at end systole. The choice of roi boundaries and the background (bgnd) roi are critical.

Figure 7.19. *Functional image of a normal MUGA study. The grey scale represents the phase of contraction of the heart chambers; the contour lines represent the amplitudes; both derived from the Fourier fitted curve (figure 7.18).*

All the data may be analysed pixel by pixel to produce two functional images, one of which shows the phase of each part of the heart motion, the other showing the amplitude. Note that in practice the curves will not be truly sinusoidal but methods of Fourier analysis may be used to introduce further refinements if necessary. Even further compression of data is possible if, say, the 30%, 50% and 80% contours are identified on the amplitude map and these are then superimposed on the phase. This has been done for data derived from a normal patient in figure 7.19. Such images are best displayed using a colour scale rather than monochrome, as the relative magnitude of the parameter in different parts of the image is more readily appreciated. However, even in grey scale it can be seen that for this normal patient, (*a*) the phase is similar throughout each chamber, (*b*) the atria are contracting out of phase with the ventricles, (*c*) the contours of equal amplitude (volume contraction) are uniform.

A major difficulty with functional imaging is the choice of a reasonably simple mathematical index that is relevant to the physiological condition being studied.

7.4.2.   Camera performance at high count rates

When a dynamic study requires rapid, sequential imaging, the gamma camera may have to function at high count rates to achieve adequate statistics. However, the system imposes a number of constraints on maximum count rate. For example, the decay time of the scintillation in the crystal has a time constant of 0.2 μs and about 0.8 μs is required for maximum light collection. The electronic signal processing time is also a major limitation. The signal from the camera pre-amplifier has a sharp rise but a tail of about 50 μs so pulses have to be truncated or 'shaped' to last no longer than about 1 μs. The pulse height analyser and pulse arithmetic circuits have minimum processing times and it may be an advantage to by-pass the circuits that correct for spatial non-linearity, if any, to reduce processing time.

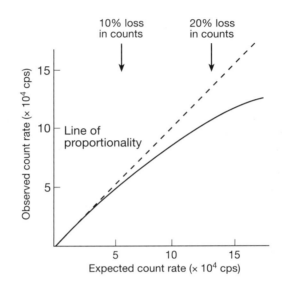

Figure 7.20. *Curve showing how the various dead times in the camera system result in count rate losses. In this example 10% and 20% losses in counts occur at about* 6 $\times$ 10^4 *and* 13 $\times$ 10^4 cps *respectively. The maximum observed count rate would be about* 30 $\times$ 10^4 cps *after which the recorded count rate would actually fall with increasing activity.*

For all these reasons, if sources of known, increasing activity are placed in front of a gamma camera under ideal conditions with no scattering material, a graph of observed against expected count rate for a modern camera might be as shown in figure 7.20. In practice, performance would be inferior to this because the camera electronics has to handle a large number of scattered photons that are subsequently rejected by the pulse height analyser. Thus the exact shape of curve is very dependent on the thickness of scattering material and the width of the pulse height analyser window. Loss of counts can occur at count rates as low as 3 $\times$ 10^4 cps and this can be a serious problem when making quantitative measurements.

The highest count rates and hence most stringent demands on both camera and radiopharmaceutical are encountered in cardiovascular studies. In a first-pass study a bolus injection is given at the highest possible radioactive concentration. One second frames are taken and the study is complete in about 30 s. Future work may see the development of ultrashort half-life radionuclides, for example gold-195m, which is generator produced and has a half-life of 30.5 s. This has the advantage of high activity during the study without a big patient dose. Also residual background activity rapidly decays to zero so repeat studies from a different angle or during exercise may be performed within a few minutes. The count rate performance of cameras may need to be improved for such studies.

7.5. QUALITY STANDARDS, QUALITY ASSURANCE AND QUALITY CONTROL

Quality standards are the standards that must be applied to the individual elements of a nuclear medicine service to ensure an agreed standard for the service overall. Quality assurance embraces all those planned and systematic actions necessary to provide confidence that a structure, system or component will perform satisfactorily in service. Quality control is the set of operations intended to maintain or improve quality. Although these are clearly separable aspects of quality, they are closely related and will be considered together in this brief overview of those aspects of the provision of a high quality nuclear medicine service where physical principles are important.

The principal aim is to ensure that the requisite diagnostic information is obtained with the minimum dose to the patient. Prime areas of concern are the accurate measurement of the administered activity and the performance of the imaging device. However, consideration must also be given to the safety of staff, other patients and the public, especially since the patient themselves becomes a source of radiation, and to the release of radioactive waste to the environment.

7.5.1. *Radionuclide calibrators and accuracy of injected doses*

The main components of a radionuclide calibrator will be a well-type ionization chamber (see section 6.3), stabilized high voltage supply, electrometer for measuring the small ionization currents, processing electronics and a display device. In the nuclear medicine department it will be used for a variety of purposes including:

(a) determination of radiopharmaceutical activities after delivery by the manufacturer;
(b) dosage of solution for injection and oral application;
(c) checking eluate activities from generators (e.g. Tc-99m and In-113m);
(d) measurement of residual activities after injections;
(e) determination of attenuation of different materials, e.g. glass and plastic containers and
(f) calibration of measuring equipment.

A protocol for establishing and maintaining the calibration of medical radionuclide calibrators and their quality control has been prepared by Parkin *et al* (1992). Drawing on experience of traceability to national standards in radiotherapy, the report gives recommended methods for calibrating reference instruments at a large regional centre and for checking field instruments. It also gives guidelines on the frequency of quality control tests and acceptable calibration tolerances for both types of instrument. A 10% limit on overall accuracy is a reasonable practical figure for a field instrument.

A number of variables can cause significant errors in radionuclide calibrators. Two are particularly important.

(a) The container size and shape, and volume of fluid can be a problem with beta and low energy gamma emitters because of self-absorption. Even the thickness of the vial can affect calibration. These variables are less of a problem for energies above about 140 keV but it is important that the gamma ray energy being used for imaging is also the one being used for calibration by the calibrator. For example the 160 keV gamma ray from I-123 is used for imaging but unless special precautions are taken to filter out low energy radiation, the calibrator will respond to the low energy characteristic X-rays at 35 keV.
(b) Contamination by other radionuclides can seriously affect calibrator accuracy if the calibrator is much more sensitive to the contaminant than to the principal product. Two examples are given in table 7.3.

For Tl-201 1.5% contamination will overestimate the activity by 13%. For Sr-85, as little as 0.2% Sr-89 will overestimate the activity by 38%.

Great precision in respect of the isotope calibrator is of little value if there is uncertainty or variation in the amount of activity actually administered to the patient. Some possible causes for the wrong activity being given would be

Table 7.3. *Effect of contaminants on accuracy of calibrator measurement of radionuclide activity.*

Radionuclide	Typical activity present (%)	Calibrator sensitivity (pA MBq^{-1})
Tl-200	0.5	2.56
Tl-201	98.5	0.86
Tl-202	1.0	5.03
Sr-89	0.2	5.26
Sr-85	99.8	0.028

(a)   variation in the volume injected,

(b)   improper mixing of the radiopharmaceutical, e.g. macro aggregates, colloid,

(c)   retention of the radiopharmaceutical in the vial (stickiness) e.g. methoxyisobutylisonitrite (MIBI), Tl-201.

It is important to check that all operators can draw up and inject a specified volume, say 1 ml, accurately. With a syringe shield in place it is possible for an inexperienced operator to draw up almost no fluid at all.

7.5.2.   Gamma camera and computer

The primary goal of quality control of the gamma camera and computer is to provide the physician and technologist with an assurance that the images produced during clinical studies accurately reflect the distribution of radiopharmaceutical in the patient.

Five elements can be identified in a good quality control programme—test to be performed, approximate frequency, accuracy and reproducibility of tests, record keeping, action thresholds.

Table 7.4 lists the more important tests and suggests an approximate frequency. There is some variation between centres. For further detail on these procedures see Hannan (1992) and Elliott (1998). Table 7.5 gives typical performance figures.

Table 7.4. *Performance measurements for a gamma camera and computer system.*

Test	Frequency	Comment
Physical inspection	Daily	
Photo-peak position	Daily	
Visual uniformity	Daily	failure of a number of functions will show as non-uniformities in a Co-57 flood image
Quantitative uniformity	Weekly	both integral and differential uniformity should be checked over the whole field of view and over the centre of the field of view
Spatial distortion	3 monthly	may be obtained from the same data set
Intrinsic resolution	3 monthly	
System spatial resolution	3 monthly	measured with the collimator in position—a weekly visual check is desirable
Energy resolution	3 monthly	
Count rate response	6 monthly	check also uniformity and spatial resolution at high count rate if appropriate
Computer	6 monthly	check analogue-to-digital converter (if appropriate) and timing

Note that QA checks must not be unduly disruptive to the work of the department. Daily checks should take no more than a few minutes and weekly checks no more than an hour.

Table 7.5. *Typical performance figures for a gamma camera.*

Parameter	Value	Conditions
Intrinsic spatial resolution	3.8 mm	FWHM over the useful field of view
System spatial resolution	7.5 mm	FWHM with high resolution collimator, without scatter at 10 cm
Intrinsic energy resolution	10%	FWHM at 140 keV
	14%	FWHM at 140 keV with collimator and scatter
Integral uniformity	2.0%	Centre of field of view
Differential uniformity	1.5%	Centre of field of view
Count rate performance	130 000 cps	20% loss of counts without scatter
	75 000 cps	Deterioration of intrinsic spatial resolution to 4.2 mm
System sensitivity	160 cps MBq^{-1}	Tc-99m and a general purpose collimator

There are many possible reasons for computer software failure and complete software evaluation is extremely difficult. Since however the basic premise of software evaluation is that application of a programme to a known data set will produce known results, new software should be tested against

(a) data collected from a physical phantom,
(b) data generated by computer simulation and
(c) validated clinical data.

Any significant variation from the expected results can then be investigated.

7.6. SUMMARY

The primary detector of gamma rays in nuclear medicine is invariably a sodium iodide crystal, doped with about 0.1% thallium. The advantages of this detector are the following.

(1) A high density and high atomic number ensure a good gamma ray stopping efficiency for a given crystal thickness.
(2) The high atomic number favours a photoelectric interaction, thus a pulse is generated which represents the full energy of the gamma ray.
(3) Thallium gives a high conversion efficiency of the order of 10%.
(4) A short 'dead time' in the crystal generally permits acceptable counting rates except for very rapid dynamic studies, when dead times both in the crystal and elsewhere in the system can be important.

The instrument of choice for a wide range of static and dynamic examinations is the gamma camera. Its mode of operation may be considered in two parts.

(1) A direct spatial relationship is established between the point of emission of a gamma ray in the patient and the point at which it strikes a large Na(Tl) crystal by collimation.
(2) An array of PMTs backed by appropriate electronic circuits is used to identify the position at which the gamma ray interacts with the crystal.

Over 90% of all nuclear medicine examinations are carried out with Tc-99m. The advantages of Tc-99m for radionuclide imaging are the following.

(1)   A monoenergetic gamma ray is emitted—this facilitates pulse height analysis.
(2)   The gamma ray energy is high enough not to be heavily absorbed in the patient, hence minimizing patient dose, but low enough to be stopped in a thin sodium iodide crystal.
(3)   No high LET radiations are emitted.
(4)   A decay product that delivers negligible dose.
(5)   A half-life that is long enough for most examinations but short enough to minimize dose to the patient.
(6)   Ready availability as an eluate from a Mo-99/Tc-99m generator.

The quality of radionuclide images is influenced by

(1)   the number of counts that can be collected for given limits on radiation dose to the patient, required resolution, and time of examination;
(2)   the ability of the radiopharmaceutical to concentrate in the region of interest;
(3)   the presence of, and ability to discriminate against, scattered radiation;
(4)   overall performance of the imaging device including spatial and temporal linearity, uniformity and system resolution.

Analogue images are produced by transmitting the signal from the positioning circuits as a light flash directly to a hard copy recording device such as film. If the signals are fed to a computer and digitized, all counts within one pixel are summed. Modern cameras collect all data digitally. For visual display they are then converted into images that look more 'analogue' by interpolation and smoothing. Digitized images can also be used to extract, under computer control, functional data for specified regions of interest. The gamma camera is now used extensively for dynamic studies where important information is obtained by numerical analysis of digitized images on a frame by frame basis.

REFERENCES

Belcher E H and Vetter H (ed) 1971 *Radioisotopes in Medical Diagnosis* (London: Butterworths)
Bremer P O 1984 The principles and practice of radiopharmaceutical production *Technical Advances in Biomedical Physics (NATO ASI series E Applied Sciences 77)* ed P P Dendy, D W Ernst and A Sengun (The Hague: Nijhoff) pp 287–318 and other relevant chapters
Britton K E 1995 Nuclear medicine, state of the art and science *Radiography* **1** 13–27
Elliott A T 1998 Quality Assurance in *Practical Nuclear Medicine* 2nd edn, ed P F Sharp, H G Gemmell and F W Smith (Oxford: IRL) pp 49–71
Faires R A and Boswell G G J 1981 *Radioactive Laboratory Techniques* (London: Butterworths)
Fenwick J D, McKenzie A L and Boddy K 1991 Intercomparison of whole body counters using a multinuclide calibration phantom *Phys. Med. Biol.* **36** 191–8
Frier 1994 Mechanisms of localization of pharmaceuticals *Text Book of Radiopharmacy Theory and Practice* 2nd edn ed C B Sampson (London: Gordon and Breach) pp 201–7
Hannan J (ed) 1992 Quality control of gamma cameras and associated computer systems *Institute of Physical Sciences in Medicine Report* 66
Parkin A, Sephton J P, Aird E G A, Hannan J, Simpson A E and Woods M J 1992 Protocol for establishing and maintaining the calibration of medical radionuclide calibrators and their quality control *Quality Standards in Nuclear Medicine (Institute of Physical Sciences in Medicine Report 65)* pp 60–77
Sharp P F, Dendy P P and Keyes W I 1985 *Radionuclide Imaging Techniques* (New York: Academic)

EXERCISES

1   What is a radionuclide generator?

2   What mass of iodine-123 will give 80 MBq of activity? (Half-life of I-123 $=$ 13 h, the Avogadro constant $= 6.02 \times 10^{23}$ mol^{-1}.)

3   A dose of Tc-99m macroaggregated human serum albumin for a lung scan had an activity of 180 MBq in a volume of 3.5 ml when it was prepared at 11.30 h. If you wished to inject 23 MBq from this dose into a patient at 16.30 h, what volume would you administer? (Half-life of Tc-99m $=$ 6 h.)

4   A radiopharmaceutical has a physical half-life of 6 h and a biological half-life of 15 h. How long will it take for the activity in the patient to drop to 10% of that injected?

5   List the main characteristics of an ideal radiopharmaceutical.

6   What are the possible disadvantages of preparing a radiopharmaceutical a long time before it is administered to the patient?

7   Why is the ideal energy for gamma rays used in clinical radionuclide imaging in the range 100–200 keV?

8   In nuclear medicine, why are interactions of Tc-99m gamma rays in the patient primarily by the Compton effect whilst those in the sodium iodide crystal are mainly photoelectric processes?

9   The sodium iodide crystal in a certain gamma camera is 9 mm thick. Calculate the fraction of the gamma rays it will absorb at (*a*) 140 keV; (*b*) 500 keV. Assume the gamma rays are incident normally on the crystal and that the linear absorption coefficient of sodium iodide is 0.4 mm^{-1} at 140 keV and 0.016 mm^{-1} at 500 keV.

10   Why is it necessary to use a collimator for imaging gamma rays but not for X-rays produced by a diagnostic set?

11   What are the differences between a collimator used to image low energy radionuclides and one used to image high energy radionuclides?

12   Why is pulse height analysis used to discriminate against scatter in nuclear medicine but not in radiology?

13   Compare and contrast the methods used to reduce the effect of scattered radiation on image quality in radiology with those used in clinical radionuclide imaging

14   How is the spatial resolution of a gamma camera measured and what is the clinical relevance of the measurement?

15   What factors affect (*a*) the sharpness, (*b*) the contrast of a clinical radionuclide image?

16   List the steps you would take in setting up a gamma camera to produce a bone image.

17   Explain, with a block diagram of the equipment, how a dynamic study is performed with a gamma camera.

18   A radiopharmaceutical labelled with Tc-99m and a gamma camera system were used for a renogram. Curves were plotted of the counts over each kidney as a function of time and although the shapes of both curves were normal, the maximum count recorded over the right kidney was higher than over the left. Suggest reasons.

19   What are functional images? Illustrate your answer by considering one application of functional images in nuclear medicine.

CHAPTER 8

ASSESSMENT AND ENHANCEMENT OF IMAGE QUALITY

8.1. INTRODUCTION

In earlier chapters the principles and practice of X-ray production were considered and the ways in which X-rays are attenuated in different body tissues were discussed. Differences in attenuation create 'contrast' on an X-ray film and differences in 'contrast' provide information about the object. Imaging systems are often described in terms of physical quantities that characterize various aspects of their performance

However, when an X-ray image, or indeed any other form of diagnostic image, is assessed subjectively, it must be appreciated that the use made of the information is dependent on the observer, in particular the performance of his or her visual response system. Therefore, it is important to consider those aspects of the visual response system that may influence the final diagnostic outcome of an investigation. Ways in which the visual process may be facilitated by image enhancement or even, in some circumstances, by-passed completely so as to provide a more objective assessment of information content in the image will also be considered. Finally, some methods of assessing image quality will be discussed.

8.2. FACTORS AFFECTING IMAGE QUALITY

A large number of factors may control or influence image quality. They may be subdivided into three general categories.

(i)   *Image parameters.*
(1)   The signal to be detected—the factors to consider here will be the size of the abnormality, the shape of the abnormality and the inherent contrast between the suspected abnormality and non-suspect areas.
(2)   The number and type of possible signals—for example the number and angular frequency of sampling in computed tomography.
(3)   The nature and performance characteristics of the image system—spatial resolution (this is important when working with an image intensifier, computed radiography systems and in nuclear medicine), sensitivity, linearity, noise (both the amplitude and character of any unwanted signal such as scatter) and imaging speed, especially in relation to patient motion.
(4)   The interplay between image quality and dose to the patient.
(5)   Non-signal structure—interference with the wanted information may arise from grid lines, overlapping structures or artefacts in CT and ultrasound.

192

(ii)   *Observation parameters.*
(1)   The display system—features that can affect the image appearance include the brightness scale, gain, offset, non-linearity (if any) and the magnification or minification.
(2)   Viewing conditions—viewing distance and ambient room brightness.
(3)   Detection requirements.
(4)   Number of observations.

(iii)   *Psychological parameters.*
(1)   *A priori* information given to the observer.
(2)   Feed-back (if any) given to the observer.
(3)   Observer experience from other given parameters—this may be divided into clinical and non-clinical factors and includes for example familiarity with the signal and with the display, especially with the types of noise artefact to be expected.
(4)   Information from other imaging modalities.

Some of these points have been considered in earlier chapters and others will be considered later in this chapter. However, it is important to realize that final interpretation of a diagnostic image, especially when it is subjective, depends on far more than a simple consideration of the way in which the radiation interacts with the body.

8.3.   ANALOGUE AND DIGITAL IMAGES

These terms would not have formed part of a radiologist's vocabulary 20–25 years ago but nowadays it is essential to appreciate the difference between them. In an **analogue** image, there is a direct spatial relationship between the X-ray photon that interacts with the recording medium and the response of that medium. The result is a continuously varying function describing the image. Developed photographic film is an example of an analogue image, providing in radiology a detailed, permanent record of the distribution of X-ray photons transmitted by the body. However this format is primarily suitable for visual inspection and it is not easy to extract quantitative data from an analogue image.

To manipulate and extract numerical information, the distribution pattern of photon interactions must be collected and stored in a computer. In principle the *x* and *y* co-ordinates of every X-ray interaction with the recording medium could be registered and stored. This is sometimes known as 'list mode' data collection and is used for a few specialized studies in nuclear medicine. However, apart from practical problems associated with the recording medium, in view of the large number of photon interactions in a conventional radiograph there would be a massive data storage problem not capable of being solved by computers used in hospitals. Therefore for the purpose of data collection and storage the image space is subdivided into a number of compartments, which are usually but not necessarily square and of equal size. In a **digitized** image the X-ray interaction is assigned to the appropriate compartment but the position of the interaction is located no more precisely than this. Thus the image has a discontinuous aspect that is absent from an analogue image.

The number and size of compartments is variable. For example in the extreme case of a 2×2 matrix illustrated in figure 8.1(*a*), all interactions would be assigned to just one of four areas. Figure 8.1(*b*) illustrates an 8×8 matrix. These compartments are frequently called **pixels**.

Matrices used for diagnostic imaging range from 64×64 to 2048×2048 or even higher and the choice of a suitable matrix size depends on a number of factors:

(1)   Resolution cannot be better than the size of an individual matrix element or pixel. For a 128×128 matrix and a 40 cm $\times$ 40 cm field of view, the pixel size is about 3 mm. Note that the pixel size is governed by the field of view. For the head, where the field of view might be only 20 cm $\times$ 20 cm, a 128×128 matrix would give a 1.5 mm pixel size.

(a) **(b)**

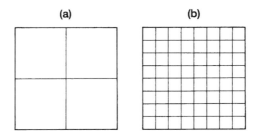

Figure 8.1. *Examples of coarse pixellation: (a) 2 × 2 matrix; (b) 8 × 8 matrix.*

(2) If pixellation is very coarse, a finer pixellation will improve resolution. Consider the extreme example of a 2 × 2 matrix in figure 8.1(*a*).

(3) There is nothing to be gained by decreasing pixel size below the resolving capability of the imaging equipment. Thus, whereas a 512 × 512 matrix (0.4 mm pixel size) might be fully justified for a high resolution image intensifier screen used in cine-angiography of the skull, larger pixels are acceptable in nuclear medicine where the system resolution of a gamma camera is no better than 5 mm.

(4) As the pixel size becomes smaller, the size of the signal becomes smaller and the ratio of signal to noise gets smaller. When counting photons, the signal size N is subject to Poisson statistical fluctuations (noise) of $N^{1/2}$ (see section 8.6). So the signal-to-noise ratio is proportional to $N^{1/2}$ and decreases as N decreases. In magnetic resonance imaging (see chapter 14), electrical and other forms of noise cannot be reduced below a certain level and thus the size of the signal, for example from protons, is a limiting factor in determining the smallest useful pixel size.

(5) Finer pixellation places a greater burden on the computer in terms of data storage and manipulation. A 256 × 256 matrix contains over 65 000 pixels and each one must be stored and examined individually.

8.4. OPERATION OF THE VISUAL SYSTEM

This is a complex subject and it would be inappropriate to attempt a detailed treatment here. However, four aspects are of particular relevance to the assessment and interpretation of diagnostic images and should be considered.

8.4.1. *Response to different light intensities*

It is a well established physiological phenomenon that the eye responds logarithmically to changes in light intensity. This fact has already been mentioned in section 4.4 and is one of the reasons for defining contrast in terms of log(intensity).

8.4.2. *Rod and cone vision*

When light intensities are low, the eye transfers from cone vision to rod vision. The latter is much more sensitive but this increased sensitivity brings a number of disadvantages for radiology.

First, the capability of the eye to detect contrast differences is very dependent on light intensity (figure 8.2). Whereas, for example, a contrast difference of 2% would probably be detectable at a light intensity or brightness of 100 millilambert (typical film viewing conditions), a contrast difference of 20% might be necessary at 10^{-3} millilambert (the light intensity that might be emitted from a very bright fluorescent screen).

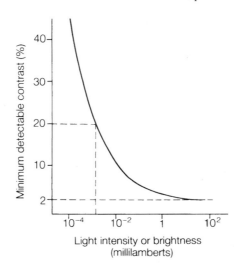

Figure 8.2. *Variation of minimum detectable contrast with light intensity, or brightness, for the human visual system.*

Second, there is a loss of resolving ability, or visual acuity, at low light intensities. With cone vision the minimum detectable separation between two objects viewed from about 25 cm is better than 1 mm. Since rods respond as bundles of fibres rather than individually, visual acuity is worse and strongly dependent on light intensity. At 10^{-4} millilambert, a more typical brightness for a traditional fluoroscopic screen, the minimum detectable separation is probably no better than 3 mm.

Loss of contrast perception and visual acuity are the two main reasons why fluoroscopic screens have been abandoned in favour of image intensifiers. Other disadvantages of rod vision are the need for dark adaptation, which may take up to 30 min, and a loss of colour sensitivity, although the latter is not a real problem in radiology.

8.4.3. Relationship of object size, contrast and perception

Even when using cone vision, it is not possible to decide whether an object of a given size will be discernible against the background unless the contrast is specified. The exact relationship between minimum perceptible contrast difference and object size will depend on a number of factors including the signal/noise ratio and the precise viewing conditions. Contrast–detail diagrams may be generated using a suitable phantom (figure 8.3(a)) in which simple visual signals (e.g. circles) are arranged in a rectangular array, such that the diameter changes monotonically vertically and the contrast changes monotonically horizontally. The observer has to select the lowest contrast signal in each row that is considered detectable in the image. A typical curve is shown in figure 8.3(b) and illustrates the general principle that the higher the object contrast, the smaller the detectable object size.

Some typical figures for a modern intravenous angiographic system will illustrate the relationship. Suppose an incident dose in air of 10 μGy per image is used. At a mean energy of 60 keV this will correspond to about 3×10^5 photons mm^{-2}. If the efficiency of the image intensifier for detecting these photons is 30%, about 10^5 interactions mm^{-2} are used per image. Assuming a typical signal/noise ratio for detectability of 5, then objects as small as 0.2 mm can be detected if the contrast is 10% but the size increases to 2.5 mm if the contrast is only 1%.

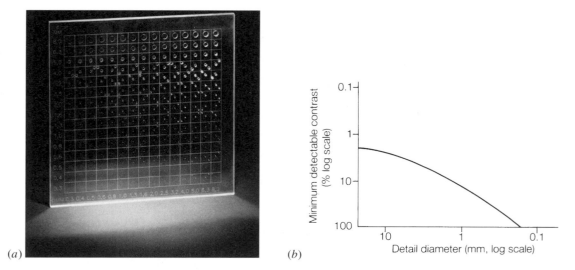

Figure 8.3. *Relationship between minimum perceptible contrast difference and object size. (a) A suitable phantom for investigating the problem. Holes of different diameter (vertical axis) are drilled in Plexiglas to different depths to simulate different amounts of contrast (horizontal axis). As the holes become smaller the contrast required to visualize them becomes greater. (Photograph kindly supplied by Nuclear Associates after a design by Thijssen* et al *(1989).) (b) Typical result for an image intensifier TV camera screening unit.*

8.4.4. Eye response to different spatial frequencies

When digitized images are displayed, the matrix of pixels will be imposed on the image and it is important to ensure that the matrix is not visually intrusive, thereby distracting the observer.

To analyse this problem, it is useful to introduce the concept of **spatial frequency**. Suppose a 128×128 matrix covers a square of side 19 cm. Each matrix element will measure 1.5 mm across, or, alternatively expressed, the frequency of elements in space is 1/0.15 or about 7 cm^{-1}. Thus the spatial frequency is 7 cm^{-1}, 700 m^{-1} or more usually expressed in radiology as 0.7 mm^{-1}.

Clearly the effect of these pixels on the eye will depend on viewing distance. The angle subtended at the eye by pixels having a spatial frequency of 0.7 mm^{-1} will be different depending on whether viewed from 1 m or 50 cm. Campbell (1980) has shown that contrast sensitivity of the eye–brain system is very dependent on spatial frequency and demonstrates a well defined maximum. This is illustrated in figure 8.4 where a viewing distance of 1.0 m has been assumed. Exactly the same curve would apply to spatial frequencies of twice these values if they were viewed from 50 cm.

Thus it may be seen that the regular pattern of matrix lines is unlikely to be intrusive when using a 512×512 matrix in digital radiology but it may be necessary to choose both image size and viewing distance carefully to avoid or minimize this effect when looking at 128×128 or 64×64 images sometimes used in nuclear medicine.

8.4.5. Limitations of a subjective definition of contrast

In chapter 4 a definition of contrast based on visual or subjective response was given. However, this definition alone is not sufficient to determine whether or not a given boundary between two structures will be visually detectable because the eye–brain will be influenced by the type of boundary as well as by the size of the boundary step. There are two reasons for this. First, the perceived contrast will depend on the sharpness of

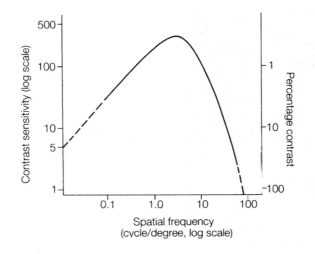

Figure 8.4. *Contrast sensitivity of the human visual system defined as the reciprocal of threshold contrast measured with a sinusoidal grating plotted against spatial frequency (after Campbell 1980).*

the boundary. Consider for example figure 8.5. Although the intensity change across the boundary is the same in both cases, the boundary illustrated in figure 8.5(*a*) would appear more contrasty on X-ray film because it is sharper. Second, if the boundary is part of the image of a small object in a rather uniform background, contrast perception will depend on the size of the object. A digital system which can artificially increase the density at the edge of a structure will subjectively increase the contrast of the structure.

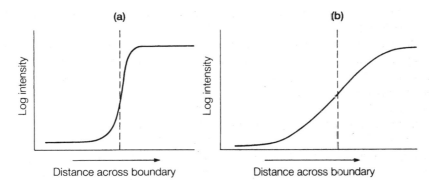

Figure 8.5. *Curves showing the difference in transmitted intensity across (a) a sharp boundary; (b) a diffuse boundary.*

8.5. PHYSICAL DEFINITION OF CONTRAST—SIGNAL-TO-NOISE RATIO

The definition of contrast given in chapter 4 is often the most appropriate for X-ray film where it is only necessary to measure the optical density of the film. However, it has limitations

(1)   because film is a non-linear device,
(2)   because it cannot readily be extended to other imaging systems.

An alternative definition uses as a starting point the signal-to-noise ratio. Imagine for example that an object consisted of a number of parallel strips of lightly attenuating material. The image would appear as in figure 8.6, with regular fluctuations in the pattern on a uniform background.

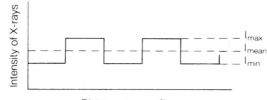

Figure 8.6. *Appearance of a perfect image of an object comprising a number of parallel strips of lightly attenuating material.*

The signal may be represented by $I_{max} - I_{min}$ and the noise is the uniform background I_{min}. Hence the signal-to-noise ratio is $(I_{max} - I_{min})/I_{min}$ or if contrast is small, to a good approximation $(I_{max} - I_{min})/I_{mean}$. Contrast is often given as $(I_{max} - I_{min})/(I_{max} + I_{min})$. Since $I_{mean} = \frac{1}{2}(I_{max} + I_{min})$ this will give values that differ by a factor of two from those obtained using I_{mean}.

This definition can be easily extended to digital systems simply by replacing 'intensity' by 'number of photon interactions in the recording medium' or any other variable that is a measure of signal. It is now easier to take account of statistical fluctuations in the signal.

8.6.   QUANTUM NOISE

When a radiographic screen–film system is exposed to X-rays with a uniform intensity distribution, the macroscopic density distribution of the developed film fluctuates around the average density. This fluctuation is called radiographic noise or radiographic mottle and the ability to detect a signal will depend on the amount of noise in the image. There are many possible sources of noise, for example noise associated with the imaging device itself, film graininess and screen structure mottle, and noise is generally difficult to analyse quantitatively. However, the effect of one type of noise, namely quantum noise, can be predicted reliably and quantitatively. Quantum noise is also referred to variously as **quantum mottle** or photon noise.

Table 8.1. *Variation in counts due to statistical fluctuations as a function of the number of counts collected (or the number of counts per pixel), N.*

N	$N^{1/2}$	$N^{1/2}/N \times 100$ (%)
10	3	30
100	10	10
1 000	30	3
10 000	100	1
100 000	300	0.3
1 000 000	1000	0.1

The interaction of a flux of photons with a detector is a random process and the number of photons detected is governed by the laws of Poisson statistics. Thus if a uniform intensity of photons were incident on an array of identical detectors as shown in figure 8.7(a), and the mean number of photons recorded per detector were N, most detectors would record either more than N or less than N as shown in figure 8.7(b).

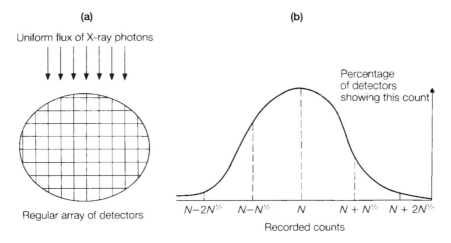

Figure 8.7. *Demonstration of the statistical variation of detected signal: (a) uniform flux of X-ray photons onto a regular array of detectors; (b) the spread of recorded counts per detector. If the mean count per detector is N, 66% of the readings lie between $N - N^{1/2}$ and $N + N^{1/2}$; 95% of the readings lie between $N - 2N^{1/2}$ and $N + 2N^{1/2}$.*

The width of the distribution is proportional to $N^{1/2}$ and thus, as shown in table 8.1, the variation in counts due to statistical fluctuations is dependent on the value of N and, expressed as a percentage, always increases as N decreases. Since the number of photons detected depends on the patient dose, quantum noise always increases as the patient dose is decreased. Whether or not a given dose reduction and the consequent increase in quantum noise will significantly affect image quality, and hence the outcome of the examination, can only be determined by more detailed assessment of the problem. For example a radiographic image might contain 3×10^5 photons mm^{-2} for an incident skin dose of 10 mGy. Thus the noise $N^{1/2} \approx 6 \times 10^2$ and the noise-to-signal ratio $N^{1/2}/N \approx 0.2\%$. In this situation changes in transmission of 1% or more will be little affected by statistical fluctuations. However, a nuclear medicine image comprising 10^5 counts *in toto* may be digitized into a 64×64 matrix (approximately 4000 pixels) giving a mean of 25 counts per pixel. Now $N^{1/2}/N = 20\%$!

Quantum noise is always a major source of image degradation in nuclear medicine. There are now also a number of situations in radiology where it must be considered as a possible cause for loss of image quality or diagnostic information. These include the following:

(1) The use of very fast intensifying screens may reduce the number of photons detected to the level where image quality is affected (see section 4.11.2).

(2) If an image intensifier is used in conjunction with a television camera, the signal may be amplified substantially by electronic means. However, the amount of information in the image will be determined at an earlier stage in the system, probably by the number of X-ray photon interactions with the input phosphor to the image intensifier. The smallest number of quanta at any stage in the imaging process determines the quantum noise and this stage is sometimes termed the **quantum sink**.

(3) Enlargement of a radiograph decreases the photon density in the image and hence increases the noise.

(4) In digital radiography (see sections 8.7 and 9.4) the smallest detectable contrast over a small area, say 1 mm^2, will be determined by quantum noise.

(5) In computed tomography (see chapter 10) the precision with which a CT number can be calculated will be affected. For example, the error on 100 000 counts is 0.3% (see table 8.1) and the CT number for uniformly attenuating material must vary accordingly.

Note that quantum mottle confuses the interpretation of low contrast images. In section 8.8.1 imager performance will be assessed in terms of modulation transfer functions which provide information on the resolution of small objects with sharp borders and high contrast. Other sources of noise may be the ultimate limiting factor in these circumstances.

8.7.   DIGITAL RADIOGRAPHY (PRINCIPLES)

A good example of the use of the signal-to-noise ratio is in digital radiography where the primary objective is to present the information from a projected X-ray image in numerical form. This idea has already been introduced in chapter 7 in the context of nuclear medicine. The principle is similar when applied to radiographic images. Instead of presenting an image that is essentially continuous, the image space is divided up into a large number of discrete picture elements or pixels and the number of photons interacting with the image receptor is recorded for each pixel. Thus the digitized image is essentially a matrix of numbers. When the image is reconstructed a certain brightness level is assigned to all numbers within a particular range and a greyscale image is generated.

To achieve a high quality digitized image, the detector system must

(1)   record X-ray quanta with a high efficiency and
(2)   be capable of providing spatial information about the distribution of X-rays.

Practical details on how this is achieved are given in section 9.4. In this chapter we shall concentrate on the many advantages of digital images for the enhancement and assessment of image quality.

When data have been stored in numerical form, the capability for manipulation provides several advantages over conventional radiology. For example if a dose in air of 10 mGy is incident on the skull, approximately 10^6 photons mm^{-2} will be transmitted. The fluctuation in this signal due to Poisson statistics is $\pm 10^3$ which is only 0.1% of the signal. Thus changes in signal of this order produced in the skull should be detectable. So far there is no difference from a conventional radiograph. However, an alternative way of stating that a change of 0.1% in the signal can be detected is to say that there are 1000 statistically distinguishable levels known to be present in the image. Unfortunately, the eye can only distinguish about 20 grey levels and for a conventional radiograph the chances of the exposure and development conditions being such that the features of radiological interest would fall exactly within those 20 levels is very small.

Numerical data can be manipulated so that the small number of grey levels to which the eye is sensitive are matched exactly to that part of the much wider dynamic range of the detector in which there is clinically useful information. This is one of the most widely used forms of image processing. It will be illustrated by comparing the much greater flexibility of a digital image with the total inflexibility of a conventional radiograph.

In order to make the comparison, the digital data will be first converted to a logarithmic scale (table 8.2). In other words, instead of relating the brightness in an image pixel to the number of X-ray photons leaving the patient it will be related to the logarithm of the number of photons. The brightness in a digital image on a TV screen then becomes closely analogous to the optical density on a radiograph (with similar advantages to those discussed in chapter 5).

Figure 8.8 shows, on the same axes, the characteristic curves for film and the digitized image. If it is assumed that the brightness range of the TV screen has been adjusted to coincide with the optical density range of film so that values between 0.25 and 2.0 produce discernible contrast, the greater dynamic range of the digital image is immediately apparent from this diagram.

However, if only about 20 brightness levels are available, this diagram shows that quite a big change in exposure will be required with the unprocessed digital image to achieve a change in brightness.

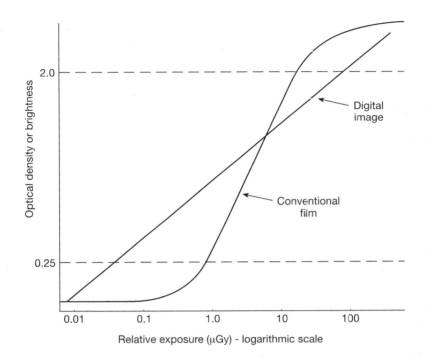

Figure 8.8. *Characteristic curves for film and a digitized image shown on the same axes.*

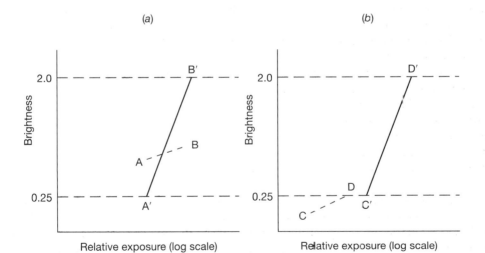

Figure 8.9. *Manipulation of digitized data to use the full range of the TV monitor. (a) Narrow window of digital data (AB), lying within the useful part of the characteristic curve amplified to use the full range of the TV camera (A'B'). This is analogous to using a film of the same speed but much bigger gamma. (b) Narrow window of digital data (CD) lying outside the useful part of the characteristic curve shifted to the right and then amplified as previously (C'D'). This is analogous to using a faster film with much bigger gamma.*

Table 8.2. *Table showing the relationship between a linear brightness scale and a logarithmic brightness scale.*

	Linear scale		Logarithmic scale	
Relative exposure	Brightness (arbitrary units)		Relative exposure	Brightness (arbitrary units)
0	0		0.1	−1
20	1		1	0
40	2		10	1
60	3		100	2
80	4		1000	3

Figures 8.9(*a*) and (*b*) show how a narrow window of digital data may be amplified and if necessary shifted into the relevant brightness range thus allowing the full range of brightness on the TV monitor to be used to display a narrow range of exposures. For detail see the figure legend.

Other forms of image processing may also be used. Consider for example figure 8.10(*a*). Here the contrast can be improved substantially by subtracting a fixed number of counts, n_0, from each pixel since

$$\text{contrast} = \log((n_1 - n_0)/(n_2 - n_0)) > \log(n_1/n_2).$$

Smoothing is another important technique widely applied to digital images of all types. The principle of smoothing is that some weight is given to the size of the signal in pixels adjacent to the pixel of interest. Consider for example a very simple form of smoothing, a nine point array of weighting factors (table 8.3).

Table 8.3. *A nine point array of smoothing factors.*

1	2	1
2	4	2
1	2	1

If the counts in a matrix of elements are as shown below

x_{i-1j-1}	x_{ij-1}	x_{i+1j-1}
x_{i-1j}	x_{ij}	x_{i+1j}
x_{i-1j+1}	x_{ij+1}	x_{i+1j+1}

then the smoothed count in pixel x_{ij} would be

$$\frac{1}{16}[4x_{ij} + 2(x_{i-1j} + x_{i+1j} + x_{ij-1} + x_{ij+1}) + (x_{i-1j-1} + x_{i+1j-1} + x_{i-1j+1} + x_{i+1j+1})].$$

The same process would be applied to every other pixel in turn. Smoothing is particularly useful for removing the intrusive visual effect of pixellation and when images are particularly noisy because of poor counting statistics.

Figure 8.10(*b*) shows another potential application of image processing where a plot of the gradient of counts per pixel is much more pronounced than the change in absolute counts across the edges.

Techniques that exploit the potential of digital radiography for high statistical accuracy without generating a wide dynamic range of data are particularly powerful. One way to do this is to use a subtraction technique on digital radiographs obtained under slightly different conditions. For example X-ray beams of different energies may be used or images can be obtained before and after injection of contrast medium. For

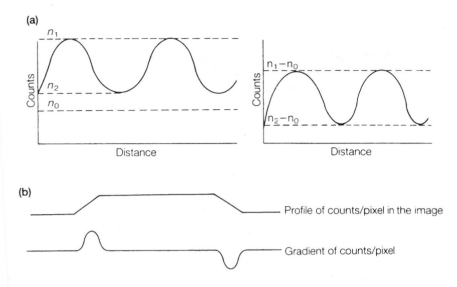

Figure 8.10. *Examples of potentially useful methods of data manipulation: (a) background subtraction; (b) edge enhancement or use of gradients.*

this purpose the absorption edge of iodine at 35 keV is particularly useful. Pullen (1981) showed that for a dose in air of 10 mGy a differential concentration of 1.5 mg ml^{-1} iodine may be sufficient to visualize 1 mm diameter vessels. Note that when the required information is dependent on the dynamic behaviour of the contrast agent, all the image space must be sampled very quickly.

Finally, there is increasing interest in combining imaging modalities. Digital images are now essential so that one set of images can be manipulated by horizontal, vertical or rotational movement, or by magnification/minification so as to match the other set. Similar techniques may also be required in subtraction studies (see section 9.5) if there is patient movement.

For further discussion on digital imaging see section 13.7.9.

8.8.   ASSESSMENT OF IMAGE QUALITY

8.8.1.   *Modulation transfer function*

It is important that the quality of an image should be assessed in relation to the imaging capability of the device that produces it. For example it is well known than an X-ray set is capable of producing better anatomical images of the skeleton than a gamma camera—although in some instances they may not be as useful diagnostically.

One way to assess performance is in terms of the resolving power of the imaging device—i.e. the closest separation of a pair of linear objects at which the images do not merge. The futility of pixelating digitized data to finer elements than the resolution capability of the system has already been mentioned. Unfortunately, diagnostic imagers are complex devices and many factors contribute to the overall resolution capability. For example, in forming a conventional X-ray image these will include the focal spot size, interaction with the patient, type of film, type of intensifying screen and other sources of unsharpness. Some of these interactions are not readily expressed in terms of resolving power and even if they were there would be no easy way to combine resolving powers.

A practical approach to this problem is to introduce the concept of modulation transfer function (MTF). This is based on the ideas of Fourier analysis, for detailed consideration of which the reader is referred elsewhere, e.g. Gonzalez and Wintz (1977). Only a very simplified treatment will be given here.

Starting from the object, at any stage in the imaging process all the available information can be expressed in terms of a spectrum of spatial frequencies. The idea of spatial frequency can be understood by considering two ways of describing a simple object consisting of a set of equally spaced parallel lines. The usual convention would be to say the lines were equally spaced 0.2 mm apart. Alternatively, one could say that the lines occur with a frequency in space (spatial frequency) of five per mm.

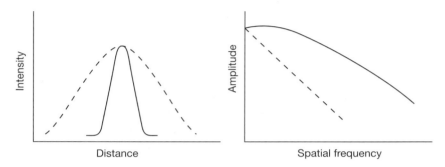

Figure 8.11. *Two objects and their corresponding spectra of spatial frequencies. Solid line is a sharp object. Dotted line is a diffuse object.*

Fourier analysis provides a mathematical method for relating the description of an object (or image) in real space to its description in frequency space. Two objects and their corresponding spectra of spatial frequencies are shown in figure 8.11. In general, the finer the detail, i.e. sharper edges in real space, the greater the intensity of high spatial frequencies in the spatial frequency spectrum (SFS). Thus, fine detail, or high resolution, is associated with high spatial frequencies.

For exact images of these objects to be reproduced, it would be necessary for the imaging device to transmit every spatial frequency in each object with 100% efficiency. However each component of the imaging device has a modulation transfer function (MTF)[1] which modifies the spatial frequency spectrum of the information transmitted by the object.

Each component of the imager can be considered in turn so

$$\frac{\text{SFS out of}}{\text{component } M} = \frac{\text{SFS into}}{\text{component } N} \rightarrow \text{component } N \rightarrow \frac{\text{SFS out of}}{\text{component } N}$$

where $(\text{SFS}_{\text{out}})_N = (\text{SFS}_{\text{in}})\, \text{MTF}_N$.

Hence $(\text{SFS})_{\text{image}} = (\text{SFS})_{\text{object}}\, (\text{MTF})_A\, (\text{MTF})_B\, (\text{MTF})_C \ldots$ where A, B, C are the different components of the imager.

No imager has an MTF of unity at all spatial frequencies. In general, the MTF decreases with increasing spatial frequency, the higher the spatial frequency at which this occurs the better the device.

The advantage of representing imaging performance in this way is that the MTF of the system at any spatial frequency, v, is simply the product of the MTFs of all the components at spatial frequency v. This is conceptually easy to understand and mathematically easy to implement. It is much more difficult to work in real space.

---

[1] Rigorous mathematical derivation of MTF assumes a sinusoidally varying object with spatial frequency expressed in cycles mm^{-1}. In the graphs which follow it has been assumed that this sinusoidal wave can be approximated to a square wave and spatial frequencies are given in line pairs mm^{-1} for ease of interpretation by the reader.

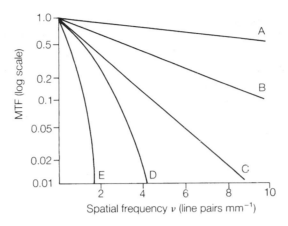

Figure 8.12. *Some typical MTFs for different imaging systems. A = non-screen film; B = film used with high definition intensifying screen; C = film with medium speed screens; D = a 150 mm Cs:Na image intensifier; E = the same intensifier with television display (adapted from Hay 1982).*

Some examples of the way in which the MTF concept may be used will now be given. First, it provides a simple pictorial representation of the overall imaging capability of a device. Figure 8.12 shows MTFs for five imaging devices. It is clear that non-screen film transfers higher spatial frequencies, and thus is inherently capable of higher resolution, than screen film. Of course this is because of the unsharpness associated with the use of screens. Note that the MTF takes no account of the dose that may have to be given to the patient.

A similar family of curves would be obtained if MTFs were measured for different film–screen combinations. The spatial frequency at which the MTF fell to 0.1 might vary from 10 line pairs mm^{-1} for a slow film–screen combination to 2.5 line pairs mm^{-1} for a faster film–screen combination. This confirms that slow film–screen combinations are capable of higher resolution than fast film–screens. The major source of this difference is in the choice of screen but recent work has shown small differences in MTF using the same intensifying screen and different films. The MTF is higher when light cross-over from one film emulsion to the other can be reduced.

Since the MTF is a continuous function, an imaging device does not have a 'resolution limit' i.e. a spatial frequency above which resolution is not possible, but curves such as those shown in figure 8.12 allow an estimate to be made of the spatial frequency at which a substantial amount of information in the object will be lost.

Second, by examining the MTFs for each component of the system, it is possible to determine the weak link in the chain, i.e. the part of the system where the greatest loss of high spatial frequencies occurs.

Figure 8.13 shows MTFs for some of the factors that will degrade image quality when using an image intensifier TV system. Since MTFs are multiplied, the overall MTF is determined by the poorest component—the vidicon camera in this example. Note that movement unsharpness will also degrade a high resolution image substantially.

As a third example, the MTF may be used to analyse the effect of varying the imaging conditions on image quality. Figure 8.14 illustrates the effect of magnification and focal spot size in magnification radiography. Curves B and C show that for a fixed focal spot size, image quality deteriorates with magnification and curves C and D show that for fixed magnification image quality deteriorates with increased focal spot size. Note that if it were possible to work with $M = 1$, then the focal spot would not affect image quality and an MTF of 1 at all spatial frequencies would be possible (curve A). All these changes could of course have been predicted in a qualitative manner from the discussions of magnification radiography in chapters 5 and 9. The point

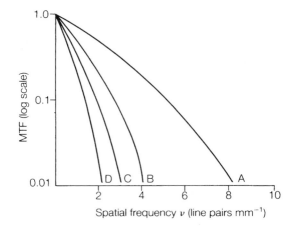

Figure 8.13. *Typical MTFs for some factors that may degrade image quality in an image intensifier TV system. A = 1 mm focal spot with 1 m focus film distance and small object–film distance; B = image intensifier; C = movement unsharpness of 0.1 mm; D = conventional vidicon camera with 800 scan lines.*

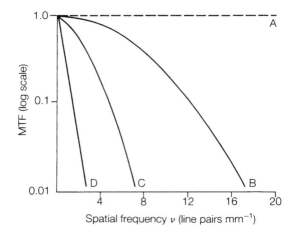

Figure 8.14. *MTF curves for magnification radiography under different imaging conditions. A = magnification of 1; B = magnification of 1.2 with a 0.3 mm focal spot size; C = magnification of 2.0 with a 0.3 mm focal spot size; D = magnification of 2.0 with a 1 mm focal spot size.*

about MTF is that it provides a quantitative measure of these effects, and one that can be extended by simple multiplication to incorporate other factors such as the effect of magnification on the screen MTF, and the effect of movement unsharpness (see for example Curry *et al* 1990). Hence it is the starting point for a logical analysis of image quality and the interactive nature of the factors that control it.

The majority of systems transmit low spatial frequencies better than high spatial frequencies. Xero-radiography (section 9.7) has an unusual MTF. This is not a good technique for visualizing large, low contrast structures and has a poor MTF at low spatial frequencies. However, the ability of xeroradiography to enhance edges means that it does transmit well the high spatial frequencies that carry information about the edges. Thus the MTF is enhanced in the region of one line pair mm^{-1}, before falling again at higher spatial frequencies because of loss of resolution caused by other problems (e.g. focal spot size and patient motion).

Physical/physiological assessment

ugh the MTF provides a useful method of assessing physical performance, its meaning in terms of
terpretation of images is obscure. For this purpose, as already noted in the discussion on contrast, it
essary to involve the observer and in this context image quality may be defined as a measure of the
effectiveness with which the image can be used for its intended purpose.

The simplest perceptual task is to detect a detail of interest, i.e. the signal, in the presence of noise. Noise
includes all those features that are irrelevant to the task of perception—e.g. quantum mottle, anatomical noise
and visual noise (inconsistencies in observer response). Three techniques that have been used to discriminate
between signal and noise will be mentioned briefly (for further details see Sharp *et al* 1985 and *ICRU Report*
54 1996).

It should be noted in passing that the visual thresholds at which objects can be (a) detected, (b) recognized
and (c) identified are not the same and this is a serious limitation when attempting to extrapolate from studies
on simple test objects to complex diagnostic images.

Method of constant stimulus

Consider the simple task of detecting a small region of increased attenuation in a background that is uniform
overall but shows local fluctuations due to the presence of noise. Because of these fluctuations, the ease with
which an abnormality can be seen will vary from one image to another even thought its contrast remains
unchanged.

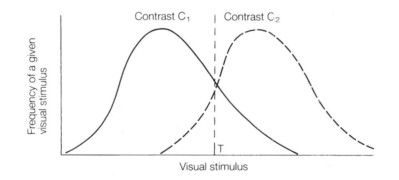

Figure 8.15. *Curves showing how for a fixed visual stimulus threshold T the proportion of stimuli detected
will increase if contrast is increased from C_1 to C_2.*

Figure 8.15 shows the probability of generating a given visual stimulus for a set of images of this type
in which the object contains very little contrast (C_1). If the observer adopts some visual threshold T, only
those images which fall to the right of T will be reported as containing the abnormality (about 25% in this
example).

If a set of images is now prepared in which the object has somewhat greater contrast (C_2), the distribution
curve will shift to the right (shown dotted) and a much greater proportion of images will be scored positive.

If this experiment is repeated several times using sets of images, each of which contains a different
contrast, visual response may be plotted against contrast as shown in figure 8.16. The contrast resulting in
50% visual response is usually taken as the value at which the signal is detectable.

Experiments of this type may be used to demonstrate, for example, that in relation to visual perception,
object size and contrast are inter-related.

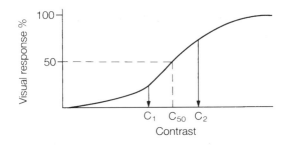

Figure 8.16. *Visual response (percentage positive identification) plotted as a function of contrast on the basis of observations similar to those described in figure 8.15.*

Signal detection theory

To apply the method of constant stimulus, contrast must be varied in a controlled manner. The method cannot therefore be applied to real images.

In signal detection theory, all images are considered to belong to one of two categories, those which contain a signal plus noise and those which contain noise only. It is further assumed that in any perception study some images of each type will be misclassified. This situation is represented in terms of the probability distribution used in the previous section in figure 8.17.

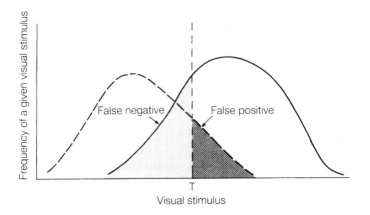

Figure 8.17. *Curves illustrating the concept of false negative and false positive based on the spread of visual stimuli for an object of fixed contrast. Dotted line = probability distribution of true negatives. Solid line = probability distribution of true positives.*

There are four possible responses: true positive, true negative, false positive and false negative. These four responses are subject to the following constraints:

true positive + false negative (all 'signal + noise' images) = constant

true negative + false positive (all 'noise only' images) = constant.

The greater the separation of the distributions, the more readily is the signal detected.

The main problem with both signal detection theory and the method of constant stimulus is that they require the visual threshold level T to remain fixed. In practice this is well nigh impossible to achieve, even for a single observer. If T is allowed to vary, for example if there are several observers, a more sophisticated approach is required (see section 8.9).

Ranking

In this method the observer is presented with a set of images in which some factor thought to influence quality has been varied. The observer is asked to arrange the images in order of preference. This approach relies on the fact that an experienced viewer is frequently able to say whether a particular image is of good quality but unable to define the criteria on which this judgement is based.

When the ranking order produced by several observers is compared, it is possible to decide whether observers agree on what constitutes a good image. Furthermore, if the image set has been produced by varying the imaging conditions in a controlled manner, for example by steadily increasing the amount of scattering medium, it is possible to decide how much scattering medium is required to cause a detectable change in image quality.

The simplest ranking experiment is to compare just two images. If one of these has been taken without scattering material, the other with scattering material, then the percentage of occasions on which the 'better' image is identified can be plotted against the amount of scattering material. The 75% level corresponds to correct identification 3/4 times, the 90% level to correct identification 9/10 times. It might be reasonable to conclude that when the better image was correctly identified 3/4 times, detectable deterioration in image quality had occurred (figure 8.18).

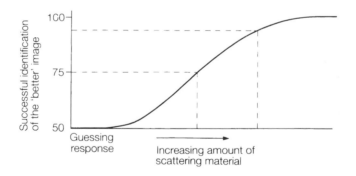

Figure 8.18. *Curve showing how the ability to detect the undegraded image might be expected to increase steadily as the amount of degradation was increased.*

The ranking approach is particularly useful for quality control and related studies. All equipment will deteriorate slowly and progressively with time and although it may be possible to assess this deterioration in terms of some physical index such as resolution capability with a test object, it is not clear what this means in terms of image quality. If deterioration with time (figure 8.19(a)) can be simulated in some controlled manner (figure 8.19(b)), then a ranking method applied to the images thus produced may indicate when remedial action should be taken. A similar approach may be used to assess the benefit of new technology, for example the effect of different film–screen combinations on the quality of tomographic images (Cohen *et al* 1976) or in mammography (Sickles *et al* 1977).

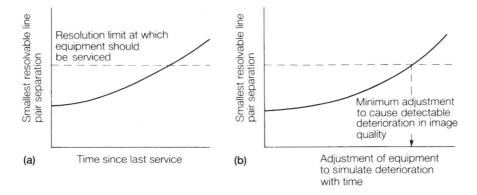

Figure 8.19. *Curves showing how a logical policy towards the frequency of servicing might be based on measurements of deterioration of imager performance.*

8.9. RECEIVER OPERATOR CHARACTERISTIC (ROC) CURVES

8.9.1. *Principles*

Images from true abnormal cases will range from very abnormal, a strong visual stimulus, to only suspicious, a weak visual stimulus. Similarly, some images from true normal cases will appear more suspicious than others and in all probability the two distributions will overlap. Note that similar distributions will be produced, perhaps with rather less overlap, if two sets of images are presented in different ways or perhaps after different pre-processing.

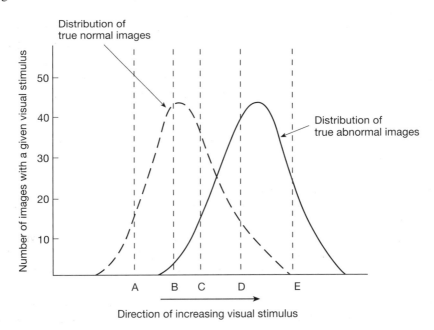

Figure 8.20. *Use of different visual thresholds with distributions of visual stimuli.*

The number of true positives identified now depends on the threshold detection level. If a very lax criterion is adopted, point A in figure 8.20, the true positives = 100% but there is also a high percentage of false positives. If the threshold detection level is higher, or the observer adopts a more stringent criterion, say point D, the result is quite different with few false positives but many true positives being missed.

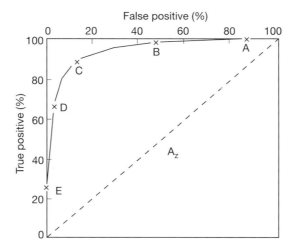

Figure 8.21. *The ROC curve that would be constructed if the two distributions shown in figure 8.20 were sampled at five discrete points.*

ROC curves overcome this problem. The observers are now encouraged to change their visual threshold and for each of the working points A to E, the false positives are plotted against the true positives (figure 8.21). Two extreme distributions can be selected to illustrate the limits of the ROC curve. First, if the distributions are completely separate a range of thresholds may be chosen which first increases the percentage of true positives without any false positives and then gradually increases the false positives until the criterion is so lax that both distributions are accepted as abnormal. In other words in the case of perfect discrimination the ROC curve tracks up the vertical axis and then along the horizontal axis. The other extreme is where the distributions overlap completely. Now each relaxation of the detection threshold which allows in a few more true positives allows an equal number of false positives and the ROC curve follows the 45 degree line (shown dotted in figure 8.21).

A useful numerical parameter is the proportion of the ROC space that lies below the ROC curve, A_Z. In the two extreme distributions just discussed $A_Z = 1.0$ and 0.5 respectively. For intermediate situations the larger the value of A_Z the greater the separation of the distributions. Note that a value of A_Z less than 0.5 indicates that the observer is performing worse than guessing!

A two-dimensional ROC curve can be converted into a three-dimensional ROC curve in which the signal intensity is the third dimension (see figure 8.22). For example when the signal intensity is high, the ROC curve will be close to the axes. When the signal intensity is zero the ROC curve will be the 45 degree guessing line. Thus a constant value for the imaging parameter produces an ROC curve whereas a profile through the surface at a constant false positive fraction yields the response curve that would be produced by the method of constant stimulus.

One limitation of the simple ROC approach is that the observer might correctly identify the abnormal image on the basis of an (erroneously) perceived abnormality in a part of the image that is actually normal. When the observer is required both to identify abnormal images and to specify correctly the position of the abnormality, localization ROC curves (LROCs) may be constructed. A further complication with clinical images is that the radiologist's decision is often based on more than one feature in the image. Clearly, there

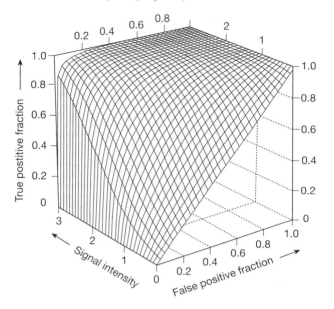

Figure 8.22. *A 'three-dimensional' graph in which two-dimensional ROC curves are plotted as a function of signal intensity, thereby generating a surface in three dimensions (from ICRU Report 54).*

will be greater confidence in the conclusion when the different features provide corroborative information. Little work has been done on the correct analytical approach when different image features provide conflicting information.

Good work can be done on systematic analysis of image quality using analogue images. However, digitized images are preferable because the data are available in numerical form. For example it is possible to investigate the interaction between, say, contrast, resolution and noise for carefully controlled, quantitative changes in the images or to look at the relative importance of structured or unstructured noise. One reason why this is desirable is that the eye–brain system is very perceptive, so there is only a narrow working region in which an observer might be uncertain or different observers disagree. Digital images provide more opportunity for fine tuning than analogue images.

Finally, under computer control it is possible to superimpose known lesions of different size, shape and contrast on normal or apparently normal clinical images that have been digitized. In the subsequent analysis the true abnormals can be unambiguously identified.

8.9.2.   *ROC curves in practice*

ROC analysis has been used to investigate numerous imaging problems. Three examples will be given to illustrate the power of the technique.

Pixel size and image quality in digitized chest radiography

Several authors have used ROC methods to address the question 'What is the largest pixel size that will generate digitized images that are indistinguishable from conventional chest radiographs?'. Clearly there are many features of the image one can examine including nodules, fine detail, mild interstitial infiltrates and subtle pneumothoraces and MacMahon *et al* (1986) were able to show, by digitizing images to different pixel sizes, that diagnostic accuracy increased significantly as pixel size was reduced, at least to 0.1 mm (figure 8.23).

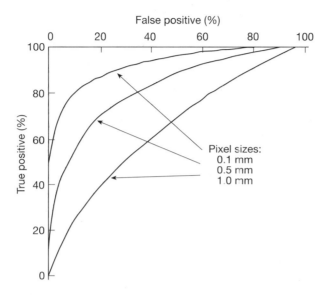

Figure 8.23. *Use of ROC curves to show how diagnostic accuracy for fine detail in the pneumothorax varies with pixel size in digitized images (adapted with permission from MacMahon et al 1986).*

Although a pixel size of 0.1 mm × 0.1 mm may be acceptable for chest radiographs, subsequent work by the same authors (Chan *et al* 1987) showed that for mammograms detection accuracy for microcalcifications was still significantly reduced if images were digitized to this level.

Assessment of competence as film readers in screening mammography

As a second example consider the question 'Is there a potential to use radiographers as film readers in screening mammography?'.

Double reading of screening mammograms has been shown to increase the cancer detection rate (Anderson *et al* 1994) but the acute shortage of radiologist resources makes this difficult. Pauli *et al* (1996) therefore set out to determine whether suitably trained radiographers could play a useful role as second readers in screening mammography. The data are suitable for ROC analysis because there are four possible outcomes to a film report—abnormal mammogram/recall (true positive), abnormal mammogram/return to routine screening (false negative), normal mammogram/recall (false positive), normal mammogram/return to routine screening (true negative).

A modified version of A_Z, represented here by A'_Z, was used to allow for the fact that distributions are very skewed because of the large percentage of normals. Results for seven radiographers were compared with those from nine radiologists and an extract from the data is shown in table 8.4.

Results showed that after suitable training radiographers achieved and maintained the same high standard of performance on training test sets as a group of radiologists.

Image quality following data compression

As a final example consider the question 'Is it possible to compress data without loss of image quality?'. Radiology departments are moving rapidly towards digital systems and film-less images. However this

Table 8.4. *Radiographer and radiologist performances in reading mammographic films by comparison of the area under an ROC curve (adapted from Pauli et al 1996).*

	A'_Z values	
	Radiographers	Radiologists
Pre-training	0.77	0.87
Post-training	0.88	0.89
After 200 screening mammograms read	0.91	—
After 5000 screening mammograms read	0.88	

will create formidable data transfer and archiving problems. An 18 cm × 24 cm mammogram digitized at 0.05 mm × 0.05 mm results in a matrix size of about 4000 × 5000, so a four-view study generates about 160 megabytes of digital data.

One way to reduce the problem is to compress the data. There is clearly redundant information in the complete image. For example the counts in pixels in the corners will be no higher than background and there will be a high degree of correlation between adjacent pixels. Data compression will then have a number of advantages, especially in respect of less archival demands, faster data transfer and faster screen build-up.

Compression algorithms are basically of two types. The first is reversible or loss-less compression. Provided that the compression/decompression schemes are known, a true copy of the original image may be recovered. Unfortunately, the maximum compression ratio that can be achieved in this way is about 4:1, often less with noisy data.

Higher compression ratios require the use of irreversible processing which results in some alteration in the information content. Note that this is not a problem *per se*. Most image data undergo some alteration in information content before they are viewed (e.g. filtering the raw projections during image reconstruction in CT—section 10.3.3). The important requirement is to be able to image a lesion optimally without introducing false readings. Provided that data compression does not influence this process negatively, a certain degree of loss of information can be tolerated.

A discussion of the relative merits of different compression algorithms is outside the scope of this book but it is clear that ROC analysis provides a useful way to select the approach which gives maximum compression with minimum loss of image quality. For an example Aberle *et al* (1993) studied 122 PA chest radiographs which had been digitized at 2000 × 2000 pixels and then compressed at an approximate compression ratio of 20:1. Five radiologists read the digitized images and the digitized/compressed images and results were analysed using ROC curves (see table 8.5).

Table 8.5. *Mean values of A_Z for five radiologists who read uncompressed and compressed chest radiographs (adapted from Aberle et al 1993).*

	A_Z	
	Digitized	Compressed
Interstitial disease	0.95 ± 0.04	0.95 ± 0.03
Lung nodules	0.87 ± 0.06	0.88 ± 0.05
Mediastinal masses	0.83 ± 0.08	0.80 ± 0.10

Although the compression process was irreversible, there was no evidence of any difference in detectability for interstitial disease, lung nodules or mediastinal masses.

8.10. DESIGN OF CLINICAL TRIALS

By concentrating on the physics of the imaging process, it is easy to overlook the fact that *in vivo* imaging is not an end in itself but only a means to an end. Ultimately the service will be judged by the quality of the diagnostic information it produces and critical objective evaluation of each type of examination is essential for three reasons. First it may be used to assess diagnostic reliability and this is important when the results of two examinations conflict. Second, objective information allows areas of weakness to be analysed and provides a basis for comparison between different imaging centres. Finally, medical ethics is not independent of economics and increasingly in the future the provision of sophisticated and expensive diagnostic facilities will have to be justified in economic terms.

It is not difficult to understand why few clinical trials have been done in the past when the constraints are listed:

(1) The evaluation must be prospective with images assessed within the reference frame of the normal routine work of the unit—not in some academic ivory tower.
(2) A typical cross section of normal images and abnormal images from different disease categories must be sampled since prevalence affects the predictive value of a positive test.
(3) A sufficient number of cases must enter the trial to ensure adequate statistics.
(4) Equivalent technologies must be compared. It is meaningless to compare the results obtained using a 1975 ultrasound scanner with those obtained using a 1996 CT whole body scanner or vice versa.
(5) Evaluations must be designed so that the skill and experience of the reporting teams do not influence the final result.
(6) Finally, and generally the most difficult to achieve, there must be adequate independent evidence on each case as to whether it should be classified as a true normal or a true abnormal.

If these constraints can be met, methods are readily available for representing the results. For example if images are simply classified as normal or abnormal, there are only four possible outcomes to an investigation, which can be expressed as a 2×2 decision matrix (table 8.6).

Table 8.6. *A 2×2 decision matrix for image classification.*

	Abnormal images	Normal images
True abnormal	a	b
True normal	c	d
Totals	$a + c$	$b + d$

$$\text{Overall diagnostic acccuracy} = \frac{\text{No of correct investigations}}{\text{total investigations}} = \frac{a + d}{a + b + c + d}.$$

Other indices sometimes quoted are

$$\text{sensitivity} = \frac{\text{abnormals detected}}{\text{total abnormals}} = \frac{a}{a + b}$$

$$\text{specificity} = \frac{\text{normals detected}}{\text{total normals}} = \frac{d}{c + d}$$

and

$$\text{predictive value of a positive test} = \frac{\text{abnormals correctly identified}}{\text{total abnormal reports}} = \frac{a}{a+c}.$$

It is well known that prevalence has an important effect on the predictive value of a positive test and to accommodate possible variations in prevalence of the disease, Bayes' theorem may be used to calculate the posterior probability of a particular condition, given the test results and assuming different *a priori* probabilities.

If the prior probability of disease, or prevalence, is $P(D_+)$ then it may be shown that the posterior probability of disease when the test is positive (T_+) is given by

$$P(D_+/T_+) = \frac{\left(\frac{a}{a+b}\right) P(D_+)}{\left(\frac{a}{a+b}\right) P(D_+) + \left(\frac{c}{c+d}\right) P(D_-)}.$$

Similarly the posterior probability of disease when the test is negative (T_-) is given by

$$P(D_+/T_-) = \frac{\left(\frac{b}{a+b}\right) P(D_+)}{\left(\frac{b}{a+b}\right) P(D_+) + \left(\frac{d}{c+d}\right) P(D_-)}.$$

For a full account of Bayes' theorem see Shea (1978).

As an alternative to simple classification of the images as normal or abnormal, the data can be prepared for ROC analysis if a confidence rating is assigned to each positive response (table 8.7).

Table 8.7. *Examples of the confidence ratings used to produce an ROC curve.*

Rating	Description
5	Abnormality definitely present
4	Abnormality almost certainly present
3	Abnormality possibly present
2	Abnormality probably not present
1	Abnormality not present

These ratings reflect a progressively less stringent criterion of abnormality and correspond to points E to A respectively in figure 8.21. Thus by comparing the ROC curves constructed from such rating data, the power of two diagnostic procedures may be compared.

Much work has been done in recent years on the design of trials for critical clinical evaluation of diagnostic tests and it is now possible to assess better the contribution of each examination within the larger framework of patient health care.

8.11.   CONCLUSIONS

The emphasis in this chapter has been on the idea that there is far more information in a radiographic image than it is possible to extract by subjective methods. Furthermore, many factors contribute to the quality of the final image, and for physiological reasons, the eye can easily be misled over what it thinks it sees. Thus there is a strong case for introducing numerical or digital methods into diagnostic imaging. Such methods allow greater manipulation of the data, more objective control of the image quality and greatly facilitate attempts to evaluate both imager performance and the overall diagnostic value of the information that has been obtained.

REFERENCES

Aberle D R, Gleeson F, Sayre J W, Brown K, Batra P, Young D A, Stewart B K, Ho B K T and Huang H K 1993 The effect of irreversible image compression on diagnostic accuracy in thoracic imaging *Invest. Radiol.* **28** 398–403

Anderson E D C, Muir B B, Walsh J S and Kirkpatrick A E 1994 The efficacy of double reading mammograms in breast screening *Clin. Radiol.* **49** 248–51

Campbell F W 1980 The physics of visual perception *Phil. Trans. R. Soc.* **290** 5–9

Chan H P, Vyborny C J, MacMahon H, Metz C E, Doi K and Sickles E A 1987 Digital mammography ROC studies of the effects of pixel size and unsharp mask filtering on the detection of subtle microcalcifications *Invest. Radiol.* **22** 581–9

Cohen G, Barnes J O and Peria P M 1976 The effects of filmscreen combination on tomographic image quality *Radiology* **129** 515

Curry T S, Downey J E and Murry R C 1990 *Introduction to the Physics of Diagnostic Radiology* 4th edn (Philadelphia, PA: Lea and Febiger)

Gonzalez R C and Wintz P 1977 *Digital Image Processing* (Reading, MA: Addison Wesley)

Hay G A 1982 Traditional x-ray imaging *Scientific Basis of Medical Imaging* ed P N T Wells (Edinburgh: Churchill Livingstone) pp 1–53

ICRU 1996 Medical imaging—the assessment of image quality *ICRU Report* 54 *International Commission on Radiation Units and Measurements*

MacMahon H, Vyborny C J, Metz C E, Doi K, Sabeti V and Solomon S L 1986 Digital radiography of subtle pulmonary abnormalities—an ROC study of the effect of pixel size on observer performance *Radiology* **158** 21–6

Pauli R, Hammond S, Cooke J and Ansell J 1996 Radiographers as film readers in screening radiography. An assessment of competence under test and screening conditions *Br. J. Radiol.* **69** 10–4

Pullan B R 1981 Digital radiology *Physical Aspects of Medical Imaging* ed B M Moores, R P Parker and B R Pullan (Chichester: Hospital Physicists' Association–Wiley) pp 275–88

Sharp P F, Dendy P P and Keyes W I 1985 *Radionuclide Imaging Techniques* (London: Academic)

Sickles E A, Genant H K and Doi K 1977 Comparison of laboratory and clinical evaluation of mammographic screen–film systems *Applications of Optical Instruments in Medicine* vol 4, ed J E Gray and W R Hendee (Boston, MA: SPIE) pp 30

Thijssen M A O, Thijssen H O M, Merx J L, Lindeijer J M and Bijkerk K R 1989 A definition of image quality: the image quality figure *Optimization of Image Quality and Patient Exposure in Diagnostic Radiology* ed B M Moores, B F Wall, H Eriskat and H Schibilla *BIR Report* 20 pp 29–34

FURTHER READING

Chester M S 1982 Perception and evaluation of images *Scientific Basis of Medical Imaging* ed P N T Wells (Edinburgh: Churchill Livingstone) pp 237–80

Dendy P P 1990 Recent technical developments in medical imaging part 1: digital radiology and evaluation of medical images *Current Imaging* **2** 226–36

Shea G 1978 An analysis of the Bayes procedure for diagnosing multistage disease *Comput. Biomed. Res.* **11** 65–75

EXERCISES

1   What factors affect: (*a*) the sharpness; (*b*) the contrast of clinical radionuclide images?

2   List the factors affecting the sharpness of a radiograph. Draw diagrams to illustrate these effects.

3   What do you understand by the 'quality' of a radiograph? What factors affect the quality?

4   Explain the terms subjective and objective definition, latitude and contrast when used in radiology.

5   Explain what is meant by the term 'perception' in the context of diagnostic imaging. Explain how perception studies may be used to assess the quality of diagnostic images.

6   Why does the MTF of an intensifying screen improve if the magnification of the system is increased?

7   Show that enlargement of an image such that the photons are spread over a larger area increases quantum noise by $(N/m^2)^{1/2}$ where N is the original number of photons mm^{-2} and m is the magnification. Is quantum noise increased by magnification radiography?

8   Explain how quantum mottle may limit the smallest detectable contrast over a small 1 mm^2 area in a digitized radiograph.

9   What is an ROC curve and how is it constructed? Give examples of the use of ROC curves in the assessment of image quality.

CHAPTER 9

SPECIAL RADIOGRAPHIC TECHNIQUES

9.1. INTRODUCTION

In chapter 2 the basic principles of X-ray production were presented and chapter 3 dealt with the origin of radiographic images in terms of the fundamental interaction processes between X-rays and the body. Chapters 4 and 5 showed how the radiographic image is converted into a form suitable for visual interpretation.

On the basis of the information already presented, it is possible to understand the physics of most simple radiological procedures. However, a number of more specialized techniques are also used in radiology and these will be drawn together in this chapter. These techniques provide an excellent opportunity to illustrate the application of principles already introduced to specific problems, and appropriate references will be made to the relevant sections in earlier chapters.

9.2. HIGH VOLTAGE RADIOGRAPHY

Increasing the generator voltage to an X-ray tube has a number of effects, some of which are desirable with respect to the resulting radiograph, some of which are undesirable. Among the desirable features are increased X-ray output per mA s, more efficient patient penetration, reduced dose to the patient and more efficient film blackening. More scattered radiation reaches the film and this is clearly a disadvantage. Finally, the fact that radiographic contrast falls with increasing tube kilovoltage is generally a disadvantage, except when it is necessary to accommodate a wide patient contrast range. These effects may all be illustrated by considering the technique used for chest radiography.

An operating kilovoltage of 60–70 kVp is a sound choice for small or medium sized patients. It gives a good balance of subject contrast, good bone definition and a sharp, clean appearance to the pulmonary vascularity. However, for larger patients, say in excess of 25 cm anterior–posterior diameter, both attenuation and scatter become appreciable. Tube current can only be increased up to the rating limit, then longer exposure times are required. Also if a scatter reducing grid is used, the dose to the patient is increased appreciably. Finally, use of a grid in this low kV region may enhance contrast excessively, resulting in areas near the chest wall and in the mediastinum being very light and central regions being too dark.

The problems encountered when a high voltage technique (say 125–150 kVp) is adopted are different. Tube output and patient penetration are good, thus for example operating conditions of 140 kVp and 200–500 mA allow short exposure times to be used. Note, however, that the X-ray tube may not tolerate consistent use at high voltage and the manufacturer should be told at the time of installation if the tube is to be used in this way. The kV of the generator output should be checked regularly since the tube will be operating near to its electrical rating limit. High tension cables may develop problems more frequently than at low kV.

Films will be of markedly lower contrast and may have an overall grey appearance. Thus it will be possible to accommodate a much wider range of object contrast but rib detail will not be distinct. If a patient's anterior–posterior diameter is markedly different over the upper and lower chest, an aluminium wedge filter may be used to compensate for differences in tissue absorption.

One consequence of using a higher kV is that the exposure latitude will be increased (see section 4.4.4). For any object, the primary radiation contrast, or subject contrast, will be less at the higher kV. Thus transmitted exposures occupy a narrower region on the characteristic curve at high kV than at low kV. Figure 9.1(*a*) shows how the narrow range of exposures AB may be shifted to A_1B_1 or to A_2B_2 without further loss of contrast. At the lower kV illustrated in figure 9.1(*b*) there is no exposure latitude since a shift either way will result in loss of contrast in either the toe or the shoulder of the characteristic curve.

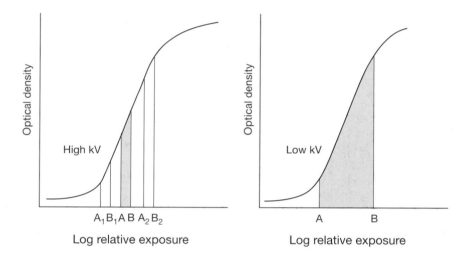

Figure 9.1. *Graphs showing how, for a given object and a given film–screen combination, there may be exposure latitude at high kV (figure 9.1(a)), but not at low kV (figure 9.1(b)).*

When using a high kV technique it is not uncommon for five times as many scattered photons as primary photons to reach the film so some form of scatter reducing technique must be used. If this is by means of a grid a 10:1 grid ratio is probably a good compromise. A higher grid ratio may improve image quality even more for a few large patients but unless the grid is changed between patients, all patients will receive a lot more radiation.

An alternative method to reduce scatter reaching the film is the **air gap technique**, the principle of which is illustrated in figure 9.2. Imagine there is a small scattering centre near the point where the X-rays leave the patient. At diagnostic energies, Compton scattered X-rays will travel almost equally in all directions (see section 3.4.3). Referring to the diagram, as the film–screen cassette is progressively moved away from the patient, the number of scattered X-ray photons intercepting the cassette decreases.

It is also clear from the diagram that the first 20–30 mm of gap will be the most important. Since there will be scattering centres throughout the body one might think that the technique would not be very effective because there is a large 'gap' between most of the scattering points and the film even when the cassette is in contact (position A). However, low energy scattered photons produced near the entrance surface are heavily absorbed within the patient, thus it is the scattered photons that originate close to the exit surface which cause most of the problem.

Note also that

(1)   whereas attenuation of scattered X-rays within the patient is an important factor, attenuation of scattered X-rays within the air gap itself is negligible;

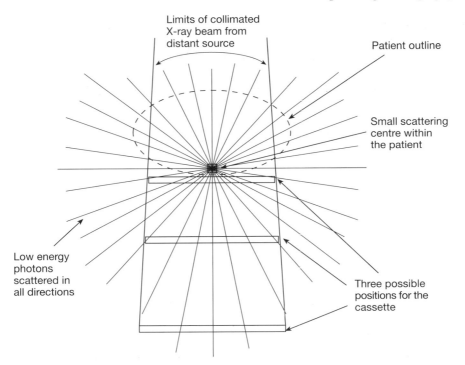

Limits of collimated
X-ray beam from
distant source

Patient outline

Small scattering
centre within
the patient

Low energy
photons
scattered in
all directions

Three possible
positions for the
cassette

Figure 9.2. *Diagram showing schematically how the number of scattered photons intercepting the cassette will decrease as the cassette is moved away from the position of contact with the patient.*

(2)   on the scale of figure 9.2, the X-ray source is distant so the margins of the collimated primary beam will be almost parallel—thus the effect of the inverse square law on the primary beam as a result of introducing the gap will be small (it is left as an exercise for the reader to confirm this for the figures given below).

A number of other features of the air gap technique—e.g. the effect of the finite focal spot size on image sharpness, the effect of the penumbra and patient movement on sharpness and the effect on patient dose are identical to those encountered in macroradiography and will be considered in the next section.

For a high kV chest technique, in the range say 125–150 kVp, a typical air gap might be 20 cm with a focus–film distance of 3 m. This would give comparable contrast to a 10:1 grid ratio. Both techniques result in an increased dose to the patient but the increase due to the inverse square law as a result of the air gap is generally less than that required to compensate for the grid.

Although the air gap technique would appear to have a number of advantages, it is not widely used, perhaps because the position of the film holder relative to the couch has to be changed.

As a final comment on high voltage radiography, this discussion has been presented in terms of low voltage (60–70 kVp) versus high voltage (125–150 kVp) techniques but intermediate voltages can of course be used, with the consequent mix of advantages and disadvantages.

9.3. MACRORADIOGRAPHY

As discussed in section 5.10.1 an object radiographed onto film is magnified in the ratio

$$\frac{\text{focus–film distance}}{\text{focus–object distance}} = \frac{\text{FFD}}{\text{FFD} - d}.$$

This geometry is reproduced in figure 9.3(a) for ease of reference. Note that magnification $M = 1$ only if the object is in contact with the film, i.e. when $d = 0$.

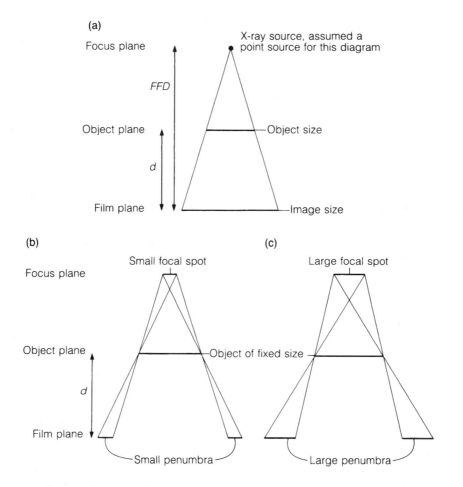

Figure 9.3. *Geometrical arrangements for magnification radiography. (a) Assuming a point source of X-rays, then by similar triangles the magnification $M = \text{FFD}/(\text{FFD} - d)$, (b) and (c) demonstrate that for an object of fixed size and fixed magnification, the size of the penumbra increases with the size of the focal spot.*

For most routine X-ray examinations, M is kept as small as possible. This is because, as discussed in section 5.9.1, if the focal spot is of finite size, which is always the case in practice, a penumbra proportional to $d/(\text{FFD} - d)$ is formed, so the penumbra is also absent only if $d = 0$. Comparison of figures 9.3(b) and (c) shows that, for a fixed value of d, the size of the penumbra depends on the size of the focal spot.

In practice structures of interest are never actually in contact with the film. Taking a typical value of $d = 10$ cm, then for an FFD of 100 cm, $M \simeq 1.1$. Sometimes, for example when looking at small bones in the extremities or in some angiographic procedures, it is useful to have a magnified image. One way to achieve this is to magnify, using optical means, a standard radiograph. However, this approach is often not satisfactory as it produces a very grainy image with increased quantum noise (see sections 4.11.2 and 8.6). The alternative is to increase the value of M, and for the purpose of the present discussion it will be assumed that this is effected by increasing d whilst keeping FFD constant. Although increasing d achieves the desired result, this has a number of other consequences as far as the radiographic process is concerned and these will now be considered.

(i) *Focal spot size.* As shown in figures 9.3(b) and (c), the size of the penumbra depends on focal spot size. If it is assumed that the penumbra is part of the magnified image, it may be shown by simple geometry that the true magnification is equal to

$$M + (M - 1)F/xy$$

where F is the focal spot size and xy the size of the object. Thus when M is large and F is of the order of xy, the penumbra contributes significantly to the image.

In order that this geometric (penumbral) unsharpness is kept to a minimum, the focal spot used for magnification radiography must be as small as possible. Spots larger than 0.3 mm are little use and 0.1 mm is preferable. This may impose rating constraints.

A focal spot of 0.3 mm or less is not easy to measure accurately and its size may vary with the tube current by as much as 50% of the expected value. A pin-hole may be used to measure the spot size (section 2.7) but to estimate the resolution it is best to use a star test pattern. The performance of a tube used for magnification radiography is very dependent on a good focal spot and careful, regular quality control checks must be carried out. It can be difficult to maintain a uniform X-ray intensity across the X-ray field using a very small focal spot. The intensity distribution may be either greater at the edge than in the centre or, conversely, higher in the middle than at the edge. Such irregularities can cause difficulties in obtaining correct exposure factors.

(ii) *Film–screen unsharpness.* Whereas magnification has a deleterious effect on unsharpness due to a finite focal spot size, screen unsharpness is in fact reduced by magnification. To understand the reason for this, consider image formation for a test object that consists of eight line pairs mm^{-1}. If the object is in contact with the screen ($M = 1$) the screen must be able to resolve eight line pairs mm^{-1}, which is beyond the capability of fast screens. Now suppose the object is moved to a point mid-way between the focal spot and screen ($d = FFD/2$ and $M = 2$). The object is now magnified at the screen to four line pairs mm^{-1}, thereby making the imaging task easier.

(iii) *Movement unsharpness.* One further important source of image degradation in magnification radiography is movement unsharpness (see section 5.9.2). If an object is moving at 5 mm s^{-1} and the exposure is 0.02 s, then the object moves 0.1 mm during the exposure. If the object is in contact with the film ($M = 1$) and the required resolution is 4 line pairs mm^{-1}, corresponding to a separation of 0.25 mm, then movement of this magnitude will not seriously affect image quality. However, as shown in figure 5.13, the effect of object movement at the film depends on d. If $d = FFD/2$, i.e. $M = 2$, the shadow of the object at the film will move 0.2 mm and this may cause significant degradation of a system attempting a resolution of 0.25 mm. Note however that the size of the penumbra has remained the same *fraction* of the object size.

A further significant contribution to movement unsharpness may come from the increased exposure time resulting from lower tube output with the smaller spot size and increased focus–film distance.

(iv) *Quantum mottle.* This is determined by the number of photons per square mm in the image (see sections 4.11.2 and 8.6) which in turn is governed by the required film blackening. As the image is being viewed under normal viewing conditions, the number of photons striking the film per square mm is exactly the same as on a normal radiograph and the quantum mottle is also the same.

(v)   *Patient dose.* If the FFD is fixed, then exposure factors are unaltered, but if the patient is positioned closer to the focal spot in order to increase magnification then the entrance dose to the patient is increased. Two factors compensate partially for this increased entrance dose. First the irradiated area on the patient can, and must, be reduced. This will require careful collimation of the X-ray beam and accurate alignment of the part of the patient to be exposed. Second, an 'air gap' has in effect been introduced (see section 9.2) so it may be possible to dispense with the use of a grid.

Some increase in the FFD may also be necessary because if the mid-plane of the patient is positioned 50 cm from the focal spot (to give $M = 2$ for an FFD of 100 cm) the upper skin surface of the patient will be very close to the focal spot and the inverse square law may result in an unacceptably high dose. Note that if the FFD is increased, exposure factors will have to be adjusted and a higher kVp may be necessary to satisfy rating requirements.

The way in which the effect on image quality of some of these factors may be analysed more quantitatively is outlined in section 8.8.1. Suffice to conclude here that resolution falls off rapidly with magnification and for a 0.3 mm focal spot the maximum usable magnification is approximately 2.0 at the object or about 1.6–1.8 at the skin surface nearest the tube.

9.4. DIGITAL RADIOGRAPHY (PRACTICAL DETAIL)

9.4.1. *Introduction*

The principles of digital radiography were discussed in section 8.7. As stated there, to generate high quality digital images, the detector system must

(1)   record X-ray quanta with a high efficiency and
(2)   be capable of providing accurate spatial information about the distribution of X-rays.

Basically, there are two ways this can be done. The first is to use one or more small, discrete detectors and to cover the area of interest by suitable detector movement if necessary. For example, a linear array of detectors will require linear movement. The alternative approach is to sample, at discrete regular intervals the analogue image produced by most image transducers used in conventional radiology. Both of these approaches will be considered.

Two factors which are important for any image transducer are:

(a)   the relationship between spatial resolution and matrix size, which essentially determines the sampling interval. According to the Nyquist sampling theorem (see section 10.3.3), if the largest spatial frequency that is to be recorded is ν_m then the linear sampling distance d must be no greater than $1/2\nu_m$. Thus for example in order to digitize a 35 cm × 42 cm chest film in which the highest recorded spatial frequency is 4 cycles mm^{-1}, the sample spacing must be no greater than 125 μm (4 cycles mm^{-1} = a repeat distance of 250 μm) and the corresponding matrix size will be 2800 × 3440 (for further detail on the sampling theorem see section 10.3.3).

(b)   the dynamic range—one of the major disadvantages of analogue images that has to be overcome is the limited dynamic range of photographic film, where only exposures that fall on the steeply rising portion of the characteristic curve can be usefully interpreted. In a high contrast chest radiograph taken at low kV the full spread of digital data may span almost five orders of magnitude. Thus ideally a dynamic range of $10^5 : 1$ will be required to record it faithfully.

9.4.2. *Methods of image formation*

For a detailed review of the methods available for digital image formation the reader is referred elsewhere (e.g. Harrison 1988, Cowan 1995a, b). Here we shall concentrate on the principles involved and the features that are relevant to the many potential advantages of digital radiography over conventional radiography.

Scintillators coupled to photodiodes or photomultiplier tubes

In theory, there are several ways in which a digital image could be produced. Assuming a point source, the possibilities are illustrated schematically in figure 9.4.

(1) A small detector scans across the area occupied by the image (figure 9.4(*a*)). This is a very simple geometry but will require a long time to acquire all the data, even in a relatively coarse 256×256 matrix.

(2) A linear array of detectors only has to make a linear movement to cover the image space (figure 9.4(*b*)). Data collection is now speeded up but may still take several seconds.

(3) A static array of detectors may be used (figure 9.4(*c*)). This will be a rapid imaging system but the number of individual detectors required becomes large.

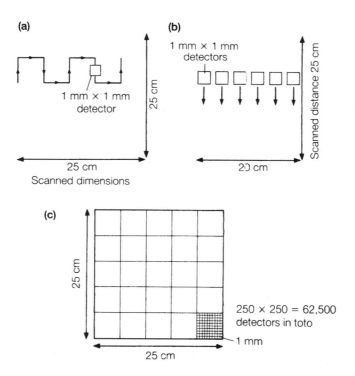

Figure 9.4. *Detector arrangements for digital radiology, (a) scanning detector, (b) linear array of detectors, (c) static array of independent detectors:* $250 \times 250 = 62\,500$ *detectors in toto.*

Most of the systems which have adopted the scanning principle have compromised on a one-dimensional array with linear motion. In one system proposed for digital radiography of the chest (Tesic *et al* 1983) a vertical fan beam of X-rays scanned traversely across the patient. An entrance slit 0.5 mm wide defined the beam, an exit slit 1.0 mm wide removed much of the scatter. The beam fell on a gadolinium oxysulphide screen backed by a vertical linear detector array consisting of 1024 photodiodes with a 0.5 mm spacing (figure 9.5). The complete system of X-ray tube, collimators and detectors scanned over the patient in 4.5 s sampling in 1024 horizontal positions. The estimated entrance dose was 250 μGy.

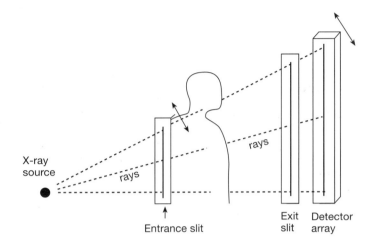

Figure 9.5. *Diagram of a scanning digital radiography system. The collimated fan beam moves horizontally across the patient. The exit slit and detectors move with the beam.*

Flat detectors are currently being developed either using an amorphous silicon array to image the light output from caesium iodide or a selenium detector (one version is 500 μm thick with 140 μm pixels) which converts X-ray photons directly into electrons.

Direct digitization of films

This has attractions as the logical extension of existing radiological practice since it permits retrospective processing of existing films. However, it is not being widely used in routine practice and the methodology is broadly similar to that described in the next section.

Image intensifier–TV system

The components of a system based on a TV camera coupled to an image intensifier are shown schematically in figure 9.6. The output from the image intensifier is fed to a TV camera which is then scanned by an electron beam. The natural line scanning motion of the beam provides digitization in one dimension and data along the scan line can be digitized by registering the accumulated signal in regular, brief intervals of time.

An important development has been the steady improvement in the performance of CsI input phosphors. Modern image intensifiers have a high detective quantum efficiency, i.e. the fraction of X-ray quanta incident on them that are actually detected (0.5 at 60 keV), and phosphor size has been increased without loss of spatial resolution. 4 lp mm^{-1} is now typical across a 23–36 cm diameter. Note that this resolution is not matched by the performance of a standard 625-line TV camera which has a limiting resolution of 1.4 lp mm^{-1} when used with a 230 mm image intensifier. Therefore 1000-line and 2000-line TV cameras have been developed. The TV camera tube will be a 25 mm Plumbicon, Saticon or similar with a diode type electron gun to give a high TV signal-to-noise ratio and an adequate operating dynamic range.

Two approaches to image acquisition are possible. With continuous low tube current exposure the TV frame can be digitized at 25 frames per second. Because of the low X-ray output, the noise in each frame

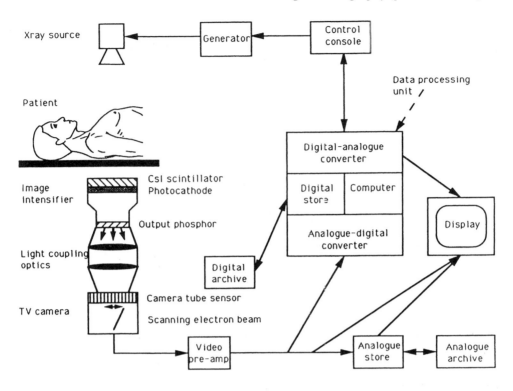

Figure 9.6. *Block diagram representation of an image intensifier–TV system for digital radiography.*

is high but this can be reduced by summing several frames, provided there is little patient movement. For certain investigations, e.g. digital cardiac imaging, acceptably low noise levels are essential at framing rates that can be anything from 12.5 to 50 frames per second. This requires pulsed operation with a much higher X-ray output during the pulse.

Charge-coupled devices/detectors

One potential application of the solid-state detector (see section 6.10) in digital radiology is the charge-coupled device or charge-coupled detector (CCD). A CCD sensor is an amorphous silicon wafer that has been etched to produce an array of elements (pixels) which are insulated from each other. A typical element would have a length of 40–50 μm and matrix sizes of 1024×1024 are already available. By applying a suitable biasing voltage, each element can be made to act as a capacitor storing charge. Thus upon exposure to radiation, electric charge proportional to the beam intensity is collected in individual pixels as in the vidicon camera (see section 4.12) to produce a charge image across the CCD. Unlike the vidicon camera however, methods of read-out are available that do not require a scanning beam of electrons. One way is to use a shift register (figure 9.7). This depends for its action on the fact that each charge sits within a potential well in the matrix (see section 1.1) but will move to an adjacent pixel if it has a deeper well. Hence by carefully controlling the depth of the potential well in each pixel, a given charge may be systematically moved around the matrix without overlapping or mixing with charge in adjacent pixels. Thus one set of control gates between the elements allows transfer of charges in one row of elements into the shift register line as shown by the vertical arrows. Once in the shift register the charges can be moved, in turn, to the right into the output gate where they can be read. The gates are reset and the next row moved down to be read. This process is now so rapid that read-out of the plate is effectively instantaneous. The analogue signal which is produced is passed through the signal wire from the plate to an analogue-to-digital converter and the digitized signal then placed in the pixel array generated in the computer to match the detector element array. Once there it can be manipulated as any other digitized image.

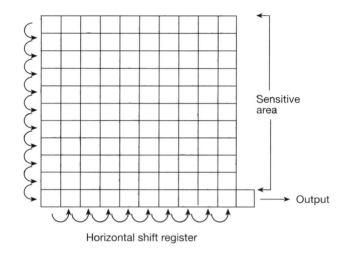

Figure 9.7. *Schematic representation of charge shifting in a CCD.*

An alternative method of read-out allows the charge on each element to be measured by using the fact that connections to a vertical line and a horizontal line uniquely define one element where they cross, allowing the charge on that element to be read.

For recording digitized images directly from X-rays, CCDs suffer from the limitation of solid-state detectors already discussed, namely a small active thickness resulting in poor detection sensitivity. In addition the CCD is susceptible to radiation damage at high fluxes.

A CCD may be used very successfully in conjunction with a luminescent screen and a photoelectric cathode surface (cf the image intensifier). The X-ray photons are converted successively to visible light photons and then into electrons which are readily captured in the potential wells. Therefore the CCD has possibilities for application in digital fluoroscopy, with the following benefits over a conventional TV camera:

(a)   the resolution is fixed by the size and interspacing of the photosensitive elements—there is no scanning electron beam to cause drifting;
(b)   the geometry of the CCD camera is precise, uniform, distortion free and stable;
(c)   the sensor is linear over a wide range of illumination and
(d)   there is evidence to suggest that the noise characteristics are more favourable (Cowan 1995).

There is intense commercial activity to produce image plates of a large enough size and uniformity for general radiography at an acceptable cost. At the time of writing, the detector area is only about 5 cm^2 so applications are limited to small area work—e.g. restricted mammography (perhaps an an aid to taking a biopsy specimen) and dental radiography.

In dental radiography the resolution is approximately 10 lp mm^{-1} and the dose required is only 10–20% of that for D speed films, the speed most commonly used. Even compared to the fastest E speed films it is approximately 50% less. One problem in dental radiography which could well affect general radiography is that some X-ray units cannot produce a short enough exposure. As the read-out is instantaneous, repeat radiographs can be made immediately, if necessary. Image manipulation can reduce the requirement for repeat radiographs, giving the opportunity to magnify images or use image enhancement.

There are few disadvantages. For dental radiography the thickness (approximately 0.5 cm) and rigidity of the image plate means that some patients cannot tolerate it in their mouth, although it is the same physical size as dental film. The availability of image processing can however, as in other situations, lead to a slower throughput as the dentist tries to obtain more information by spending time on image manipulation.

Photo-stimulable phosphor computed radiography

This is a technique that is finding increasing application as a system based on discrete sampling of the analogue image produced by the transducer.

The principles of photo-stimulable luminescence were discussed in section 4.9. The latent X-ray is recorded on a re-usable plate (typically 35 cm × 35 cm) coated with crystals of an appropriate photo-stimulable phosphor. The plate can be read after X-ray exposure by scanning a laser beam in raster motion over it and measuring the light emitted with a photomultiplier tube and light guide. The laser scanning spot will be typically 100 μm and will sample at between 5 and 10 pixels mm^{-1}. The final image may be recorded on film or stored in digital form.

Desirable properties of the photo-stimulable phosphor are (a) high emission sensitivity at the wavelength of a readily available laser, e.g. He/Ne at 633 mm, (b) light emission in the range 300–500 mm where photo-multiplier tubes have a high quantum efficiency. A class of europium activated barium fluorohalide crystals (Ba FX:Eu where X may be Cl/Br/I) appears to be most suitable. The wavelength range for photostimulation is 500–700 nm and light emission is in the region of 400 nm.

The relative intensity of light emitted is proportional to X-ray exposure at the plate over approximately four decades and this wide dynamic range can be exploited in chest radiography by using a windowing technique (section 8.7) or dual energy subtraction (section 9.5).

Since the laser beam is very small, it does not limit spatial resolution, which is governed more by read-out time and the light intensity required for stimulation. A variety of plates has been constructed for general radiology, mammography, tomography and subtraction techniques, with typical spatial frequencies of 2–3 lp mm^{-1}.

9.4.3. Resolution requirements

Widespread acceptance of digital images by radiologists will depend on two criteria:

(a) confidence that the 'new' images result in no loss of clinically relevant data and
(b) presentation of the images in a way that the radiologist finds acceptable—i.e. the images must not look too different from the more familiar analogue images.

This has led to a number of studies on maximum acceptable pixel size (recall that this is a more relevant criterion than matrix size which will depend on the field of view), and many of these studies have used the methods of assessment of image quality discussed in chapter 8. Results have not been entirely consistent, some reports suggesting that there is no loss of observer accuracy until the pixel size increases above 1 mm, others suggesting that diagnostic accuracy continues to increase with reduced pixel size down to 0.1 mm. For more detailed discussion on this point see Dendy (1990).

An effective pixel size of 0.2 mm requires a 2048 × 2048 digital array to image a 40 cm × 40 cm field of view and is likely to suffice for many applications. Ultra high definition cameras operating onto a 2048 × 2048 matrix are feasible and smoothing techniques (see section 8.7) can make such images virtually indistinguishable from analogue images. Also 4096 × 4096 systems, with an effective pixel size of 0.1 mm, are being developed. They will probably eliminate any remaining concerns over image quality in erect examination of the chest but even smaller pixel sizes may be necessary for digital mammography. Note that finer pixellation carries a penalty in terms of additional data storage.

9.4.4. *Patient doses*

This is still a difficult area. For digital radiology to become widely adopted, it should ideally result in doses that are no greater than, and preferably smaller than, those used in conventional radiology.

In some respects this will almost certainly be true. For example the wide dynamic range of digital radiology ensures that it is almost always possible to generate a diagnostic quality image by suitable image processing (see section 8.7). Thus repeat images are largely eliminated. However care is required to check that exposure factors which result in an unnecessarily high dose to the patient are not being selected. This problem would be immediately recognized by an over-exposed film but may be obscured by post-processing of a digital radiograph.

Perhaps a more reasonable question is 'Does digital radiology result in lower doses when optimal settings are used for comparable image quality?'. At the time of writing the jury is still out. For example Marshall *et al* (1994) have recently investigated doses associated with five different imaging systems for chest radiography but not all of their results are in agreement with those of other workers.

One set of experiments will be described briefly to illustrate some of the physics problems involved in attempting to answer the question. Dobbins *et al* (1992) compared the imaging performance of photo-stimulable phosphor computed radiography (PPCR) with a medium speed screen–film combination using a contrast–detail phantom. Under conditions appropriate to high kV chest radiography PPCR needed a significant (75–100%) increase in exposure to achieve parity of image quality. Alternatively, comparable exposures resulted in a 100% fall in contrast detectability with PPCR.

However, some of this effect can be attributed to the fact that the mean atomic number of barium fluorohalide in the PPCR is less than that of gadolinium oxysulphide in the screen. Thus the X-ray absorption, and hence efficiency of the barium fluorohalide, will deteriorate more rapidly at high kV. This explanation was largely supported by the fact that PPCR performed relatively better at lower kilovoltages (80 kV).

Another possible contribution to relative image quality might be the responses of the two different detectors to scattered radiation. Recall that materials absorb very well at energies just above their K absorption edges. For barium and gadolinium these are at 37.4 keV and 50.2 keV respectively. Thus the PPCR screen would absorb rather too well scattered radiation in the 40–50 keV range.

These results illustrate that the relative patient doses for a given image quality may be very dependent on the precise technique chosen. Much more work remains to be done and, until there is a broad consensus on the doses associated with different techniques, the key question is likely to remain unresolved.

9.5. SUBTRACTION TECHNIQUES

These are techniques whereby unwanted information is eliminated from an image, thereby making the diagnostically important information much easier to visualize. They are particularly useful where sequential images differ in a small amount of detail which one wishes to highlight. A typical example would be during angiography where one requires to see changes in the position and amount of contrast medium in blood vessels between two images separated by a short time interval.

The basic principle of the technique is quite simple. A radiograph is a negative of the object data. If a negative of this negative is prepared (a positive of the object data) and positive and negative are then superimposed., the transmitted light will be of uniform intensity. This is because regions that were black on the original negative are white on the positive and vice versa, the two compensating exactly. The positive of the original image is often called the 'mask'. When a second radiograph is taken of the patient, with one or two details slightly different, e.g. following the injection of contrast medium, superimposition of the mask and the second radiograph will result in all the unchanged areas transmitting uniformly, but the parts where the first and second radiograph differ will be visualized.

For the technique to be successful, the two images must superimpose exactly and no patient movement must take place between exposures. The exposure factors for the radiograph and the tube output must remain the same to ensure an exact match of optical density which should be in the range 0.3–1.7. The copy film making the mask must have a gamma equal to 1.0. When the films are viewed together the combined optical density is approximately 2.0 and a special viewing box with a high illumination is required.

Many of these limitations have been overcome in digital radiology and subtraction techniques are one of the major applications of digital imaging.

9.5.1. Digital subtraction angiography

The technique is broadly similar to that adopted for simple mask subtraction in film–screen angiography. A series of individual X-ray exposures is made, typically at a rate of one per second before, during and after injection either intra-venously or intra-arterially of X-ray contrast material containing a high atomic number element such as iodine. However, every X-ray exposure can now be stored as a digitized image.

All the technical details considered in section 9.4 for good digital images are relevant. For example to achieve high resolution, small focal spot sizes (0.5 mm) are necessary. Coupled with repeat pulsed exposures of short duration, especially in cardiac work, this places new demands on X-ray tube rating (see section 2.5). For optimum contrast scatter must be greatly reduced or eliminated completely. Several rather complex procedures, beyond the scope of this book, for achieving this have been suggested. One idea, whereby grids placed both in front of and behind the patient and moved in synchrony might remove scatter more effectively than a single grid, is shown in figure 9.8. Note that the digital image processor may also be used to control the X-ray tube generator, thereby helping to ensure that all exposures are identical except in those regions affected by contrast agent. Also it is important to realize that the requirement for a modern intra-venous angiographic system to be capable of recording 60 images per second of 512×256 pixels causes major problems for data collection, manipulation, storage and display.

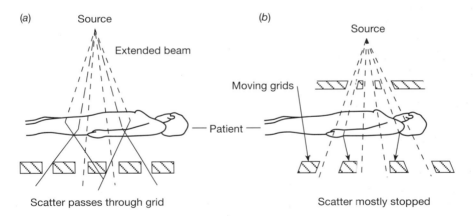

Figure 9.8. *(a) Single grid. (b) Use of grids both in front of the patient and between patient and detector and moved in synchrony to remove more scattered radiation than a single grid.*

Many of the advantages of digital imaging can now be exploited. For example any pair of frames may be subtracted, one from the other, to form a new image. The wide dynamic range of the digitized image and the techniques of image processing in section 8.7 may be used to bring out features of interest. Because of the method of formation, subtracted images are rather noisy. Imagine for example that each frame contains

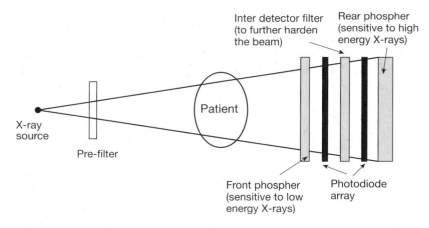

Figure 9.9. *Schematic diagram of a single exposure dual energy system.*

on average $10\,000$ photons mm^{-2}. The Poisson error ($n^{1/2}$) is 100 photons and this represents the noise. If a small object attenuates 1% of the photons, this is also 100 photons and coincidentally in this specific example the signal and noise are equal. However, at least two frames are required to form the subtracted image so the noise is now $(20\,000)^{1/2}$ whilst the signal has remained unchanged.

Digital techniques may be used to sum frames retrospectively—sometimes known as frame integration—and it is left as an exercise to the reader to show that if, say, four pre-injection frames and four post-injection frames are summed the signal-to-noise ratio in the final image will be improved. Note that frame integration reduces the effects of both quantum noise and electronic noise, as a fraction of the signal, whereas increasing the dose only reduces the effect of quantum noise.

Finally, manipulation of the digital image can, in theory, overcome problems of patient movement after injection of the contrast agent but such techniques are sometimes difficult to apply.

9.5.2.   *Dual energy subtraction*

Another subtraction technique that is particularly amenable to digital methods makes use of the fact that the attenuating properties of different materials in the body are kV dependent. Thus for example very low kVs are required to generate soft tissue contrast, whereas bone will show reasonable contrast even at high kV.

Individual approaches differ somewhat but figure 9.9 illustrates the general principle. The first part of this detector comprises an yttrium oxysulphide phosphor screen coupled to a photodiode array. Since the K absorption edge for yttrium is at 17 keV, a signal corresponding to the detection of low energy photons, i.e. a bone + soft tissue image is generated. The X-ray beam now passes through a copper filter for further beam hardening before falling on a thick gadolinium oxysulphide phosphor screen. The K edge for gadolinium is at 50 keV so a signal corresponding to the detection of high energy photons is generated. If the image recorded in the second screen is suitably weighted and then subtracted from the image obtained in the first screen, an image virtually devoid of bony structures is produced (figure 9.10). Note there is no movement blurring because the images are produced simultaneously.

9.5.3.   *Time interval differencing*

The third application of digital subtraction techniques uses a property of digital images that has already been discussed, namely that any pair of frames can be subtracted, retrospectively one from the other. Thus it is not necessary to use the same frame as the mask for each subtraction.

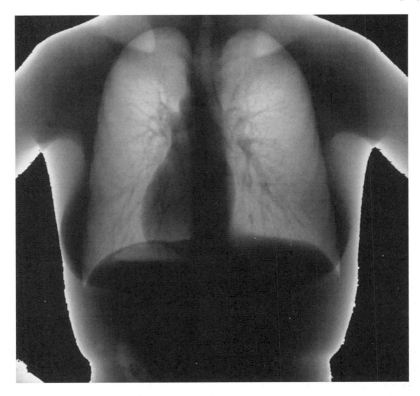

Figure 9.10. *A soft tissue chest image obtained by digital subtraction of images obtained with a prototype dual energy chest unit (image courtesy of Professor Gary T Barnes).*

Imagine that 30 frames per second have been collected for several seconds. In time interval differencing a new series of subtracted images is constructed by using frame 1 as the mask for say frame 11, frame 2 as the mask for frame 12, frame 3 as the mask for frame 13 and so on. Thus each subtracted image is the difference between images separated by some fixed interval of time—one third of a second in this example. Such processing can be effective in situations where an organ is undergoing regular cyclical patterns of behaviour, for example in cardiology. Note that the statistics on a single frame may be poor but the technique is still applicable to a fixed time difference between two groups of frames. Both the time difference and the number of frames in the group can now be adjusted to give the best images.

9.6. MAMMOGRAPHY

There are several difficulties associated with imaging the breast to determine whether a carcinoma or precancerous condition exists. First, there is no physical density difference between suspect areas and normal breast tissue and only a small difference in atomic number. Any radiological differentiation is therefore very dependent on photoelectric attenuation. Secondly, one of the prime objectives of mammography is to identify areas of microcalcification, even as small as 0.1 mm in diameter. To achieve the necessary geometric resolution, a small focal spot size is required and problems of X-ray tube output and rating must be considered. Finally, breast tissue is very sensitive to the induction of breast cancer by ionizing radiation (especially for women between the ages of 14 years and menopause). High regard must therefore be paid to the radiation dose received during mammographic examinations (see for example Law 1997). This section will identify a number of technical developments which have improved image quality and at the same time reduced patient dose.

9.6.1. Optimum kilovoltage and tube design

In order that maximum contrast may be achieved, a low kV must be used since both attenuation coefficients themselves and their difference (e.g. $\mu_{\text{water}} - \mu_{\text{fat}}$) decrease with increasing energy (see section 5.2). The choice of kV is however a compromise. Too low a kV results in insufficient penetration and a high radiation dose to the breast. X-ray photons below 12–15 keV contribute very little to the radiograph and must if possible be excluded.

Excellent contrast for typical compressed breast thicknesses of 3–5 cm is obtained with X-rays in the 17–22 keV range. Theoretical work based on signal-to-noise ratios (see section 8.5) has indicated that for thicker breasts the optimum voltage is higher, perhaps 21–25 keV (Jennings *et al* 1981) and some experimental data to support this conclusion have been reported (Beaman *et al* 1983).

Although this difference in photon energy is apparently very small, it has a fundamental effect on tube design and construction. Most, if not all, mammography units designed to produce the majority of their X-ray photons in the 15–20 keV region use a molybdenum anode in a tube with a beryllium window. Additional filtration is provided by a molybdenum filter. A tube designed to produce X-rays in the 21–25 keV region uses a tungsten anode tube with special filters.

9.6.2. Molybdenum anode tubes

A typical spectrum for a molybdenum anode tube operating at 35 kV constant potential and using a 0.05 mm molybdenum filter was shown in figure 3.16. A significant proportion of the X-ray photons are in fact Kα (17.4 keV) and Kβ (19.6 keV) characteristic X-rays from molybdenum. These lines lie just below the K shell energy (absorption edge) of molybdenum at 20.0 keV. Figure 9.11(*a*) shows a similar spectrum at 28 kVp using a slightly different quantity on the *y* axis and showing the characteristic lines resolved. An X-ray tube window of low atomic number (beryllium with $Z = 4$) is used so that wanted X-rays are not attenuated. The total filtration on such a tube should be equivalent to at least 0.3 mm of aluminium to remove low energy radiation.

For reasons of geometric resolution, discussed in the general remarks on mammography, a small focal spot size must be used. For contact work the source to image distance is typically 60–65 cm and the object-to-film distance is about 6 cm giving a magnification of about 1.1. To achieve the necessary resolution of about 13 lp mm^{-1} with this magnification, a focal spot size of about 0.3–0.4 mm is required.

The quality of the X-ray beam depends to some extent on the ripple on the kV wave form. The evolution from single phase to three phase and recently to high frequency generators reduced the kV ripple from 100% (one phase) to 25% (three phase) and typically less than 4% (high frequency). Most commercially available mammography units use high frequency generators and rotating anodes. Tube currents are typically 100 mA on broad focus and, because of rating considerations, 30 mA on fine focus. Exposure times are typically about 2 s because the efficiency of output of the low atomic number molybdenum anode is poor. Note that the orientation of the anode–cathode axis allows the heel effect to be used to compensate for thickness of breast.

Although the characteristic lines in the molybdenum spectrum are ideal for breast imaging, recent work has shown that the higher energy photons in the continuous spectrum still make a significant contribution to the image. Thus the kVp has to be controlled rather more precisely than would be necessary if the characteristic lines (which are independent of kVp) totally dominated the spectrum. Also there is some merit in increasing the kVp for thicker breasts (perhaps from 28 to 30 kVp for breasts greater than 7 cm) although this results in some loss of contrast.

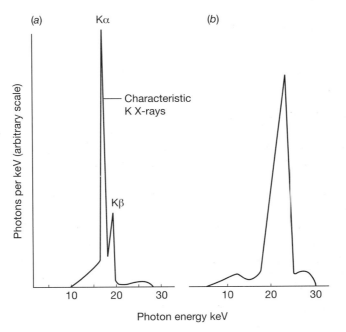

Figure 9.11. *(a) Spectral output of a molybdenum anode X-ray tube operating at 28 kVp with a 0.05 mm molybdenum filter. (b) Spectral output of a tungsten anode X-ray tube operating at 30 kVp with a 0.05 mm palladium filter.*

To counteract the effect of body attenuation as much as possible, whilst retaining a low kV, mammography units always use special applicators to compress the breast to as small a thickness as possible. Applying compression also minimizes or even eliminates movement unsharpness during the relatively long exposures, reduces geometric unsharpness because the breast is closer to the receptor and improves contrast by reducing scatter (see section 5.7).

9.6.3. Tungsten anode tubes

The effect of a 0.05 mm palladium filter on the output spectrum of a tungsten anode tube operating at 30 kVp is shown in figure 9.11(b). The K absorption edge of palladium is at 24.3 keV, so below this energy the attenuation of the filter is very much lower than at higher energies (see section 3.6). Thus a window of energies is transmitted rather readily and such a filter is sometimes known as a K edge filter. The spectrum transmitted by palladium most closely matches that which theory predicts will be most suitable for mammography.

Since a tungsten anode tube gives a good output, a small focal spot (0.2 mm) can be used and this can be reduced to an effective focal spot of 0.1 mm by mounting the tube at an angle of 5°. Note that greater care is required in setting and checking the generator kilovoltage since there are no characteristic lines in the spectrum as in the case of molybdenum.

9.6.4. Film–screen combinations

Many different films and intensifying screens are offered by manufacturers exclusively for mammography. The films are single sided, thus eliminating parallax, and are used with a single screen which is often much thinner than standard screens. The screen is positioned behind the film as this causes less loss of resolution and is also more efficient for blackening the film. The explanation for these effects is shown in figure 9.12. In the upper diagram, using high energy X-rays, the loss of X-ray intensity in the screen is small, so the amount of light produced in successive layers of the screen is similar. Both the amount of light reaching the film (which determines blackening) and the mean distance of the light source from the film (which affects resolution) will also be similar. For mammography X-rays, in the lower diagram, the situation is different. Now there is appreciable attenuation of the X-rays so it is important that as much light as possible is produced close to the film. Clearly this occurs when the screen is behind the film not in front of it.

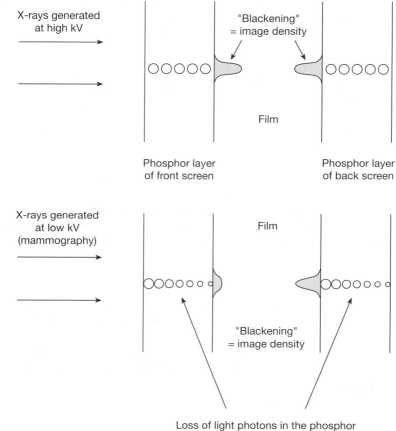

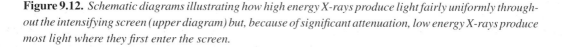

Loss of light photons in the phosphor
layer due to attenuation of X-rays

Figure 9.12. *Schematic diagrams illustrating how high energy X-rays produce light fairly uniformly through-out the intensifying screen (upper diagram) but, because of significant attenuation, low energy X-rays produce most light where they first enter the screen.*

The screens are generally rare earth, e.g. lanthanium bromide or gadolinium oxysulphide, and screen and film are pulled into very close contact, for example by using a flexible plastic cassette that can be vacuum evacuated. The screens are only about one tenth as fast as ordinary screens in order to ensure that very high resolution can be achieved—typically 20–22 lp mm^{-1}, whilst quantum mottle is also eliminated.

The optimum film–screen combination for a given mammography unit is generally found by trial and error and may involve combining the film of one manufacturer with the screen of another. To enhance contrast differences the film has a high gamma and hence a small film latitude—see section 4.4.4. Therefore automatic exposure control (AEC) is used routinely to provide consistent film density, typically centred on an optical density of about 1.5, over the clinically useful range of breast thicknesses and X-ray tube potentials. Note that if the automatic exposure control is not properly set then film density may decrease with increasing breast thickness, beam hardening being the main cause.

9.6.5. Patient doses

As mentioned in the introduction to this section, skin doses must be kept very low because the breast is very radiosensitive. As techniques have improved, the mean glandular dose (MGD) has fallen from several tens of mGy per exposure 15 years ago to less than 1 mGy using the best current techniques with molybdenum anodes (see table 9.1).

Table 9.1. *Results of 1994 survey of 23 mammography units in East Anglia using a standard breast phantom.*

Average MGD (mGy)	Range of MGD (mGy)	Range of optical density
1.17	0.72–1.68	1.30–1.69

Courtesy Mr D Goodman.

The patient dose using a tungsten anode tube and a K edge filter is even less than that using a molybdenum anode. At operating voltages of about 30 kVp the dose reduction factor for a thin breast is approximately two. This dose advantage is partially offset by a slight reduction in image quality since the molybdenum spectrum is better than the tungsten spectrum for radiographing thin breasts. For thicker breasts not only can the dose reduction factor rise to as much as 5 but the image quality produced by the tungsten tube is also superior.

9.6.6. Contrast improvement

In mammography contrast is reduced by scatter by as much as 50% for a thicker breast. The use of compression to reduce scatter has already been mentioned. Contrast may also be improved by using a grid, usually of the moving type, to blur grid lines and with a typical grid ratio of 4.0–5.0. The grid will be placed above the screen–film cassette and must have high transmission for low energy photons. It may improve contrast by a factor of between 1.2 (2 cm breast) and 1.7 (8 cm breast). The dose to the breast will of course be increased—typically by a factor of between two and three for the same optical density. Movement of the grid must not cause unsharpness due to vibration.

9.6.7. Quality control

Asymptomatic mammography screening is an excellent example of the principle that the medical benefit, i.e. the number of unsuspected breast lesions detected, must clearly outweigh the number of radiation-induced cancers caused. Such an analysis requires a knowledge of the excess lifetime risk of mortality in females from breast cancer, a precise knowledge of the MGD and accurate information on the extra cancers detected. Both the radiation risk (table 9.2) and the pick-up rate are very age dependent.

Factors that can affect MGD are kept under critical review. These include:

(a)   X-ray beam spectrum (kVp, anode filter),
(b)   compression force and compressed breast thickness—the MGD increases rapidly with breast thickness,
(c)   the effect of the age of women at exposure on tissue density in the compressed breast,
(d)   use of grids,
(e)   magnification techniques,

Table 9.2. *Excess lifetime risk of mortality in females from breast cancer (from ICRP 60).*

Age	Excess risk per Sv ($\times 10^{-2}$)
15	2.95
25	0.52
35	0.43
45	0.20
55	0.06
65	—

(f)   film–screen combinations/processing/optical density and

(g)   number of exposures and total mA s.

Strict attention to image quality is also required to ensure a high pick-up rate. Optical density must be checked—a low dose mammogram that produces too light a film is diagnostically useless. Test objects are available to check spatial resolution, threshold resolution and granularity and focal spot size must be checked carefully. Note that because of the line focus principle this may be different in different directions.

Typical tolerance limits on some of the more important variables are shown in table 9.3.

Table 9.3. *Typical tolerance limits on some important variables in mammography.*

Measure	Tolerance limit
Resolution	10 lp mm^{-1}
Threshold contrast	1%
MGD	2 mGy with grid
kV	± 1 kV of intended value
Filtration	0.3 mm A1
Output consistency/kV dependence/ variation with the tube current and focus	5%
Film density (measured with 4 cm Perspex block under AEC)	± 0.2 of target value and in OD range 1.4–1.8
AEC breast thickness compensation	± 0.2 in OD
AEC variation with the tube kV, tube current	10% of mean

9.7.   XERORADIOGRAPHY

This method of recording X-ray images has been investigated with particular reference to mammography but has also been used for other high definition work particularly imaging bones and extremities. For an early account see Boag *et al* (1972).

Instead of using a film as the recording medium, xeroradiography uses a photoconductor, which is a special class of semiconductor, as the surface on which the image is recorded. As discussed in sections 2.3.4 and 6.10, a semiconductor can be converted from an insulator into a conductor under certain conditions. In a photoconductor the energy gap between the valence and conduction bands is so small that even the energy provided by the absorption of a photon of visible light (for green light of wavelength 540 mm, the quantum energy is about 2.3 eV) is sufficient to cause this change. Thus if a photoconducting surface is covered uniformly with charge and then exposed to visible or X-ray photons, exposed regions will become conducting and the charge will flow away. Hence the pattern of charge free regions after exposure is an image of the irradiated area.

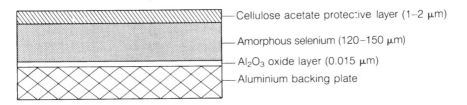

Figure 9.13. *Construction of a selenium plate for xeroradiography.*

The photoconductor used in xeroradiography is amorphous selenium and the construction of a xeroradiography plate is shown in figure 9.13. The aluminium plate must be very smooth and the selenium layer laid down upon it must be very uniform in thickness. The cellulose acetate layer, as well as protecting and thus extending the life of the selenium layer, also prevents the lateral conduction of a charge across the surface.

9.7.1. Mode of use

The selenium is charged to a potential difference of about 1500 V. The charge, which is normally positive, must be very uniformly distributed on the plate surface. The plate acts as a capacitor so a negative charge is induced on the aluminium. This charge is prevented from leaking into the selenium by the aluminium oxide layer. The plate is then placed in a light tight cassette, since the photoconductor is sensitive to light. During an exposure, X-ray photons interact with the selenium plate and these interactions differ from the interaction of light photons in two ways. First, the X-ray photons may penetrate some distance into the selenium plate before they interact. Secondly, each primary interaction results in the release of secondary electrons that have the capability of causing many more conduction centres in the selenium plate. Where there are many X-ray photon interactions most of the charge leaks away. The X-ray photon image is thus converted to a charge image, known as the electrostatic latent image, on the plate.

This charge image is converted to a visual image by first coating the surface of the plate with a very fine (<5 μm diameter) charged powder. This powder is attracted to the surface by the residual charge. Excess powder is removed. The plate is then covered with a special paper and the powder image is transferred into it electrostatically. Finally the paper is heated to fuse the powder into it creating a permanent record. By altering the charge on the powder, either a positive or negative can be produced.

Any powder remaining on the selenium plate can be cleaned off and the plate is then stored, uncharged, ready for reuse.

9.7.2. Formation of powder image

The electric field across a charge distribution is shown in figure 9.14. This figure illustrates the unique advantage of xeroradiography over conventional radiography. At the edge of a body of charge the field does not fall to zero in a simple manner. Instead, on moving away from the charged region there is first a sharp rise in the field strength, immediately followed by a sharp negative field pulse, before the zero-field-strength position is reached. Charged powder near the edge is either very strongly attracted or repulsed. This gives the characteristic 'outlining' effect of a xeroradiograph making it particularly suitable for visualizing edges where there may be only a small contrast difference. As shown in figure 9.15, a difference in potential of only some 10–20 V at an edge is enough to cause differences in electric field that will enable the edge to be sharply outlined in the resulting image.

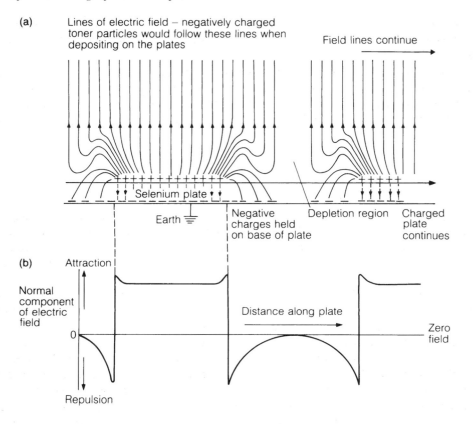

Figure 9.14. *The effect of charge distribution on (a) the electric field lines, (b) the component of electric field normal to the charged surface.*

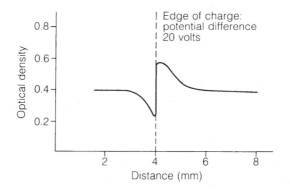

Figure 9.15. *Variation in optical density on a xeroradiography plate across an edge that represents a potential difference of only 20 V.*

Conversely xeroradiography produces poor images of large, homogenous structures, because the field lines are weak and uniformly spaced in these regions. Thus a xeroradiograph has poor broad area contrast, i.e. when the exposure is uniform the optical density changes very little with plate voltage and hence radiation exposure. This has the advantage that xeroradiography exhibits a very wide exposure latitude. For a more quantitative treatment of these effects on image formation in terms of modulation transfer functions see section 8.8.1.

If very fine particles (<5 μm in diameter) are used, resolution of approximately 50 line pairs mm^{-1} can be achieved on a xeroradiograph. However, this is better than the resolution limit imposed by the tube focal spot size and/or patient movement. Thus in practice the resolution limit of the system is closer to 10 line pairs mm^{-1}. However, this does mean that microcalcifications can be seen easily, cysts of 1–2 mm can be identified and carcinomas of 2–5 mm can be found.

9.7.3. *Advantages and disadvantages of xeroradiography*

In addition to the imaging advantages already discussed, xeroradiography has the cost advantage of not being based on an imaging system involving silver and, in theory at least, a plate is reusable indefinitely. A tungsten anode operating at about 45 kVp is used for xeroradiography so it is not necessary to have a dedicated tube. A xeroradiograph has a short processing time and can be read quickly. The nature of the edge enhancement process also means that there is a very broad exposure latitude in the system allowing good results to be achieved with less than optimum generator settings.

Possible disadvantages of the technique are that great care must be taken to ensure a uniform charge distribution prior to exposure. Mechanical damage to the selenium surface and exposure to light must be avoided. The charging process must be performed immediately prior to use because a certain amount of charge leakage is inevitable if charged plates are stored. A further limitation is that the selenium plate cannot be reused too quickly as it carries a 'memory' and may produce a 'ghost' of the original image.

The major disadvantage, however, is that most reports to date indicate a mean glandular breast dose for a xeroradiograph that is some three to 10 times higher than for a conventional radiograph. Given the sensitivity of the breast to induction of cancer, this is sufficient reason for the technique not to have been widely introduced. Recent technical developments, including the use of a thicker photoreceptor layer to increase X-ray absorption, a higher conversion of absorbed X-ray signals to charge and a more sensitive developer have brought mean glandular breast doses down to within a factor of two of conventional mammography. This may reawaken interest in the technique, a fuller account of which can be found in Curry *et al* (1990).

9.8. PAEDIATRIC RADIOLOGY

The most recent reports on the long term effects of radiation (see chapter 11) have confirmed that the risk from X-rays to children is greater than the risk to adults. This is partly because the radiation risk itself depends strongly on the age at which exposure occurs, and partly because the opportunity for expression, i.e. the life expectancy at the time of exposure, is greater. Thus the numerous factors that contribute to the complex relationship between image quality and dose must be re-examined to ensure that, for the imaging process as a whole, the principle of optimization still applies.

There are many reasons why imaging criteria applicable to the standard adult may not be appropriate for paediatric imaging—for example the smaller body size, age dependent body composition, lack of co-operation and functional differences (e.g. higher heart rate, faster respiration).

In this brief review, which will be restricted to factors directly related to the physics of radiology, attention will focus on those aspects where the optimum technique for children may differ from that for adults.

(i) *Exposure times.* These will generally be short, and indeed should be short to minimize movement problems. However, many generators are incapable of delivering the exposures in the 1–4 ms range that are frequently required. There may be problems associated with the rise time to maximum output, which in older units can be longer than the required exposure and if the cable between the transformer and the tube has a significant capacitance, there may be a post-exposure surge of radiation.

To achieve a rectangular configuration of kV with minimal ripple and hence a uniform X-ray output over a very short exposure time (say 1 ms) requires a grid-controlled, 12-pulse multipurpose, or constant potential generator. The shorter the exposure, the more stringent the requirements. This leads to the rather paradoxical conclusion that the smallest patients require the most powerful machines.

Note that it is generally not acceptable to achieve more 'convenient', i.e. longer, exposure times by reducing the kV. This will increase exposure times, by the converse of the processes discussed in section 2.5.4, and will also increase contrast but the penalty in terms of increased dose is likely to be high.

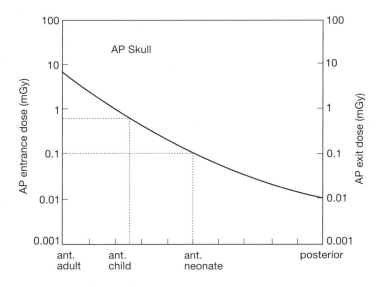

Figure 9.16. *Schematic representation of the difference in entrance dose for an adult, a child and a neonate. The initial attenuation in the adult must be introduced by added filtration for paediatric radiology.*

(ii)   *Added filtration.* An optimized radiograph of a child cannot be obtained simply by reducing the exposure time in proportion to body thickness. This can be appreciated from figure 9.16. At the depth in an adult corresponding to the entrance surface of the child there will be not only beam attenuation but also beam hardening. This beam hardening must be introduced externally for paediatric measurements. Additional filtration of 1 mm aluminium plus 0.1–0.2 mm copper (equivalent to about 3–6 mm aluminium at standard diagnostic radiographic voltages) is appropriate. Note that in general-purpose equipment it may not be easy to change the filtration frequently, nor to avoid the risk of using the wrong filtration for a given patient.

Attempts have been made to use filter materials with K shell absorption edges (see section 3.6) at specific wavelengths, for example rhodium, silver and cadmium, with K edges at 23.2, 25.5 and 26.9 keV respectively. Considerable dose reduction can be achieved with these filters.

(iii) *Low attenuation materials.* In view of the increased radiation sensitivity of the patient, the use of carbon fibre or some of the newer plastics in materials for table tops, grids (if used), front plates of film changers and cassettes is strongly recommended. In the voltage range used for paediatric patients, dose reduction of up to 40% may be achieved.

(iv)   *Field size and collimation.* Inappropriate field size is a frequent fault in paediatric radiology. Too small a field may result in important anatomical detail being missed. Too large a field will not only impair image contrast and resolution by increasing the amount of scattered radiation but will also result in unnecessary ionizing radiation energy being deposited in the body outside the region of interest. In the neonatal period the tolerance for maximum field size should be no more than 1 cm greater than the minimum value. After the neonatal period this may be relaxed to 2 cm.

(v) *Effective immobilization.* Incorrect positioning is another frequent cause of inadequate image quality and no exposure should be allowed unless there is a high probability that the exact positioning will be maintained. In paediatric work appropriate immobilization devices may be required to ensure that (i) the patient does not move, (ii) the beam can be centred correctly, (iii) the film is obtained in the correct projection, (iv) accurate collimation limits the beam to the required area and (v) shielding of the remainder of the body is possible.

In many situations physical restraint will be a poor substitute for a properly designed device.

(vi) *Use of grids.* There is much less scattered radiation from an infant than from an adult and grids are frequently unnecessary especially for fluoroscopic examinations and spot films. Therefore equipment in which the grid can easily be removed should be used. If a grid is necessary, the grid ratio should not exceed 1:8 and the line number should not exceed 40 cm^{-1}. Moving grids may cause problems at very short exposure times.

(vii) *Film–screen combination.* Table 9.4 shows some typical values for the dose required at the image receptor and the lower limit of visual resolution for different speed classes.

Note that these figures, which were obtained at 80 kV with a suitable phantom, should be used only as a guide. Exact values will vary with kV and filtration, especially that caused by the object.

Reference to table 9.4 shows that a high speed film–screen combination gives a big dose saving. Furthermore it results in a shorter exposure which minimizes motion unsharpness, the most important cause of blurring in paediatric imaging. The slight loss in resolution is rarely a problem although it may be a reason for choosing a lower speed class for selected examinations.

(ix) *Automatic exposure control (AEC).* There are a number of potential problems with using AEC in paediatric work: (i) the system may not be able to compensate for the very large variation in patient size; (ii) the detector may be too big for the critical region of interest; (iii) it may not be possible to move the detector to the critical region; (iv) the usual ionization chamber of an AEC is built behind the grid which is frequently not necessary; (v) a variety of film–screen combinations may be required for different examinations and the AEC system will have to be calibrated for each of them and (vi) the AEC may require a longer exposure time than the radiographic examination.

An AEC specifically designed for paediatric work will have a small mobile detector that can be positioned very precisely, mounted behind a lead free cassette and must respond to an absorbed dose that is considerably less than 1 μGy. The technical problems are such that it may be preferable to use predetermined exposure charts based on the infant's size and weight.

(x) *Patient dose and quality criteria for images.* Recent work on patient doses in paediatric radiology was discussed in chapter 6 and the CEC Radiation Protection Programme has made recommendations on quality criteria for diagnostic radiographic images in paediatrics. Although not all these criteria are necessarily universally accepted, figure 9.17 adapted from the work of Schneider *et al* (1992) shows a clear inverse correlation between the number of technique criteria met and mean surface entrance dose.

This is strong evidence for the need for customized facilities, if not dedicated facilities, for paediatric radiology.

9.9. DENTAL RADIOLOGY

In dentistry, as in other specialties, accurate diagnosis is a fundamental pre-requisite for correct treatment and radiology provides one of the more valuable diagnostic aids. Furthermore, in the late 1980s the annual

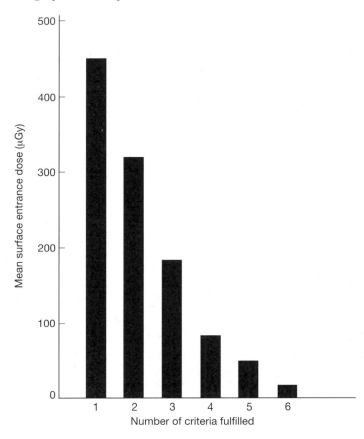

Figure 9.17. *Correlation between fulfilment of X-ray technique criteria and dose for chest radiography in 10 month old infants (adapted from Schneider et al 1992).*

Table 9.4. *Relationship between speed class, radiation dose and resolution for different film–screen combinations.*

Speed class	Dose requirement at image receptor (μGy)	Lower limit of resolution (lp mm^{-1})
25	40	4.8
50	20	4.0
100	10	3.4
200	5	2.8
400	2.5	2.4
800	1.25	2.0

expenditure in the UK on dental radiology was comparable with that on many other aspects of dental care (e.g. extractions). Thus the emphasis given to dental radiology in education and training and the standards expected in dental radiology practice should be at least comparable with those in other areas of dental care.

9.9.1. Intra-oral radiography

A major source of unnecessary dose to the patient is too low an operating potential. Many older sets operate at 45–50 kV and, whereas they give excellent contrast, the dose is high. With potentials in the range 60–70 kV contrast is still adequate and the dose is greatly reduced. Total beam filtration should be at least 1.5 mm aluminium up to 70 kVp and 2.5 mm aluminium above 70 kVp. The added filtration should not be so thick that exposure times are greater than one second.

As in other applications, X-ray units incorporating high frequency circuitry and giving an effectively constant potential output have a number of advantages. For example, the higher X-ray output rate permits techniques involving longer focus-to-skin distances whilst maintaining short exposure times; accurate, very short exposure times are possible for digital intra-oral radiographic systems, which are just beginning to be introduced.

Exposure times must provide increments that are small enough for accurate selection of the required exposure. Timers must be carefully checked; errors in excess of 0.2 s are unacceptable.

One of the problems with current dental radiology is that the image receptor, a rectangular film, is not the same shape as the collimated X-ray beam, which is circular. To minimize unnecessary exposure the beam diameter should not exceed 60 mm at the patient end of the cone. However, this is still a major source of inefficiency since either the X-ray beam will be greater than its useful area or the field of view will be impractically small. The introduction of rectangular collimation would be a major source of dose reduction. A closely related problem is the need to introduce suitable film holders and alignment devices which would improve the consistency of set-up and overall diagnostic quality of the films.

When the output of the X-ray set is low, another consequence of low kV, it may be necessary to work at short focus-to-skin distance to keep the exposure time short. However, this results in high entrance skin doses because of the inverse square law effect. For dental radiology a minimum focal–skin distance of 200 mm is recommended.

9.9.2. Panoramic radiography

Panoramic radiographs are obtained by collimating the X-ray output with a slit and rotating the source and slit in a horizontal plane. A complementary slit in front of the cassette holder collimates the transmitted beam onto the film. The beam size should not exceed 10 mm × 150 mm and the area of the beam should not exceed that of the receiving slit by more than 20%. It is important to check that these slits remain aligned as they rotate. Note that narrower beams and shorter cycle times require an effectively constant output. Any significant fluctuation, for example caused by the kV dropping with the mains frequency, would cause banding on the film.

Many panoramic radiographs are unacceptable because of poor patient positioning and alignment. Equipment must be provided with effective positioning aids, preferably of the light beam type. There should be sufficient variation on exposure times to take advantage of the fastest film–screen combinations.

9.9.3. Receptors

The fastest film consistent with satisfactory clinical results should be used. There is no justification for using slower than D speed film and current opinion is that good diagnostic films can even be obtained with E speed films if care is taken over processing.

Increased fog levels on dental films due to poor storage conditions prior to use is a common problem. Films should be stored in a cool, dry place and must be far enough away from, or adequately shielded from scattered X-rays. They should not be used after their expiry date.

Fast films and rare earth intensifying screens should be used for extra-oral and vertex occlusal views where practicable. It is important to ensure that the properties of the film and screen are compatible. Digital systems for intra-oral radiography have been referred to briefly. They incorporate an intra-oral sensor to replace the film. Currently the area of the sensor is rather small but exposure times are lower than for films.

9.9.4. Protection

Careful attention to dose reduction techniques is the best method of patient protection. In particular there is no case for using a lead apron, except perhaps for the infrequent vertical occlusal projection. The apron will not protect against radiation scattered internally in the body and direct measurements have shown that under optimal conditions external scatter results in a dose of no more than 0.01 μSv/exposure to the gonads.

If the thyroid gland is in the primary beam, there may be a case for using a thyroid collar. The introduction of rectangular collimation and longer focal–skin distances would eliminate this requirement almost entirely.

Staff will be adequately protected by distance if they are at least two metres from the set. If this condition can be satisfied there is no need to leave the room. Note that the beam is not fully stopped by the intra-oral film, nor is it fully absorbed by the patient. Therefore it should be considered as extending beyond the patient until attenuated by distance or stopped by suitable shielding e.g. a solid wall.

9.9.5. Quality assurance

Quality assurance procedures in dental radiology show no essential differences from those used in conventional radiology. Checks should be made on the equipment, films, processing and dark room—a light leak or the wrong safe light are not unknown sources of fog. The dose at the cone tip should not exceed 2.5 mGy for a 70 kV set when using D speed film, half this value for E speed film.

Table 9.5. *Typical effective doses for dental examinations under different conditions.*

Examination and conditions	Effective dose (mSv)
Two dental bitewings, 70 kV set, 200 mm fsd, rectangular collimation, E speed film	0.002
Two dental bitewings, 70 kV set, 200 mm fsd, round collimation, E speed film	0.004
Two dental bitewings, 50–60 kV set, 100 mm fsd, round collimation, E speed film	0.008
Two dental bitewings, 50–60 kV set, 100 mm fsd, round collimation, D speed film	0.016

Image quality is a good test of the entire quality assurance programme. Routine films can be checked quickly and simply by comparison with good quality reference radiographs obtained under carefully controlled conditions. To obtain such films a suitable phantom consisting of human teeth and facial bones embedded in plastic or other tissue equivalent material might be useful. For absolute evaluation a phantom containing calibrated step wedges and wired meshes to test resolution will be required.

Guidelines on radiology standards for primary health care (NRPB 1994) indicate that current failure to use optimal technique is resulting in significant unnecessary extra exposure to radiation (table 9.5).

It has been estimated that in the UK, where over 16 million intra-oral dental radiographs are taken each year, universal introduction of the conditions in the top line of table 9.5 would result in a saving in collective effective dose (see section 11.6) of about 125 man Sv.

REFERENCES

Beaman S, Lillicrap S C and Price J L 1983 Tungsten anode tubes with K-edge filters for mammography *Br. J. Radiol.* **56** 721–7
Boag J W, Stacey A J and Davis R 1972 Xerographic recording of mammograms *Br. J. Radiol.* **45** 633–40

Cowan A R 1995a A review of digital x-ray imaging acquisition systems part 1: Advances in established techniques *Imaging* **7** 204–18

——1995b A review of digital x-ray imaging acquisition systems part 2: New and emerging technologies *Imaging* **7** 259–67

Curry T S, Dowdey J E and Murry R C Jr 1990 *Christensen's Introduction to the Physics of Diagnostic Radiology* 4th edn (Philadephia, PA: Lea and Febiger)

Dendy P P 1990 Recent technical developments in medical imaging part 1: Digital radiology and evaluation of medical images *Current Imaging* **2** 226–36

Dobbins J T *et al* 1992 Threshold perception performance with computed and screen–film radiography. Implications for chest radiography *Radiology* **183** 179–87

Harrison R M 1988 Digital radiography *Phys. Med. Biol.* **3** 751–84

Heath M 1995 Digital radiography—the future? *Radiography* **1** 49–60

ICRP 1991 1990 recommendations of the International Commission on Radiological Protection, ICRP 60 *Ann. ICRP* **21** Nos 1–3

Law J 1997 Cancers detected and induced in mammographic screening; new screening schedules and younger women with family history *Br. J. Radiol.* **70** 62–9

Marshall N W, Faulkner K, Busch H P, Marsh D M and Pfenning H 1994 An investigation into the radiation dose associated with different imaging systems for chest radiology *Br. J. Radiol.* **67** 353–9

NRPB 1994 Guidelines on radiology standards for primary dental care *Report by the Royal College of Radiologists and the National Radiological Protection Board (Documents of the NRPB vol 5 No 3)* (Chilton: NRPB)

Schneider K *et al* 1992 Results of a dosimetry study in the European Community on frequent x-ray examinations in infants *Radiat. Protect. Dosim.* **43** 31–6

FURTHER READING

CEC 1991 Quality criteria for diagnostic images in paediatrics *Documents of the CEC* XII/307/91

Jennings R J, Eastgate R J, Siedband M P and Ergun D L 1982 Optimal x-ray spectra for screen film mammography *Med. Phys.* **8** 629–39

Law J, Dance D R, Faulkner K, Fitzgerald M G, Ramsdale M L and Robinson A 1994 The commissioning and routine testing of mammographic x-ray systems *Institute of Physical Sciences in Medicine Report* 59, 2nd edn

Starritt H C, Faulkner K, Wankling P F, Cranley K, Robertson J and Young K 1994 Quality assurance in dental radiology *Institute of Physical Sciences in Medicine Report* 67

Tesic M M, Mattson R A, Barnes G T, Jones R A and Stickney J B 1983 Digital radiography of the chest: design features and considerations for a prototype unit *Radiology* **148** 259–64

EXERCISES

1   Since a low kVp is required for high contrast, why should the use of a high kVp sometimes be advantageous?

2   Outline the principles of macroradiography and indicate its limitations.

3   What is the relationship between brightness and contrast in (*a*) a radiographic image, (*b*) a digital image. Explain how this relationship may be affected by windowing for the digital image.

4   Explain how quantum mottle may limit the smallest detectable contrast over a small, 1 mm^2 area in a digitized radiograph.

5   Suggest some situations in which a subtraction technique might be used and indicate how the required information would be obtained.

6   What is the effect of digital image subtraction on noise and how is this overcome in digital subtraction angiography?

7   What are the basic requirements of an X-ray tube that is to be used for mammography? How can these be achieved with (*a*) a tungsten anode tube, (*b*) a molybdenum anode tube.

8   What steps would you take to obtain the best results when using rare earth screens for mammography?

9   Discuss the relative advantages and disadvantage of magnification mammography.

10   What are the relative merits of xeroradiography and normal mammography techniques for investigation of the female breast?

11   Discuss the need for effective immobilization in paediatric radiology.

12   Explain the need for good quality assurance procedures in dental radiology and discuss the potential for reducing the collective dose to the population.

CHAPTER 10

TOMOGRAPHIC IMAGING

10.1. INTRODUCTION

Three fundamental limitations that apply equally to imaging with both X-rays and with gamma rays from radionuclides can be identified.

(i) *Superimposition.* The final image is a two-dimensional representation of an inhomogeneous three-dimensional object with many planes superimposed. In radiology the data relate to the distribution of attenuation coefficients in the different planes through which the X-ray beam passes. In nuclear medicine the data relate primarily to the distribution of radioactivity, although the final result is also affected by attenuation of the gamma rays as they emerge from the patient.

The confusion of overlapping planes results in a marked loss of contrast making detection of subtle anomalies difficult or impossible. A simplified example of the consequence of superimposition taken from nuclear medicine is shown in figure 10.1.

(ii) *Geometrical effects.* The viewer can be confused about the shapes and relative positions of various structures displayed in a conventional X-ray picture and care must be taken before drawing conclusions about the spatial distribution of objects even when an orthogonal pair of radiographs is available.

(iii) *Attenuation effects.* When the intensity of X-rays striking a film is described by the equation $I = I_0 e^{-\mu x}$ it must be appreciated that beam attenuation, I/I_0, depends on both μ and x. This can cause ambiguity since an observed difference in attenuation can be due to changes in thickness alone (x), in composition alone (μ) or a combination of these factors.

Two methods have been used to overcome the effects of superimposition of information in different planes. The first is longitudinal or blurring tomography. This relies on the principle that if the object is viewed from different angles, such that the plane of interest is always in the same orientation relative to the detector or film, the information in this plane will superimpose when the images are superimposed, but information in other planes will be blurred.

Longitudinal tomography or body section radiography with X-rays is considered briefly in section 10.2, although it is being used less and less in modern radiology departments. Longitudinal tomography with gamma ray emissions from radionuclides is also possible using either a slant hole collimator or a multiple pin-hole collimator, both of which can be used to take views from different angles. However, longitudinal tomography is not widely used in nuclear medicine.

In longitudinal tomography, the angle between the normal to the plane of interest and the axis of the detector, ϕ, can be anything from $1°$ to $20°$. Since the object space is not completely sampled, the in-focus plane is not completely separated from its neighbouring planes, but the tomographic effect increases with increasing ϕ. The second method, transverse tomography, can be considered as a limiting form of longitudinal tomography with $\phi = 90°$. The object is now considered as a series of thin slices and each

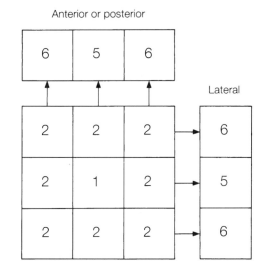

Figure 10.1. *Illustrating how superimposition of signals from different planes reduces contrast. If the numbers in the nine squares represent radioactive concentrations in nine compartments, the theoretical contrast is 2:1. If, however, the activity is viewed along any of the projection lines shown, the contrast is only 6:5.*

slice is examined from many different angles. Isolation of the plane of interest from other planes is now complete and is determined by the width of the interrogating X-ray beam, collimator design and the amount of scatter. Transverse tomography provides an answer to geometrical and attenuation problems as well as that of superimposition and the majority of this chapter is devoted to consideration of the many physical factors that contribute to the production of high quality transverse tomographic images.

10.2. LONGITUDINAL TOMOGRAPHY

10.2.1. The linear tomograph

This is the simplest form of tomography and historically is a good way to introduce the principles. As shown in figure 10.2, the X-ray tube is constrained to move along the line SS' so that the beam is always directed at the region of interest P in the subject. As the X-ray source moves in the direction $S \to S'$, shadows of an object P will move in the direction $F' \to F$. Hence if a film placed in the plane FF' is arranged to move parallel to that plane at such a rate that when S_1 has moved to S_2, the point P_1 on the film has moved to P_2, all the images of the object at P will be superimposed.

Consider however the behaviour of a point in a different plane (figure 10.3). As S_1 moves to S_2, the film will move a distance $P_1 P_2$ as before, so the point F_1, which coincides with X_1, will move to F_2. However, the shadow of X will have moved to X_2. Hence the shadows cast by X for all source positions between S_1 and S_2 will be blurred out on the film between F_2 and X_2, in a linear fashion. A similar argument will apply to points below P. Thus only the image of the point P, about which the tube focus and film have pivoted, will remain sharp.

Consideration by similar triangles of other points in the same horizontal plane as P, for example Q, shows that they also remain sharply imaged during movement. The horizontal plane through P is known as the **plane of cut** or sometimes the **focal plane**.

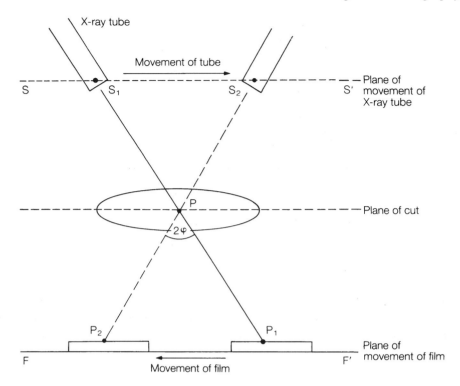

Figure 10.2. *Diagram illustrating the movement of the X-ray tube and the film in linear longitudinal tomography.*

10.2.2. Thickness of plane of cut

Blurring is the loss of definition of objects outside the focal plane. Although the plane of cut is not completely separated from adjacent planes, there is a minimum amount of blurring that the eye can discern and this effectively determines the 'thickness' of the plane that is in focus.

It is instructive to calculate the thickness of the plane in terms of the amount of blurring, B say, and the geometry of the system. Referring again to figure 10.3, suppose the point X suffers the minimum amount of blurring that is just detectable. From the previous discussion:

$$\text{film blurring} = X_2F_2.$$

By geometry this is equal to $X_2X_1 - X_1F_2 = X_2X_1 - P_2P_1$. This must be set equal to B.

Now

$$P_1P_2 = S\frac{d}{D}$$

where S is the distance moved by the tube, and

$$X_1X_2 = S\frac{(d+t)}{(D-t)}.$$

Hence

$$B = S\left[\frac{(d+t)}{(D-t)} - \frac{d}{D}\right] \simeq S\frac{(D+d)}{D^2}t \qquad (\text{since } t \ll D).$$

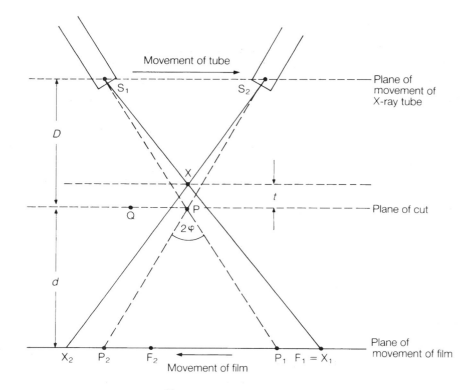

Figure 10.3. *Demonstration of the blurring effect that will occur in longitudinal tomography for points that are not in the plane of cut.*

Since

$$\frac{S}{D} \simeq 2\phi$$

$$B = 2t\left(\frac{D+d}{D}\right)\phi.$$

The same argument can be applied to points below P so the thickness of cut,

$$2t = B\frac{1}{\phi}\frac{D}{(D+d)}. \tag{10.1}$$

Thus the thickness of cut increases if the value assumed for minimum detectable blurring, B, or the focus–plane of cut distance D increases. However, it decreases if the angle of swing 2ϕ or the focus–film distance $(D+d)$ increases.

Assuming typical values:

angle of swing	$2\phi = 10° = 10 \times 2\pi/360$ radians
focus–film distance	$(D+d) = 90$ cm
focus–plane of cut	$D = 75$ cm.

Taking $B = 0.7$ mm, a reasonable figure for the blurring that will not detract significantly from the interpretation of most images

$$\text{thickness of cut, } 2t = 6.6 \text{ mm.}$$

Note the following additional points:

(1)   The plane of cut may be changed by altering the level of the pivot.
(2)   Since the tube focus and film move in parallel planes, the magnification of any structure that remains unblurred throughout the movement remains constant. This is important because the relatively large distance between the plane of cut and the film means that the image is magnified (in this worked example by $90/75 = 1.2$).
(3)   For two objects that are equidistant from the plane of cut, one above and one below, the blurring is greater for the one that is further from the film. This may influence patient orientation if it is more important to blur one object than the other.
(4)   Care is required to ensure that the whole exposure takes place whilst the tube and film are moving
(5)   The tilt of the tube head must change during motion—this minimizes reduction in exposure rate at the ends of the swings due to obliquity factors but there is still an inverse square law effect.

10.2.3.   Miscellaneous aspects of longitudinal tomography

There is insufficient space to discuss in detail other aspects of longitudinal tomography but they will be summarized briefly. For a fuller account see for example Curry *et al* (1990).

Alternative linear movements

When ϕ is large it may be difficult to compensate adequately for inverse square law effects using the movement discussed in section 10.2.1. If both the tube and film move along the arcs of circles that are centred on a point in the plane of cut, the focus–film distance remains fixed. This movement allows easier engineering design for the X-ray tube but is more difficult to achieve for the film.

Non-linear movements

An alternative way to overcome inverse square law problems is to use non-linear movement. For example, if both tube and film execute a circular motion (figure 10.4), the desired blurring will be achieved in a circular fashion but the source–film distance will not change. Such a movement also eliminates another major problem of linear tomography, namely that structures lying parallel to the line of motion of the tube are not well blurred because the blurred image of one part of the object is simply superimposed on another part of the object. Although any planar movement is theoretically possible, circular and elliptical movements illustrate adequately the realistic alternatives.

Choice of angle swing

As shown in equation (10.1), the angle of swing will determine slice thickness. Whereas a thin slice may be desirable, contrast between two structures will be proportional to $(\mu_2 - \mu_1)t$. Thus large angle tomography giving thin slices may be used when there are large differences in atomic number and/or density between structures of interest, but if $(\mu_2 - \mu_1)$ is small, a somewhat larger value of t may be essential to give detectable contrast.

Zonography

This is the term given to very small angle tomography ($2\phi \sim 1°$–$5°$) usually done with circular motion. The slice is now very thick and structures within it appear almost as on a normal radiograph. The technique may be useful if very thick structures are being examined or if the difference $(\mu_2 - \mu_1)$ is very small.

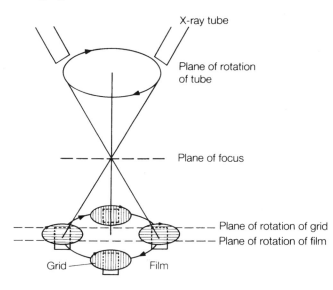

Figure 10.4. *Cyclic movement in longitudinal tomography ensures a fixed focus–film distance. Note that the grid rotates with the X-ray tube so that it maintains the same orientation relative to the target.*

10.3.   COMPUTED AXIAL TRANSMISSION TOMOGRAPHY

This full terminology is rarely used, having been progressively shortened to computed axial tomography and computed tomography (CT). It is however important not to overlook the words omitted since they give a more accurate description of the technique.

10.3.1.   Principles

As discussed in section 8.4.3, there is an inverse relationship between the contrast of an object and its diameter at threshold perceptibility. Thus the smaller the object the larger the contrast required for its perception, and loss of contrast resulting from superimposition of different planes has an adverse effect on detectability. Hence, although the average film radiograph is capable of displaying objects with dimensions as small as 0.25 mm, when due allowance is made for film resolution, geometric movement blur and scattered radiation, a contrast difference of about 10% is required to achieve this.

The objective of CT scanning is to take a large number of one-dimensional views of a two-dimensional transverse axial slice from many different directions and reconstruct the detail within the slice. Digital techniques are normally used so the object plane can be considered as a slice of variable width subdivided into a matrix of attenuating elements with linear attenuation coefficients $\mu(x, y)$ (figure 10.5). A typical matrix size will be 512×512 corresponding to a pixel size of 0.5 mm for a 25 cm diameter image. The image will often be interpolated to 1024×1024 for viewing. Slice thickness z may vary from 1 mm to 10 mm.

When data are collected at one particular angle, θ, the intensity of the transmitted beam I_θ will be related to the incident intensity I_0 by

$$P_\theta = \ln(I_0/I_\theta)$$

where P_θ is the projection of all the attenuation along the line at angle θ.

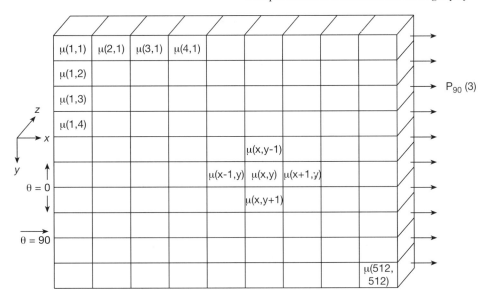

Figure 10.5. *A slice through the patient considered as an array of attenuation coefficients. If $\theta = 0$ is chosen arbitrarily to be along the y axis, the arrows show the P_{90} projections.*

For example for $\theta = 90°$, *one* of the values of P_{90}, say $P_{90}(3)$ to indicate that it is the projection through the third pixel in the y direction, is given by

$$P_{90}(3) = \mu(1, 3) + \mu(2, 3) + \mu(3, 3) + \cdots + \mu(512, 3).$$

Note that, strictly, each μ value should be multiplied by the x dimension of the pixel in the direction of X-ray travel. All values of x are the same in this projection. Pixels contribute different amounts to different projections.

The problem now is to obtain sufficient values of P to be able to solve the equation for the $(512)^2 \simeq$ 260 000 values of $\mu(x, y)$. The importance of computer technology to this development now becomes apparent, since correlating and analysing all this information is beyond the capability of the human brain but is ideal for a computer, especially since it is a highly repetitive numerical exercise.

10.3.2. Data collection

The process of data collection can best be understood in terms of a simplified system. Referring to figure 10.6 which shows a single source of X-rays and a single detector, all the projections $P_{90}(y)$, i.e. from $P_{90}(1)$ to $P_{90}(512)$, can be obtained by traversing both the X-ray source and detector in unison across the section. Both source and detector now rotate through a small angle $\delta\theta$ and the linear traverse is repeated. The whole process of rotate and traverse is repeated many times such that the total rotation is at least $180°$. If $\delta\theta = 1°$ and there is a $360°$ rotation, 360 projections will be formed.

Although this procedure is easy to understand and was the basis of the original or first generation systems, it is slow to execute, requires many moving parts and requires a scanning time of at least 3 min. Thus a major technological effort has been to collect data faster and hence reduce scan time, thereby reducing patient motion artefacts. Historically scan times were reduced by:

(1)   using several pencil beams and an array of detectors;
(2)   using a fan-shaped beam, wide enough to cover the whole body section, and an array of detectors.

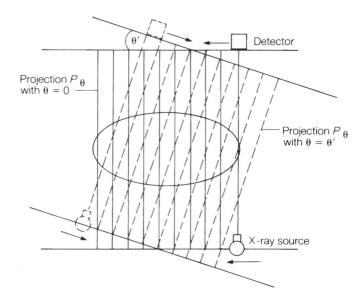

Figure 10.6. *Illustrating the collection of all the projections using a single source and detector and a translate–rotate movement.*

However, modern scanners achieve even faster scanning times either by arranging for source and detectors to rotate continuously, maintaining a fixed geometrical relationship between them, or by utilizing a complete ring of detectors and arranging for the source only to rotate continuously (figure 10.7). Modern equipment may contain multi-row detectors (total number in excess of 2000) and achieve scan times of 0.5–1.0 s. To achieve even faster scan times, 'scanners' with no moving parts have been built. In one approach, termed electron beam CT, there is a semicircle of tungsten targets below the patient and a complementary semicircle of detectors above the patient. Note that reconstruction is possible from profiles collected over angles of a little more than 180°. Collection over a full 360° is not necessary. The beam of electrons is deflected electromagnetically so that the tungsten targets are bombarded in sequence. Very high tube currents can be used since each part of the anode is only used very briefly and scan times as short as 50 ms have been achieved but 250 ms is more normal.

CT places a number of new demands on the design and construction of X-ray tubes. Data reconstruction procedures assume an exact geometrical relationship between the X-ray source and the detectors and a very stable X-ray output. Voltage fluctuations should not exceed about 0.01%. High resolution requires a small focal spot size, typically of the order of 1.0 mm, and the tube must be arranged with its long axis perpendicular to the fan beam to avoid heel effect asymmetry.

Tube design has been improved greatly in recent years with larger, faster rotating, ceramic mounted anodes for better tube rating and improved heat storage capacity. The typical capacity of a modern anode is about 3.5×10^6 heat units with a cooling rate of 820 kilo heat units per minute and this permits continuous exposures for up to 60 seconds at tube currents of over 200 mA.

A further advance has been the development of slip-ring technology to permit continuous rotation providing up to 60 reconstructed 1 second images. Rotation may be in a fixed plane (cine mode) to obtain information about the way in which contrast medium reaches an organ or fills blood vessels, alternatively the patient may be moved through the gantry aperture during such continuous data acquisition. This provides a spiral or helical data set (see section 10.3.7).

Choice of X-ray energy is a compromise between high detection efficiency and good image contrast

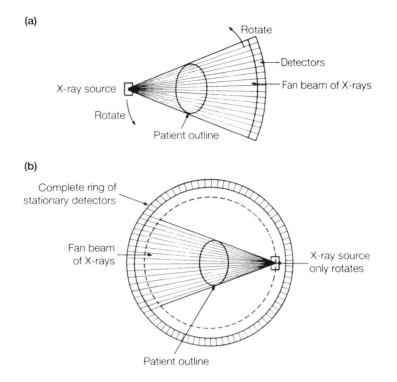

Figure 10.7. *Design of different generations of CT imagers. (a) Rotate–rotate (sometimes known as a third generation scanner). (b) Source only rotates (fourth generation scanner).*

on the one hand, low patient dose and high tube output on the other. Calculation of a unique matrix of linear attenuation coefficients assumes a monoenergetic beam. Heavy filtration is required to approximate this condition—for example 0.5 mm Cu might be used with a 120 kVp beam to give a mean energy of about 70 keV. When corrections are made for differences in mass attenuation coefficient and density, 0.5 mm Cu is approximately equivalent to 8 mm Al. Note however that the characteristic radiation that would be emitted from copper ($Z = 29$, K shell energy = 9.0 keV) as a result of photoelectric interactions would be sufficiently energetic to reach the patient and increase the skin dose. Thus the copper filter is backed by 1 mm Al to remove this component.

Modern scanners are tending to use less filtration, perhaps 3 mm Al. The beam is now more heterochromatic but the tube output is higher. The effects of beam hardening are largely corrected by software in the reconstruction programme.

Beam size is restricted by a pair of slit-like collimators, one near the tube, the other near the detectors. Perfect alignment is essential. These collimators will define the slice thickness and the collimator at the detector is essential to control Compton scatter, especially at the higher energies. Normally only a few per cent of scattered radiation is detected.

Note that scattered radiation is more of a problem with a fan beam than with a pencil beam. The reader may find the discussion of broad beam and narrow beam attenuation in section 3.7 helpful in understanding the reason.

The requirements of radiation detectors for CT are: a high detection efficiency, high dynamic range, fast response, linearity, stability, reliability and, because many detectors have to be accommodated in a limited volume, small physical size and low cost.

This has been an area of intensive commercial development. Originally thallium-doped sodium io-dide NaI(Tl) crystals and photomultiplier tubes (PMTs) were used. One problem with NaI(Tl) crystals is afterglow—the emission of light after exposure to X-rays has ceased. This effect is particularly bad when the beam has passed through the edge of the patient and suffered little attenuation because the flux of X-ray photons incident on the detector is high. Solid-state scintillation detectors are now either cadmium tungstate or doped rare earth ceramics and PMTs have now been replaced by high purity temperature-stabilized silicon photodiodes. These detectors have a wide dynamic range, close detector spacing, high efficiency and no high voltage power supply is required. A modern solid-state detector has about 98% X-ray quantum detection efficiency and 80% geometric efficiency. The detector element spacing is about 1 mm and some designs incorporate a quarter detector shift to double the spatial sampling density.

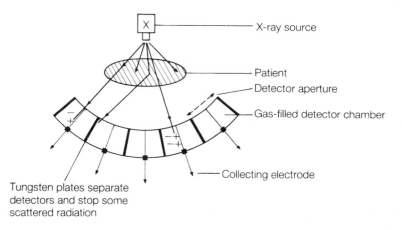

Figure 10.8. *Design of an array of gas-filled detectors for CT.*

In spite of the inherently low sensitivity of gas-filled ionization chambers, with suitable design (figure 10.8) they provide a realistic alternative for CT scanning. Use of high atomic number gases (Xe/Kr) at pressures of up to 30 atmospheres in ionization chambers several cm deep gives a sensitivity of about 50% and the detectors are closely spaced. There is little variation in sensitivity and since the current produced is independent of the voltage across the ion chamber, they have excellent stability.

Commercial machines have been built with all four possible combinations of fixed detector/moving detector and solid-state detector/xenon detector but the trend is towards solid-state because of the high quantum detection efficiency.

10.3.3. Data reconstruction

The problem of reconstructing two-dimensional sections of an object from a set of one-dimensional projections is not unique to radiology and has been solved more or less independently in a number of different branches of science. As stated in the introduction, in diagnostic radiology the input is a large number of projections, or values of the transmitted radiation intensity as a function of the incident intensity, and the solution is a two-dimensional map of X-ray linear attenuation coefficients.

Mathematical methods for solving problems are known as algorithms and a wide variety of algorithms has been proposed for solving this problem. Many are variants on the same theme so only two, which are both fundamental and are quite different in concept, will be presented. The first is filtered back-projection, sometimes called convolution and back-projection, and the second is an iterative method.

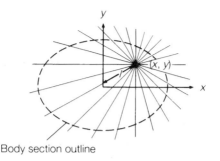

Body section outline

Figure 10.9. *The projections that will contribute to the calculation of linear attenuation coefficient at some arbitrary point* (x, y).

(i) *Filtered back-projection.* Figure 10.9 shows several projections passing through the point (x, y) which is at some distance l from the centre of the slice. The philosophy of back-projection is that the attenuation coefficient $\mu(x, y)$ associated with this element contributes to all these projections, whereas other elements contribute to very few. Hence if the attenuation in any one direction is assumed to be the result of equal attenuation in all pixels along that projection, then the total of all the projections should be due to the element that is common to all of them. Hence one might expect an equation of the form

$$\mu(x, y) = \sum_{\substack{\text{over all} \\ \text{projections}}} P_\theta(l)$$

where $P_\theta(l)$ is the projection at angle θ through the point (x, y) and $l = x \cos\theta + y \sin\theta$, to give an estimate of the required value of $\mu(x, y)$.

Note that because the beam has finite width, the pixel (x, y) may contribute fully to some projections (figure 10.10(*a*)) but only partially to others (figure 10.10(*b*)). Due allowance must be made for this in the mathematical algorithm. Similar corrections must be applied if a fan beam geometry is being used (see section 10.3.8).

If this procedure is applied to a uniform object that has a single element of higher linear attenuation coefficient at the centre of the slice, it can be shown to be inadequate. For such an object each projection will be a top-hat function, with constant value except over one pixel width (figure 10.11(*a*)). These functions will back-project into a series of strips as shown in figure 10.11(*b*). Thus simple back-projection creates an image corresponding to the object but it also creates a spurious image. For the special case of an infinite number of projections, the spurious image density is inversely proportional to radial distance, r, from the point under consideration.

Further rigorous mathematical treatment is beyond the scope of this book and the reader is referred to one of the numerous texts on the subject. Suffice to say that the back-projected image in fact represents a blurring of the true image with a known function which tends to $1/r$ in the limiting case of an infinite number of profiles. Furthermore, correction for this effect can best be understood by transforming the data into spatial frequencies (see section 8.8.1). In terms of the discussion given there, the blurred image can be thought of as the result of poor transmission of high spatial frequencies and enhancement of low spatial frequencies. Any process that tends to reverse this effect helps to sharpen the image since good resolution is associated with high spatial frequencies and it may be shown that in frequency space correction is achieved by multiplying the data by a function that increases linearity with spatial frequency.

An important factor determining the upper limit to the spatial frequency (ν_m) is the amount of noise that can be accepted in the image. Also this upper limit is related to the size of the detector aperture,

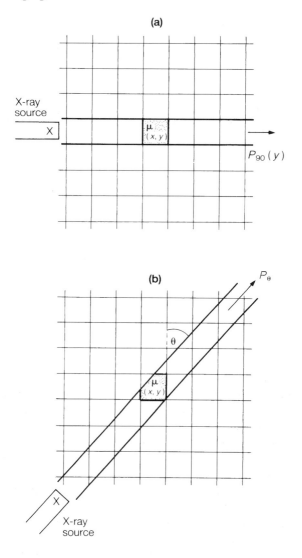

Figure 10.10. *Illustration of the difference between the matrix size and the size and shape of the scanning beam for most angles. Assuming, for convenience, that the beam width is equal to the pixel width, then for the P_{90} projection (figure 10.10(a)) it can be arranged for the beam to match the pixel exactly. However this is not possible for other projections. Only a fraction of the pixel will contribute to the projection P_θ (figure 10.10(b)) and allowance must be made for this during mathematical reconstruction. Any variation in μ across a pixel, or since the slice also has depth, the volume element (voxel), will be averaged out by this process.*

the effective spot size of the X-ray tube and the frequency of sampling. The value of v_m imposes a constraint on the system resolution since for an object of diameter D reconstructed from N profiles, the cut-off frequency v_m should be of the order of $N/(\pi D)$ and ensures spatial resolution of $\frac{1}{2}v_m$. For example if $N = 720$ and $D = 40$ cm, the limit placed on spatial resolution by finite sampling is about 1 mm. Note that reference to figure 10.11 shows that sampling is higher close to the centre of rotation

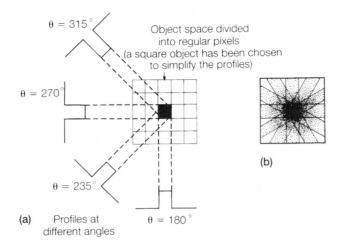

$\theta = 315°$

Object space divided
into regular pixels
(a square object has been chosen
to simplify the profiles)

$\theta = 270°$

(b)

$\theta = 235°$

$\theta = 180°$

(a)   Profiles at
different angles

Figure 10.11. *(a) The projections for a single pixel of higher linear attenuation coefficient at the centre of the field of view. (b) Back-projection shows maximum attenuation at the centre of the field of view but also a star-shaped artefact.*

than at the periphery. Thus if only the centre of the field of view is to be reconstructed, higher spatial frequencies may contribute and better resolution may be achieved.

(ii) *Aliasing.* Many texts state that the generation of spurious spatial frequencies resulting in multiple low intensity images owing to inadequate sampling of data is known as aliasing. However, relatively few give a detailed explanation in simple terms.

A good starting point is the approach of Sorenson and Phelps (1987). Scan profiles are not continuous functions but collections of discrete point by point samples of the scan projection profile (in digital and CT work the pixel). The distance between these points is the linear sampling distance. Additionally, profiles are obtained in CT scans only at a finite number of angular sampling intervals around the object. The choice of linear and angular sampling intervals and the maximum spatial frequency of the cut-off filter (the cut-off frequency), in conjunction with the detector resolution, determine the reconstruction image resolution.

The sampling theorem (Oppenheim and Wilsky 1983) states that to recover spatial frequencies in a signal up to a maximum frequency v_m requires a linear sampling distance d given by $d \leq 1/2v_m$. Alternatively, if the value of d is fixed, there is a limit on v_m of $1/2d$. If frequencies higher than v_m are transmitted (e.g. by the MTF of the detector) or amplified by the filter function, aliasing will result.

To understand how under-sampling can introduce spurious spatial frequencies, consider figure 10.12 which shows a simple sinusoid with repeat distance x. If it is sampled at five equally spaced points A B C D E, the curve will be uniquely defined. No other sinusoid of lower frequency will have the same amplitude values at these five points. Note that the data have been sampled five times in the distance $2x$ so the sampling theorem is satisfied.

If the data are only sampled at A, C and E, an alternative sinusoid (rising as it passes through A and E) of lower spatial frequency may also be fitted.

Finally, the correction function cannot rise linearly to v_m and then stop suddenly because this will introduce another source of image artefacts. Figure 10.13 shows how a sharp discontinuity in spatial frequency translates into a function varying as $(\cos x)/x$ in real space—the effect is very similar to the diffraction of light by a narrow slit.

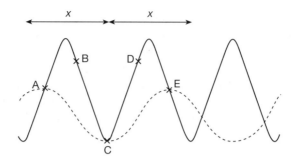

Figure 10.12. *An example of aliasing. If the sine wave is sampled at A B C D E, it is uniquely determined. If it is only sampled at A C E, the values may be fitted by a lower frequency curve shown dotted.*

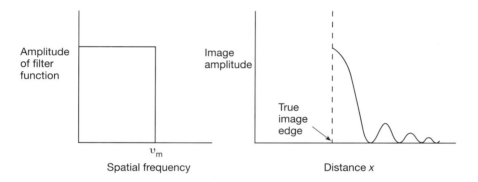

Figure 10.13. *Effect of a sharp discontinuity in the filter function (left) on the sharpness of the edge of an image (right).*

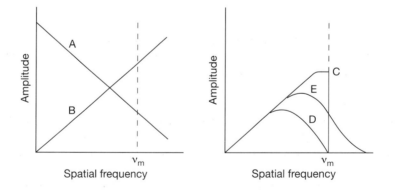

Figure 10.14. *Typical functions associated with image reconstruction by filtered back-projection. For explanation see text.*

Figure 10.14 summarizes the position with respect to filter functions. On the left, curve A shows the effect of simple back-projection on the amplitudes of different spatial frequencies; curve B shows the (theoretically) ideal correction function. However this function will cause amplification of high spatial frequency noise and aliasing. On the right filter C will give good resolution but will cause artefacts because of the sharp edge in frequency space. Filter D will cause loss of resolution because high spatial frequencies near to but less than v_m are inadequately amplified. Filter E will give good resolution without too much noise amplification but there will be some aliasing.

Clearly there is no ideal filter function and the final choice must be a compromise. However the exact shape of the filter function has a marked effect on image quality in radiology and manufacturers generally offer a range of filter options. The performance of a filter function with respect to resolution can be well represented by its modulation transfer function. MTFs for a standard and a high resolution filter are shown in figure 10.15. These would correspond to spatial resolution limits of about 0.7 mm, and 0.4 mm respectively.

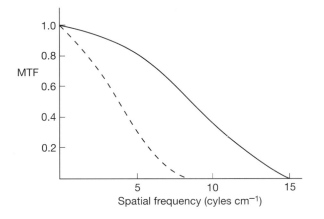

Figure 10.15. *Modulation transfer functions for two CT filter functions: – – – standard; —— high resolution.*

For a standard body scan one would normally use short scan times because of body movement with a wide slice and standard convolution filter. For a high resolution scan, say of the inner ear, a sharper filter, i.e. one which falls to zero more steeply, and thinner slices will be selected. Both of these changes will increase the relative importance of noise. If this is unacceptable it can be counteracted by increasing the mA s, i.e. the dose.

(iii) *Iterative methods.* The starting point for such a method is to guess a distribution of values for $\mu(x, y)$. This is quite arbitrary, for example the μ for each pixel could be set equal to the average for the whole slice. One projection to be expected from the assumed values of μ is now compared with the actual value and the difference between observed and expected projections is found. Values of μ along this projection are then adjusted so that the discrepancy is attributed equally to all pixels. This process is repeated for all projections and when all μ values have been adjusted in this way, a revised set of assumed projections is calculated and the whole process is repeated.

Clearly this sequence can be repeated indefinitely but there is a high cost in computing time and a procedure that requires large numbers of iterations is probably unsatisfactory. Also care is necessary to ensure that the iterative process converges to a unique solution. If μ values are changed too much, the revised image may be less similar to the object than the original image. A further disadvantage of iterative methods is that all data collection must be completed before reconstruction can begin.

No one reconstruction method holds absolute supremacy over all others and it is essential to assess the efficiency of a particular algorithm for a particular application.

10.3.4. Data presentation and storage

For each pixel in the reconstructed image a CT number is calculated which relates the linear attenuation coefficient for that pixel $\mu(x, y)$ to the linear attenuation coefficient for water μ_w according to the equation:

$$\text{CT number} = K \frac{\mu(x, y) - \mu_w}{\mu_w}$$

where K is a constant equal to 1000 on the Hounsfield scale. Some typical values for CT numbers are given in table 10.1.

Table 10.1. *Typical CT values for different biological tissues.*

Tissue	Range (Hounsfield units)
Air	−1000
Lung	−200 to −500
Fat	−50 to −200
Water	0
Muscle	+25 to +40
Bone	+200 to +1000

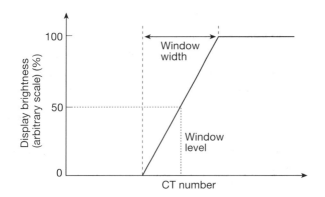

Figure 10.16. *Illustrating the relationship between window width and window level when manipulating CT numbers.*

An interactive display is normally used so that full use can be made of the wide range of CT numbers. Since only about 30 distinguishable grey levels can be shown on the screen between 'black' and 'white', adjustment of both the mean CT number, the **window level**, and the range of CT numbers covered by the grey levels, the **window width**, must be possible quickly and easily (figure 10.16). A wide window is used when comparing structures widely different in μ value, but a narrow window must be used when variations in μ are small. Consider for example a 5% change in μ, which represents a range of about 25 CT numbers. With a window from +500 to −500 this may show as the same shade of grey. If a narrow window, ranging from say, −10 to +50, is selected, 25 represents almost half the range and several shades of grey will be displayed. Data manipulation in the manner discussed for digitized images in section 8.7 is possible and by setting the window width equal to 1 and changing the level, the precise CT number of a pixel may be obtained. All systems contain sophisticated region of interest (ROI) packages. Finding the average CT number in an ROI may sometimes help to decide the exact nature of the imaged tissues.

Data storage and retrieval present a formidable problem in CT. Clearly the final images must be stored and these can consist of several slices with 512 pixels across a diameter ($\pi D^2/4 =$ about 200 000 pixels), together with information on more than 1000 grey levels. If each attenuation value is stored with a 12-bit accuracy, each image requires about 300 kbytes of storage (1 byte = 8 bits). Furthermore, it may be necessary to store the original projections if several reconstruction algorithms are available and the radiologist is unsure which one will give the best image.

Access to the data must be rapid so that reconstructions can be completed whilst the patient is still in the department and so that the radiologist may quickly recall earlier studies. Finally, the long term storage medium must be reliable, physically compact and not too expensive.

For short term storage, floppy disks are frequently used. They allow ready transfer between the scanner and remote viewing console, but have limited storage capacity—as little as six images for body scans. The viewing console itself has limited storage.

For longer term storage most new systems now offer optical disks. The most important advantage of a disk over old-fashioned magnetic tape is that it permits random access since the video disks are numbered and the operator can go straight to any desired number.

Optical disks are being used increasingly and operate on the write once read many times (WORM) principle. An optical disk comprises a rigid base covered by a photoactive material into which a laser beam etches a pattern. Read-out depends on scattering of light from these contours in the active layer. Disk technology is still a rapidly developing area in the search for ever greater and cheaper data storage capacity.

10.3.5. *System performance and quantum mottle*

For computed tomography it is difficult to define image quality. However, as with conventional radiology, image quality, noise and patient exposure or radiation dose are all inter-related. Two forms of resolution can be considered: (a) the ability to display as separate images two objects that are very close, (b) the ability to display as distinct images areas that differ in contrast by a small amount.

For a high contrast object the system resolution is determined by the focal spot size, the width of detector aperture, the separation of measurement points or data sampling frequency and the display pixel size. The focal spot can affect resolution as in any other form of X-ray imaging although in practice with the 1.0 mm focal spots now used in CT this is not a problem. However, the width of the detector aperture is important and if resolution is considered in terms of the closest line pairs that can be resolved, simple ray optics show that the line pair separation at the detector must be greater than the detector aperture.

Note however that since the patient is at the centre of rotation, not close to the detectors, there is always some magnification in CT, typically a factor of 1.5–2.0. Thus the resolution in the patient will be better than the resolution at the detectors.

The pixel size must match the resolving capability of the remainder of the system. For example a 512×512 matrix across a 50 cm field of view corresponds to a pixel size of just less than 1 mm. This would be capable of displaying an image in which the resolution was 5 lp cm^{-1} (i.e. a resolution of an object in which there are five lines 1 mm wide separated by five spaces 1 mm wide). However, it could not display adequately an image in which the resolution was 10 lp cm^{-1}. Note that by restricting the field of view to the central 25 cm, this higher resolution image can be adequately displayed on a 512×512 matrix.

The effect of sampling frequency on spatial resolution was discussed in section 10.3.3.

Resolution becomes a function of contrast and dose at low contrast. When a CT imager is used in the normal 'fast scan' mode, a typical relationship between resolution and contrast will be as shown in figure 10.17. The minimum detectable contrast decreases as the detail diameter increases. Note however, that for objects as small as 2–3 mm, the minimum detectable contrast is less than 1%. This compares very favourably with a minimum detectable contrast of about 10% in conventional radiology and illustrates well one of the principles of tomographic imaging discussed in the introduction to this chapter.

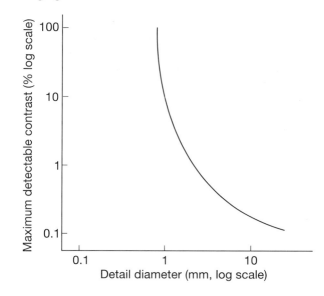

Figure 10.17. *Relationship between minimum detectable contrast and detail size for a CT scanner.*

At low contrast, *quantum mottle* may become a problem. If a uniform water phantom were imaged, then even assuming perfect imager performance, not all pixels would show the same CT number. This is because of the statistical nature of the X-ray detection process which gives rise to quantum mottle as discussed in section 8.6. CT in fact operates close to the limit set by quantum mottle. A 0.5% change in linear attenuation coefficient, corresponding to a change in CT number of about 5, can only be detected if the statistical fluctuation on the number of counts collected ($n^{1/2}$) is less than 0.5%. Hence

$$\frac{n^{1/2}}{n} < \frac{5}{1000}$$

which gives a value for n of about 4×10^4 photons. This is close to the collection figure for a single detector in one projection.

Any attempt to reduce the dose will increase the standard deviation in the attenuation coefficient due to this statistical noise. When measurements are photon limited, statistical noise increases if the patient attenuation increases. However it decreases if the slice thickness increases or the pixel width increases because more photons contribute to a given attenuation value.

Quantum mottle is still a considerable problem in the very obese patient especially for a central anatomical site such as the spine.

10.3.6. *Patient doses*

CT is a relatively high dose examination. A study in the mid-1980s showed that in the UK CT accounted for only 2% of the patient examinations but over 20% of the patient dose (NRPB 1992). It is therefore an important area in the radiation protection of patients.

Furthermore, the number of CT examinations per annum per million of the population is increasing sharply in the developing countries. During the 6 years from 1983 to 1989 there was almost a fourfold increase in the UK.

Since CT examinations involve the irradiation of thin transverse slices of the body by a rotating beam of X-rays, conditions of exposure are very different from those used in conventional radiology and special methods are required to calculate effective doses. To assess the risk correctly it is necessary to know the dose to all the radiosensitive organs of the patient arising from the various highly localized patterns of exposure. For a detailed discussion of the techniques involved the reader is referred to specialized texts, e.g. Shrimpton *et al* (1991), only a very brief outline of one possible approach is given below.

First some basic information will be required for each examination performed on the scanner—examination type, number of slices, slice thickness, couch increment and scan mode.

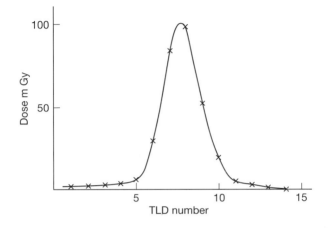

Figure 10.18. *A typical dose profile acquired with 255 mA s and measured with a stack of thermoluminescent dosimeter chips each 0.8 mm thick. The area under the curve is 83 mGy mm and the nominal slice width 2 mm. Hence the CTDI is 0.16 mGy (mA s)*$^{-1}$.

Second it is important to realize that each slice will not have a dose profile that is an ideal rectangular shape but will have a more spread distribution (figure 10.18). Moreover the amount of spreading will be variable from one scanner to another. Therefore a method is required to find the equivalent peak dose for a perfectly collimated CT slice. This can be done for example by placing a row of thermoluminescent dosimeters in a suitable position. Several suggestions have been made for the most appropriate place to make a measurement. The profile in figure 10.18 was obtained in air on the axis of rotation. **The computed tomographic dose index** (CTDI) may then be calculated as the area under the curve divided by the nominal slice width and the mA s.

For a given scanner the CTDI should be approximately independent of slice width (table 10.2). Note however that for this particular scanner the CTDI is approximately double for the 1 mm slices. This is because only the detector collimators collimate to 1 mm, the source collimators remaining at 2 mm.

Table 10.2. *Results of CTDI measurements with a Siemens Somatom Plus.*

Slice width (mm)	CTDI (mGy (mA s)$^{-1}$)
1	0.29
2	0.16
5	0.14
10	0.14

CTDI values will vary considerably from one scanner to another; 0.15–0.4 mGy (mA s)$^{-1}$ would be a typical range. The CTDI may then be multiplied by the appropriate mA s to find the axial in air dose.

Finally organ doses may be estimated from axial air doses using Monte Carlo techniques developed by the NRPB (Jones and Shrimpton 1991) and the effective dose may then be calculated using the methods described in section 11.7.1.

Some typical effective doses published by the NRPB (Shrimpton *et al* 1991) are shown in table 10.3. For fixed mA s, slice width and number of slices, the effective dose will be proportional to the CTDI.

Table 10.3. *Mean levels of effective dose from CT in the UK.*

Procedure	NRPB results (1991)	East Anglian results (1996)
Routine head	1.8	1.0
Sinuses	0.3	0.7
Routine chest	7.8	5.2
Mediastinum	7.6	5.8
Pancreas	5.2	4.8
Routine pelvis	7.1	6.0
Lumbar spine	3.3	3.8

Since the NRPB survey was conducted, average effective doses have come down. Results of a recent local audit are also shown in table 10.3. This is partly the result of technical improvements, e.g. high efficiency detectors, new beam shaping absorbers, rapid X-ray rise and fall times; partly greater awareness amongst radiologists that CT is a high dose technique. One new dose reduction idea is an auto exposure function with tube mA varied under computer control in response to changes in patient cross section.

Some protocols still give doses above these mean values. They generally involve a large number of slices. Thus CT examinations should be made with the minimum settings of the tube current and scan time to give adequate image quality and with the minimum number of slices and minimum irradiated volume, i.e. region of interest, required to give the necessary information. Particular care is required with high resolution modes because these are also high dose modes. Large focal spots result in a wider slice profile and higher doses whereas higher kVp results in lower doses as in conventional radiology.

Note that scan projection radiography or 'scout views' is a relatively low dose technique and may frequently be used to determine the level at which CT slices are required. Modern scanners can traverse 700 mm in about 10 seconds with a 1 mm aperture and generate good quality images. This is an acceptable alternative to conventional radiography with fast screens if radiological examination of the abdomen during pregnancy is unavoidable.

In contrast, dynamic CT investigations are usually high dose studies. Dynamic CT, which allows measurements of temporal changes in contrast density, for example in blood vessels and soft tissues, after administration of contrast material, has much in common with dynamic studies in nuclear medicine. For example a sequence of temporal images will be collected and analysed to map changes in CT number (cf changes in activity) as a function of time. Displayed images may be functional images, for example a map of time to maximum contrast, and selection of suitable regions of interest to act as controls can be difficult.

The requirements for dynamic studies are a short scan time, short scan intervals, a high scan frequency and a large number of total scans. Furthermore resolution must be high and noise levels must be low so that small changes in density can be registered. These two conditions are only met if high mA s values are selected and, combined with the need for multiple images of the same slice, the net result may be a high effective dose. The effective dose can be reduced considerably by careful selection of the correct scan plane initially.

10.3.7. Spiral CT

Spiral scanning is achieved by continuously acquiring CT data while the patient is moved through the gantry. It is known variously as **spiral CT, helical CT or volumetric CT**. A prerequisite is a continuously rotating X-ray tube and detector system for third generation scanners or a continuously rotating X-ray tube for fourth generation scanners.

Clearly, for this mode of data collection it is necessary for the X-ray tube to emit radiation for a long period of time. An initial restriction on spiral CT was set by the rating limit on the tube, limiting the method to low tube currents and poor counting statistics. The developments in anode design discussed in section 10.3.2 have been important. A top of the range modern scanner will operate at over 200 mA for 60 s and movement at 10 mm s^{-1} will provide data from a volume 60 cm long.

A second early limitation was that image quality was not comparable with slice by slice single shot acquisition. This was partly due to the noise associated with the low counts, partly to the fact that data collected at different angles are not in the same orthogonal plane relative to the patient (figure 10.19). Some degradation of the slice sensitivity profile is inevitable and this will be greater the higher the rate of table feed.

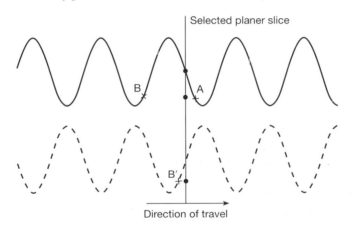

Figure 10.19. *Graph showing that because of patient movement during data acquisition the X-ray focus traces out a spiral relative to the patient. To reconstruct a plane of interest, data not actually collected in that plane must be used. Interpolation between two positions separated by* 180° *(AB′) produces a better result than interpolation between positions separated by* 360° *(AB).*

One approach to this problem is to use linear interpolation between two adjacent data points obtained at the same angular position—for a simple account see Miller (1996). In figure 10.19 measured values at A and B for the same arbitrary angle θ may be used to deduce the actual value in the plane of interest. However B is rather a long way from this plane so an improved algorithm effectively shifts the phase of this sinusoid by 180° and then interpolates between the measured values A and B′ for a more accurate estimate of the required value.

An important term in relation to spiral CT is the table **pitch**. This relates the volume traced out by the spiral to the nominal slice thickness. The table **pitch** is 1.0 if the distance travelled by the couch during one full rotation is equal to the nominal slice thickness (e.g. nominal slice thickness = 10 mm, duration of 360° rotation 1 s, rate of movement = 10 mm s^{-1}). If the rate of movement is doubled, the table pitch becomes 2.0. Slices of nominal thickness 10 mm can still be reconstructed but there will be degradation of the slice sensitivity profile.

To store 60 s of acquisition, a large random access memory (RAM) capable of holding several hundred raw data sets, together with images, is necessary.

There are a number of advantages to spiral CT.

- Slice thickness, slice interval and slice starting point may be chosen retrospectively rather than prospectively. The simple example of a single 5 mm lesion being detected with 5 mm slices illustrates this advantage. In conventional CT, unless the slice starts exactly at the edge of the lesion, it will contribute to two slices with corresponding loss of contrast.
- Reconstruction is not limited to transverse sections. They can also be coronal or sagittal, pathology can be viewed from any angle and vessel tracking techniques may be used to follow tortuous vessels as they pass through the body.
- The faster scan permits a full study during a single breath hold. This is important to eliminate respiration where motion artefacts can be troublesome—for example when visualizing small lesions in the thorax.
- The faster scan also means that high levels of contrast are maintained in vessels for the duration of the study and spiral CT angiography is claimed to have some advantages over MRI.

10.3.8. Artefacts

CT systems are inherently more prone to artefacts than conventional radiography and these will be summarized briefly.

Mechanical misalignment and patient movement

These will generate streak artefacts, intense straight lines, not always parallel and bright or dark in the direction of motion. Examples of mechanical imperfections are—non-rigid gantry, misalignment, X-ray tube wobble. In each case the X-ray beam position will deviate from that assumed by the reconstruction algorithm. Patient movement is more serious than in conventional imaging because the superimposed profiles have not passed through the same part of the body. Movement may be either voluntary (respiration) or involuntary (peristalsis, cardiac).

Detector non-uniformities

These cause ring artefacts which result from errors in projection over an extended range of views. In X-ray CT, variations in X-ray output or detector sensitivity of as little as 1 in 5000 can cause problems. For third generation scanners detector balancing is essential so gas-filled detectors are an advantage. In fourth generation scanners there is continuous real-time calibration of the detectors using the leading edge of the rotating fan beam. Since rings in image space are equivalent to vertical lines in projection space, the latter can be searched for such lines which may then be removed. Unfortunately these lines have much lower contrast than the rings they reconstruct.

Non-uniformities are a particular problem when a gamma camera is used for tomography in nuclear medicine (see section 10.4), with much better uniformity being required for this purpose than for conventional gamma camera imaging.

Partial volume effects

These were illustrated in figure 10.10. Anatomical structures do not intersect the section at right angles so a long thin voxel could have one end in bone, the other in soft tissue. An average value of μ will then be computed. Note that with a diverging beam, the object may be in the scanning plane for one projection but not for another. Partial volume effects are best overcome by using thinner slices.

Beam hardening

Even if the X-ray beam is heavily filtered, it is still not truly monochromatic. A certain amount of beam hardening will occur on passing through the patient owing to selective attenuation of the lower energy X-rays. This will affect the measured linear attenuation coefficients. For example there may be a false reduction in density at the centre of a uniform object, it appears to have a lower value of μ, and false detail near bone–soft tissue interfaces. Although beam hardening can cause streaks, it is more likely to cause shading artefacts, bright or dark unpredictable CT number shifts, often near objects of high density. For example apparent areas of low attenuation can develop within hepatic parenchyma adjacent to ribs and these 'rib artefacts' can simulate lesions.

Aliasing

This phenomenon, whereby high frequency noise generated at sharp, high contrast boundaries appears as low frequency detail in the image, was discussed in section 10.3.3 in connection with the shape of the filter function. It is caused by undersampling the highest spatial frequencies. The effect is less marked when iterative techniques are used but cannot be eliminated entirely.

Noise

If the photon flux reaching the detectors is severely reduced, for example by heavy attenuation in the patient, there will be statistical fluctuations (quantum noise) resulting in severe streaks.

Scatter

The logarithmic step in the reconstruction process makes the effect of scatter significantly non-linear and the distribution becomes very object dependent. Shading artefacts result. Post-patient collimation (third generation) or software algorithms (fourth generation) are the main methods used to remove scatter.

10.3.9. Quality control

As with other forms of X-ray equipment, specification of performance characteristics and their measurement for CT scanners can be considered under three headings: (a) type testing which is a once-only measurement of all relevant parameters associated with the function and use of a particular equipment, (b) acceptance testing when a CT scanner is installed in a hospital, (c) quality assurance measurements which check that the machine performance is maintaining an acceptable level throughout its daily use.

The measurements that relate to type testing or acceptance such as tube output, kilovoltage calibration, size of focal spot, alignment of scanner geometry, pixel size, image distortion and image recording devices will probably be made by specialists and are outside the scope of this book.

However, there are a number of tests of performance that are both specific to the CT scanner and also required to be done on a regular basis. Therefore they should be rapid, so that they take minimal scanner time; uncomplicated, so that all staff are capable of performing them; quantitative, so that as many tests as possible produce objective, numerical answers. Six basic measurements give a great deal of information about the performance of the scanner:

(i)   *Quantum noise.* An image is taken of a large region of interest in a water-filled phantom. As explained in section 10.3.5, the CT number will not be constant throughout the image. The actual variation in CT numbers will depend on the radiation dose, i.e. the mA s.

(ii)   *CT number constancy.* This may be checked by imaging a test object containing inserts to simulate various tissues—for example rib bone (500), brain (20), breast (−60), adipose tissue (−100). Typical CT numbers are given in brackets.

If the exact composition and density of the materials is known, these measurements may also be used to check that the CT number is varying linearly with the linear attenuation coefficient. Note however that for this measurement to be successful it is necessary to know the effective energy of the X-ray beam at all points in the phantom because of beam hardening.

(iii) *High contrast resolution.* Several rows of high contrast rods are inserted in a phantom. Typical diameters might range from 2.0 mm to 0.5 mm at 0.2 mm intervals, rods being placed a diameter apart. Visual inspection will show the separation at which rods are no longer resolved.

(iv) *Low contrast resolution.* Now the rods are larger, perhaps ranging from 1 to 10 mm, but the contrast relative to the surroundings is relatively small, say 2%. Again the diameter at which visual contrast is lost is noted.

(v)   *Slice thickness.* This may be obtained as part of a CTDI measurement as discussed in section 10.3.6.

(vi) *Axial dose.* This has also been discussed in section 10.3.6.

For all these features, careful measurement at acceptance testing will establish a baseline and subsequent deterioration during routine checking, perhaps weekly, can immediately be investigated.

10.4.   SINGLE PHOTON EMISSION COMPUTED TOMOGRAPHY (SPECT)

Although many of the principles of SPECT are similar to those of CT, the objective is somewhat different, namely to recover from a series of projections a map of the concentration of radionuclide which is varying continuously throughout the volume of interest.

If $C(x, y)$ is the number of counts per unit time recorded in a normal gamma camera image at an arbitrary point (x, y, z), this is related to the concentration of radionuclide (activity per unit volume $A(x', y', z)$) at some arbitrary point (x', y', z) in the same slice by the equation

$$C(x, y) = \int_{\substack{\text{all the volume} \\ \text{occupied by activity}}} A(x', y', z)S(x - x', y - y', z)e^{-\mu t}dx'dy'dz$$

where $S(x - x', y - y', z)$ represents the response of the detector (at x, y, z) to a point source of activity (at x', y', z), μ is the linear attenuation coefficient of the medium and t is the thickness of attenuating medium traversed by the gamma rays. The recovery of the function $A(x', y', z)$ for the whole slice from the available data $C(x, y)$ represents a complete solution to the problem.

Use of a single value of μ is of course an approximation. Ideally it should be replaced by a matrix of values for the linear attenuation coefficient in different parts of the slice, so one could say that the end point of CT is the starting point for SPECT!

Three fundamental limitations on emission tomography can be mentioned. The first is collection efficiency. Gamma rays are emitted in all directions but only those which enter the detector are used. Thus collection efficiency is severely limited unless the patient can be surrounded by detectors.

The second is attenuation of gamma rays in the patient. Allowances can be made and corrections simplified by adding counts registered in opposite detectors. As shown in figure 10.20, for a uniformly distributed source a correction factor $\mu L/(1 - e^{-\mu L})$ where L is the patient thickness can be applied. However, experimental work indicates that the correct value of μ is neither that for narrow beam attenuation, nor that for broad beam attenuation, but somewhere between the two. Note that accurate attenuation correction is only essential when SPECT is used quantitatively. Uncorrected images may be acceptable for qualitative interpretation. The third problem is common to all nuclear medicine studies, namely that the time of collection is only a small fraction of the time for which gamma rays are emitted. Hence the images are seriously photon limited.

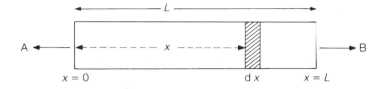

Figure 10.20. *Correction factor to be applied for gamma ray attenuation in the patient when the source of radioactivity is uniformly distributed. If the total activity is I, then the activity per unit length is I/L and the activity in the strip dx is $I\,dx/L$. The signal recorded at A is $\frac{I}{L}\int_0^L e^{-\mu x}\,dx$ where μ is the linear attenuation coefficient of the medium. The signal recorded at B is $\frac{I}{L}\int_0^L e^{-\mu(L-x)}\,dx$. Both expressions work out to $\frac{I}{L}\frac{(1-e^{-\mu L})}{\mu}$ and since, in the absence of attenuation, the signal recorded at A and B should be I, the total activity in the strip, the required correction factor is $\frac{\mu L}{(1-e^{\mu L})}$.*

The required projections are normally collected by rotating a gamma camera around the patient, although a few machines have been built which surround the patient with banks of NaI(Tl) crystals to improve detection efficiency and operate on the translate–rotate principle. Both systems will either collect data at 3° intervals for 15–20 min or during continuous rotation for the same time. The scanner has greater sensitivity for a single slice, but since several slices can be reconstructed from the gamma camera projections, camera sensitivity is higher for multisection images.

Tomography places more stringent demands on the design and performance of gamma cameras than conventional imaging. For example, multiple views must be obtained at precisely known angles and the centre of rotation of the camera must not move, for example under its own weight, during data collection. The face of the camera must remain accurately parallel to the long axis of the patient and the mechanical and electronic axes of the camera must be accurately aligned. Camera non-uniformities are more serious than in conventional imaging since they frequently reconstruct as 'ring' artefacts. If views are corrected with a non-uniformity correction matrix collected at a fixed angle, care must be taken to ensure that the pattern of non-uniformity does not change with camera angle. Such changes could occur, for example, as a result of changes in PM tube gain due to stray magnetic fields.

For all these reasons, especially very poor counting statistics, resolution is inferior in SPECT to that in conventional gamma camera imaging and much inferior to CT. Resolution decreases with the radius of rotation of the camera. This is because the circumference is $2\pi r$ and for N profiles the sampling frequency at the edge is $N/2\pi r$. For body sections the resolution is about 8–10 mm so 3° sampling ($N = 60$) is quite adequate. For example with objects 20 cm in diameter $N/2\pi r \simeq 0.1$ mm^{-1}. If this is the minimum sampling frequency, then by the Nyquist theorem (see section 10.3.3) the resolution limit set by sampling is $\simeq 1/2v_m \simeq 5$ mm which is less than the limit set by other factors.

Furthermore, matrices finer than 64×64 (about 6 mm) are not really necessary for reconstruction. Slices are usually two or three pixels (12–18 mm) thick to improve statistics.

Single head rotating gamma camera systems are now in use for routine SPECT throughout the world. A state of the art gamma camera that is well maintained should produce high quality SPECT images consistently.

Relatively recently, dual headed gamma cameras have been introduced to double the sensitivity for a given administered activity and imaging time. With SPECT applications particularly in mind, they incorporate slip-ring technology to allow continuous data collection and the separation of the detectors follows body contours to improve resolution. A range of collimators is available to allow different combinations of resolution and sensitivity and standard data processing packages are available for tomographic myocardial perfusion, tomographic HMPAO brain studies and gated tomography.

10.5. POSITRON EMISSION TOMOGRAPHY

Since radionuclide imaging provides functional or physiological information, it would be highly desirable to image elements such as carbon, oxygen and nitrogen which have a high abundance in the body. The only radioisotopes of these elements that are suitable for imaging are short half-life positron emitters (carbon-11 with a half-life of 20.5 min, nitrogen-13 with a half-life of 9.9 min and oxygen-15 with a half-life of 2.0 min).

For positron emitters, the origin of the detected radiation is the gamma rays released when the positron comes to rest and annihilates with an electron.

$$\beta^+ + \beta^- \rightarrow 2\gamma.$$

The gamma rays have a well defined energy of 0.51 MeV (the energy equivalence of the rest mass of each particle according to $E = mc^2$) and are released simultaneously in nearly opposite directions. Thus coincident detection of these two 'annihilation photons' in a pair of opposed detectors establishes the line on which the positron came to rest. The coincidence gate is ultimately limited by the time of flight of the photons.

For tomographic imaging, positron emitters have some advantages. For example in tomographic projections it is only the line of origin, not the point of origin that is important. Second, since coincidence detection eliminates stray and scattered radiation, conventional lead collimators are not necessary, leading to increased geometric efficiency and sensitivity. Finally, the total path travelled by the two gamma rays in the attenuating medium is always equal to L, the thickness of the patient in that projection, irrespective of the point of origin of the gamma rays (figure 10.21(a)). So for coincident detection the attenuation correction is $e^{\mu L}$ for all coincident events arising along that projection.

One advantage of PET is that micromolar quantities of radiopharmaceutical can be used to measure metabolic function rates, receptor densities, blood flow and statistically significant changes in function. By comparison millimolar quantities are required for MRI. A number of interesting studies have been reported using positron emitters. For example $C^{15}O$ will bind to red cells and acts as a very good marker for them. $H_2{}^{15}O$ can be used for a variety of studies. Another important positron emitter is fluorine-18 (half-life 110 min). This may be synthesized into fluorodeoxyglucose which transports and phosphorylates like glucose. It may thus be used to study sugar metabolism—for example the pattern of glucose demand in the brain or active tumour regrowth after radiotherapy. A further strength of PET is that it may be used for strictly quantitative work. However, this requires an attenuation correction which is measured on an individual basis using gamma rays from outside the body, for example from germanium-68[1].

There are a number of technological problems. For example to achieve adequate sensitivity a modern scanner may contain more than 10 000 bismuth germanate crystal detectors arranged in several rings around the patient (figure 10.21(b)). Individual, fairly thick crystals are used to improve sensitivity of detection with the higher energy gamma radiation. Spatial resolution is about 5 mm at the centre but as with all tomographic systems increases with radial position.

Random coincidences and scattered radiation give a high background unless some collimation is used, for example 1 mm tungsten interplane septa. When the beams are collimated each image is a two-dimensional slice. Recently, successful imaging without collimation has been achieved. This gives a three-dimensional data set and requires only one quarter of the administered activity.

Although the short physical half-lives might suggest that high activities could be used, the emitted positrons have a high kinetic energy which is deposited locally before annihilation takes place, thereby contributing a significant dose to the patient. Finally, because of the short half-lives, the positron emitters considered here can only be used close to the cyclotron where they are produced. Compact cyclotrons have now been developed that are capable of producing large beam currents (20–50 μA) of high energy protons

---

[1] Germanium-68 (half-life 280 d) decays by **orbital electron capture**—a decay process in which the nucleus traps an electron from the K shell. The daughter product, gallium-68, has a half-life of only 68 min so it is in secular equilibrium with the germanium-68 (see section 1.7). Gallium-68 is a positron emitter and the positron annihilates to give 0.51 MeV gamma rays.

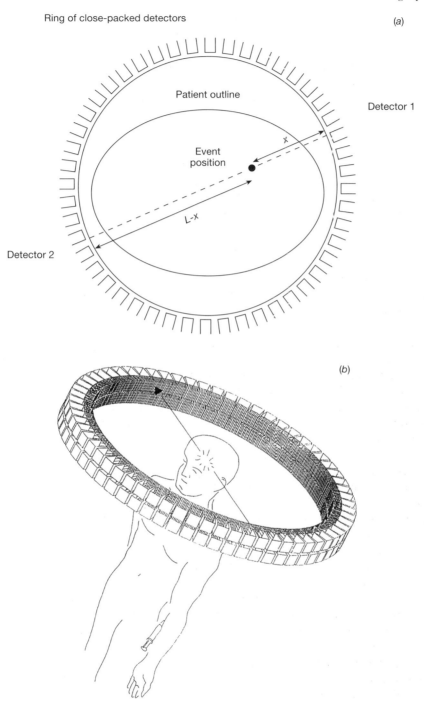

Figure 10.21. *(a) Demonstration that the total path travelled by two coincident gamma rays is equal to the length of the patient in that projection and that a fixed attenuation correction can be applied. If one photon travels a distance x, the other one travels $L - x$. If $I_1 = I_0 e^{-\mu x}$, then $I_2 = I_0 e^{-\mu(L-x)}$ and $I_1 I_2 = I_0^2 e^{-\mu L}$ which is independent of x. (b) Illustration of the arrangement of detectors in a PET scanner (courtesy Dr J Clark).*

or deuterons. By reactions such as $^{14}N(d, n)^{15}O$ when deuterons bombard nitrogen gas, 3–4 GBq per minute of oxygen-15-labelled products can be produced. Alternatively 60 min bombardment of oxygen-18-enriched water with protons can produce 30 GBq of fluorine-18 by the $^{18}O(p, n)^{18}F$ reaction. Besides being very compact, modern cyclotrons cause very little unwanted radioactivity and the stray radiation during operation can be adequately shielded to acceptable levels for radiation protection. Thus these cyclotrons are suitable for installation in a hospital environment.

Largely as a result of relatively rapid technical developments, the technique of positron emission tomography is now beginning to extend beyond rather specialized applications in research centres.

10.6.   CONCLUSIONS

A radiographic image is a two-dimensional display of a three-dimensional structure and in a conventional image the required detail is always partially obscured by the superimposition of information from underlying and overlying planes. The overall result is a marked loss of contrast.

Tomographic imaging provides a method for eliminating, either partially or totally, contributions from adjacent planes. Longitudinal tomography essentially relies on the blurring of structures in planes above and below the region of interest. It is a well established technique and the main consideration is choice of the thickness of plane of cut. If the focus–plane of cut distance and focus–film distance are fixed, the thickness is determined by the angle of swing, decreasing with increasing angle. Large angle tomography may be used when there are large differences in atomic number and/or density between the structures of interest, but if differences in attenuation are small, somewhat larger values of slice thickness may be desirable.

In CT imaging a large number of views are taken of a transverse slice of the patient from different angles. Mathematical methods, using a computer to handle the large amount of data, are then used to map the linear attenuation coefficients $\mu(x, y)$ for an array of small elements in the slice of interest. Values of $\mu(x, y)$ are expressed as CT numbers, and a grey scale which can be adjusted to cover different ranges of CT numbers is used to display the data for visual interpretation.

Several generations of CT scanner have been introduced, each designed primarily to give a faster scan time, which is now of the order of 1–2 s per slice. Contrast, resolution and dose are inter-related. At high contrast, resolution of the order of 1 mm or better can be achieved for a thin slice. At low contrast typical approximate figures are that a 0.4% contrast might permit a resolution of about 4 mm for a skin dose at the entrance surface of the patient of about 50 mGy.

The frequency of CT examinations has increased rapidly in recent years, no doubt reflecting the value of the clinical data obtained. However, CT is a relatively high dose technique so care should be exercised in its use. Several simple precautions, such as taking the minimum number of slices and using the lowest mA s that will give diagnostically useful images, should be routinely adopted. A careful quality assurance programme will also help to ensure that image quality is optimized.

New applications for CT scanning continue to be introduced. Spiral CT has generated valuable three-dimensional data sets and the latest generation of X-ray tubes and generators are opening up new possibilities in dynamic CT.

Resolution with SPECT is markedly inferior to that with CT for a number of reasons. For example radionuclides produce a low photon flux for a long time whereas X-ray generators produce a high photon flux for a short time. Furthermore, emission from radionuclides is isotropic. Many of the photons are wasted and use of collimators means that only about 1 in 2000 of the photons are detected. An X-ray beam is highly collimated and hence for equal patient dose very many more photons can be collected in the CT image.

The time of examination is much longer for SPECT, so movement of the patient is a greater problem and there is even time for movement of activity within the patient. Scattering of the uncollimated beam of gamma photons from the radionuclide means that the appropriate attenuation coefficient is difficult to determine. Indeed, the end point for CT, calculation of a matrix of μ values, is the starting point for calculating attenuation corrections that are a necessary part of high precision SPECT. Hence complex iterative reconstruction techniques are required.

For all these reasons the limiting resolution of SPECT is only about 10 mm for the body, 7 mm for the head. Nevertheless it has established a key role in selected nuclear medicine examinations.

There are a number of theoretical points in favour of positron emission tomography. For example the biologically important radionuclides carbon-11, nitrogen-13 and oxygen-15 are all positron emitters and the fact that two gamma rays are emitted simultaneously makes coincidence detection possible. However the requirement for on-site cyclotron production of most physiologically useful radionuclides and the high cost of the imaging equipment ensure that, at present, this technique is still mainly restricted to major research centres.

REFERENCES

Curry T S, Dowdey J E and Murry R C 1990 *Christensen's Physics of Diagnostic Radiology* 4th edn (Philadelphia, PA: Lea and Febiger)

Jones D G and Shrimpton P C 1993 *Normalised Organ Doses for X-ray Computed Tomography Calculated using Monte Carlo Techniques* NRPB SR250 (London: HMSO)

Miller D 1996 Principles of spiral CT—practical and theoretical considerations *Rad. Mag.* Feb. 28–9

NRPB 1992 Protection of the patient in X-ray computed tomography. *Doc. NRPB* vol 3, no 4

Oppenheim A V and Wilsky A S 1983 *Signals and Systems* (Englewood Cliffs, NJ: Prentice-Hall) ch 8

Shrimpton P C, Jones D G, Hillier M C, Wall B F, Letteron J C and Faulkner K 1991 *Survey of CT Practice in the UK Part 2 Dosimetric Aspects* NRPB R249 (London: HMSO)

Sorenson J A and Phelps M E 1987 *Physics in Nuclear Medicine* 2nd edn (Orlando, FL: Grune and Stratton) p 401

FURTHER READING

Boyd D P and Parker D L 1983 Basic principles of computed tomography *Computed Tomography of the Body* ed A A Moss, G Gamsu and H Genant (Philadelphia, PA: Saunders) pp 1–22

Claussen C and Lochner B 1985 *Dynamic Computed Tomography (Basic Principles and Clinical Applications)* (Berlin: Springer)

Davison M 1982 X-ray computed tomography *Scientific Basis of Medical Imaging* ed P M T Wells (Edinburgh: Churchill Livingstone) pp 54–92

Gemmell H G and Staff R T 1998 Single photon emission computed tomography (SPECT) *Practical Nuclear Medicine* ed P F Sharp, H G Gemmell and F W Smith 2nd edn (Oxford: Oxford University Press) pp 13–24

Gordon R, Herman G T and Johnson S A 1975 Image reconstruction from projections *Sci. Am.* **233** 56–68

Pullan B R 1979 The scientific basis of computerized tomography *Recent Advances in Radiology and Medical Imaging* vol 6, ed T Lodge and R E Steiner (Edinburgh: Churchill Livingstone)

Swindell W and Webb S 1988 X-ray transmission computed tomography *The Physics of Medical Imaging* ed S Webb (Bristol: Hilger) pp 98–127

EXERCISES

1   Explain briefly, with the aid of a diagram, why an X-ray tomographic cut is in focus.
2   List the factors that determine the thickness of cut of a longitudinal X-ray tomograph and explain how the thickness will change as each factor is varied.
3   Describe and explain the appearance of objects that are in the plane of cut of a longitudinal tomogram but are not parallel to it.
4   Explain the meaning of the terms pixel and CT number and discuss the factors that will cause a variation in CT numbers between pixels when a uniform water phantom is imaged.
5   Explain why the use of a fan beam geometry in CT without collimators in front of the detector would produce an underestimate of the μ values for each pixel.
6   Discuss the factors that would make a radiation detector ideal for CT imaging and indicate briefly the extent to which actual detectors match this ideal.
7   Explain why the technique of tomography can eliminate shadows cast by overlying structures. Suggest reasons why the dose to parts of the patient might be appreciably higher than in many other radiographic examinations.
8   Describe the production of a tomogram using an X-ray CT scanner and explain how the production of a tomogram using a radionuclide and a gamma camera differs.
9   Figures 8.3 and 10.17 show the relationship between contrast and resolution for an image intensifier and CT scanner respectively. Explain the differences.
10 Describe and explain how the CT number for a tissue might be expected to change with kVp.
11 Discuss the relative merits of CT, SPECT and PET.

CHAPTER 11

RADIOBIOLOGY AND RADIATION RISKS

11.1. INTRODUCTION

This chapter deals with a problem that is central to the theme of the book - namely that ionizing radiation, even at very low doses, is potentially capable of causing serious and lasting biological damage. If this were not so, steps that are taken to reduce patient doses, for example the use of rare earth screens, would be unnecessary and generally undesirable. Furthermore, the amount of physics a radiologist would need to know would be greatly reduced and this book might not be necessary!

Medical exposures are not the source of the highest radiation dose to the population. Some 200 million gamma rays pass through the average individual each hour from soil and building materials and about 15 million potassium-40 atoms disintegrate within us each hour emitting high energy beta particles and some gamma rays.

However, as shown in table 11.1, medical exposures contribute a far greater proportion of the average annual dose to the UK population than any other form of man-made radiation and the majority of this can be attributed to diagnostic radiology. Since this radiation can cause deleterious effects, it is essential for the radiologist to know what these effects are and to be aware of the risks when a radiological examination is undertaken.

Table 11.1. *Annual contribution to the dose to the UK population from different sources of radiation (NRPB 1994).*

Source	Percentage (%)
Natural	
Cosmic rays	10
Gamma rays from ground and buildings	14
Internal from food and drink	11.5
Radon and thoron	50
Man made	
Medical	14
Other (nuclear discharges, occupational, fall-out)	0.5

11.2.   RADIATION SENSITIVITY OF BIOLOGICAL MATERIALS

11.2.1.   Evidence for high radiosensitivity

Evidence on the lethal dose to humans is rather limited. Some information comes from the effects of the Japanese bombs, some from medical exposures and some from criticality accidents such as Los Alamos in 1958 and Chernobyl in 1986. However, there are often confounding factors. For example many of the firemen at Chernobyl suffered severe but superficial beta 'burns' (see section 11.2.4) so the precise cause of death is difficult to determine.

Extrapolation from experiments with animals, combined with the limited data available on humans, suggests that the dose of acute whole body radiation required to kill about 50% of a human population within the first 30 days (LD 50/30), without medical intervention, would be about 4 Gy to the bone marrow. This would correspond to a much higher skin dose of 80 kV X-rays owing to self-absorption in the body.

Several aspects of this statement need to be discussed, but first consider its meaning in terms of absorbed energy. When X-rays are absorbed in living material this energy is deposited unevenly in discrete packets or at ionization events (see chapter 1). The critical difference between ionizing and non-ionizing radiation is the size of the individual packets of energy, not the total energy involved. Diagnostic X-rays with photons in the energy range 30–100 keV create large packets of energy each of which is big enough to break a chemical bond and initiate the chain of events that culminates in biological damage.

Because of this, in terms of absorbed energy ionizing radiation is by far the most potent agent known to man. A lethal dose of 4 Gy corresponds to absorbed energy of 280 J in a 70 kg man and is about the same as the amount of energy in a small quantity of warm tea. The rise in body temperature may be calculated as energy absorbed/mass $\times$ specific heat. Using the specific heat capacity of water (4.2×10^3 J kg^{-1} K^{-1}), the temperature rise is $280/(70 \times 4.2 \times 10^3) = 10^{-3}$ K. This would be virtually undetectable.

It is against this extreme sensitivity of cells and tissues to ionizing radiation that the use of X-rays in diagnosis must be assessed.

11.2.2.   Cells particularly at risk; 'law' of Bergonié and Tribondeau

Cell populations may be sub-divided into several categories. A few of the more important ones are

(a)   closed static—composed entirely of fully differentiated cells with no mitotic activity,
(b)   transit—a steady state is maintained by balanced in-flow and out-flow (some cell division may occur during transit),
(c)   stem cells—a self-maintaining system with significant mitotic activity producing cells for another population.

There is a long established 'law' in radiobiology, first proposed by Bergonié and Tribondeau (1906), which, in modern parlance, states that 'the more rapidly a cell is dividing, the greater its sensitivity'. From this it follows that the lower the degree of functional and morphological differentiation, the higher the radiosensitivity.

The law applies well to rapidly dividing cells such as spermatogonia, haematopoetic stem cells, intestinal crypt cells and lymphoma cells, which are all very radiosensitive. Differentiated cells are generally relatively radioresistant but three important exceptions should be noted. Small lymphocytes, primary oocytes, especially just before release, and neuroblasts are all radiosensitive.

At the lowest dose at which radiation-induced death is likely to occur, the primary effect will be severe depletion of the bone marrow stem cells. Hence dose to bone marrow is the most meaningful LD 50/30 to quote.

11.2.3. *Time course of radiation-induced death*

After a potentially lethal dose of about 4 Gy to the whole body a typical time sequence might be the following.

(a) *0–48 h.* Loss of appetite, nausea and fatigue.
(b) *48 h to 2–3 weeks.* Latent period of apparent well-being.
(c) *2–3 weeks to 4–5 weeks.* The manifest illness stage, which may include fever, loss of hair, extreme susceptibility to infection, haemorrhage and other symptoms.
(d) *5 weeks onwards.* The situation will have resolved itself one way or the other.

This timescale reflects the changes taking place at the molecular and cellular level.

After exposure to ionizing radiation, physical processes of absorption of photons of energy hf, ionization and excitation will be complete within about 10^{-15} s. With any form of ionizing radiation there is a possibility that this interaction will be directly with critical targets in the cell. Experimental irradiation with microbeams has shown that the cell nucleus, containing DNA and the chromosomal apparatus, is the most sensitive to radiation injury so this is where the targets are likely to be located.

For diagnostic X-rays however, it is more likely that the action will be indirect, the X-rays first interacting with other atoms or molecules in the cell to produce free radicals that are able to diffuse far enough to reach and damage the critical targets.

Since 80% of a cell is composed of water, indirect effects of radiation are most likely to involve water molecules. As a result of interaction with a photon of X-rays, the water molecule may become ionized

$$H_2O \rightarrow H_2O^+ + e^-.$$

H_2O^+ is an ion radical with an extremely short lifetime, about 10^{-10} s. It decays to a free radical which is uncharged but still has an unpaired electron and hence is highly reactive. Further reactions may now occur; e.g. if this free radical reacts with another water molecule, the highly reactive hydroxyl radical OH^- may be formed

$$H_2O^+ + H_2O \rightarrow H_3O + OH^\bullet.$$

Reactive radicals can diffuse over distances of a few nanometres, thus reaching and damaging DNA. Alternatively, they may react together to form toxic products such as hydrogen peroxide

$$OH^\bullet + OH^\bullet \rightarrow H_2O_2.$$

Hence the body is flooded with toxins and this is why the general feeling of malaise results.

The steps leading from these initial physico-chemical changes to the observation of cell death (see section 11.3) are still poorly understood. Nevertheless, according to the law of Bergonié and Tribondeau, the differentiated cells will have resisted the radiation well and will continue to fulfil their specialized functions, so a period of relative well-being should ensue. However, as these cells die, their replacement from the stem cell pool will have been severely depleted or will have stopped completely. The patient becomes manifestly ill when there is a marked loss of a wide range of differentiated, mature cells particularly in the circulating blood. Ultimately the cause of death will usually be failure to control infection or failure to prevent haemorrhage.

The time scale of events for sub-lethal or just lethal radiation damage, together with details of the techniques available to study the various stages, is shown in figure 11.1.

11.2.4. *Other mechanisms of radiation-induced death*

Above 4 Gy, damage to the gastrointestinal stem cells becomes increasingly important and this effect dominates in the dose range 10–100 Gy with loss of body water and body salts into the gut being the major cause of death. The time scale is now much shorter and death will occur in about 3–4 days.

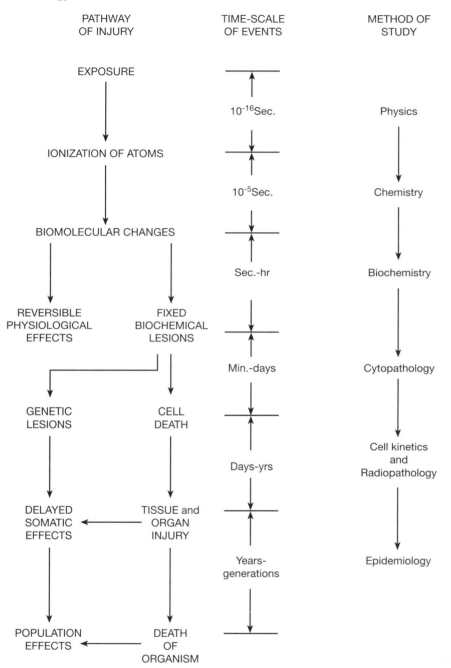

Figure 11.1. *Timescale of events for radiation-induced damage.*

At doses of about 10 Gy to the skin, radiation 'burns' will occur as a result of damage to the single basal layer of rapidly dividing stem cells in the epidermis. Note that these are not 'burns' in the thermal sense but an acute radiation reaction.

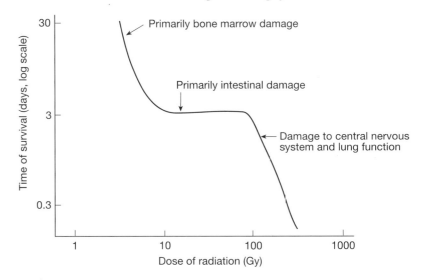

Figure 11.2. *Approximate representation of survival time plotted as a function of dose for acute exposure to whole body radiation.*

At even higher doses, death may be caused by disturbance of the central nervous system (100–150 Gy), loss of lung function (150 Gy) and ultimately simultaneous disruption of the total body chemistry (above 200 Gy).

Survival time plotted as a function of dose might be as shown in figure 11.2, but note that doses shown are only orders of magnitude and in any given situation death will probably be due to a combination of causes.

11.3.   EVIDENCE ON RADIOBIOLOGICAL DAMAGE FROM CELL SURVIVAL CURVE WORK

About 40 years ago, the first reports appeared of successful attempts to use a clonogenic assay of cell survival following irradiation. Basically the technique is as follows. A small number of single cells is placed in a Petri dish with growth medium and incubated for 10–14 days. The cells settle on the base of the dish and, if they are capable of cell division (reproductive integrity), they develop into submacroscopic colonies which may be counted. Not all cells are capable of growing into colonies, even if unirradiated. Suppose that 100 cells are seeded and 90 colonies grow. If now a second sample of the same cells is irradiated before incubation, from 1000 cells only about 180 colonies might develop. The expected colonies from 1000 cells would be 900 and therefore 180/900 or 20% of the irradiated cells have survived. By repeating this experiment at different doses, a survival curve may be obtained and for mammalian cells exposed to X-rays it would resemble figure 11.3.

For further details of the extensive literature on survival curves, including the evidence that qualitatively similar curves are obtained *in vivo*, the reader is referred to more specialized texts (e.g. Coggle 1983, Hall 1988). However, two aspects that are of great relevance to this chapter will be discussed.

11.3.1.   *Cellular recovery and dose rate effects*

After a dose of 1–2 Gy, some cells are killed and others damaged. It may be demonstrated convincingly that some damaged cells will recover by performing a 'split dose' experiment. In figure 11.4 the dotted curve is reproduced from figure 11.3. The solid curve would be obtained if a dose of say 4 Gy were given but a time interval of 10 h elapsed before any further irradiation. The 'shoulder' to the curve has reappeared and

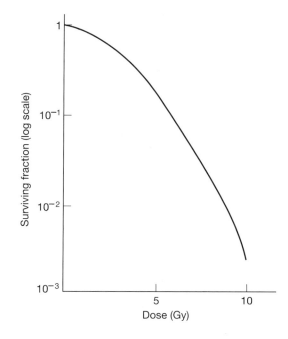

Figure 11.3. *A typical clonogenic survival curve for mammalian cells irradiated in vitro with X-rays.*

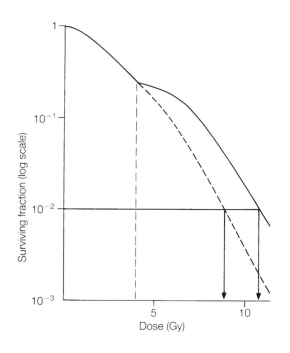

Figure 11.4. *Typical results for a 'split dose' experiment to demonstrate radiation recovery. If an initial dose of 4 Gy is delivered but there is a time delay of a few hours before any further irradiation, the survival curve follows the solid line rather than the dotted line. Note that the shoulder to the curve reappears and the total dose to produce a given surviving fraction is now higher.*

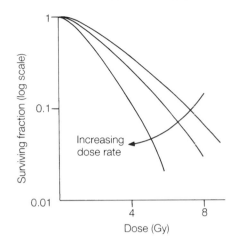

Figure 11.5. *Graphs demonstrating that because of recovery the killing effect of X-rays may be dose rate dependent.*

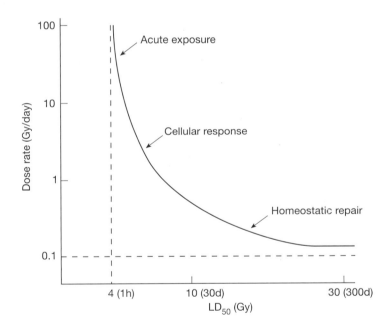

Figure 11.6. *Curve to demonstrate that the LD_{50} in vivo is very dependent on dose rate because of cellular and homeostatic recovery mechanisms. The LD_{50} was measured under conditions of continuous irradiation. The duration of irradiation is shown in brackets.*

the total dose to achieve a given surviving fraction is higher for the split dose than for the single dose. The recovery effect is detectable by 2 h and reaches a maximum by 24 h.

One consequence of recovery is that when radiation exposure is protracted, the effect may be dose rate dependent (see figure 11.5). A simple explanation is that recovery is occurring during radiation exposure. Conversely, dose rate effects are evidence of recovery. They indicate that cell killing is occurring as a result of a sequence of events following interaction of more than one X-ray photon with the cell.

Cellular recovery is observed *in vivo* but now another recovery mechanism is also observed. This has a much longer timescale and is primarily a result of homeostatic mechanisms involving stem cells in the whole animal. One consequence is shown in figure 11.6. The LD 50/30 for acute exposure is about 4 Gy (4 Gy at 100 Gy per day corresponds to an exposure time of about 1 h), but the LD 50/30 increases when the radiation is protracted, first because of cellular repair and subsequently because of homeostatic repair. The results suggest that animals could tolerate 0.1 Gy per day for a long time.

11.3.2.   *Radiobiological effectiveness (RBE)*

If the survival curve experiment is repeated with neutrons, the result shown in figure 11.7 will be obtained. A smaller dose of radiation is required to produce a given killing effect and the curve has a smaller shoulder, indicating less capacity to repair sub-lethal damage.

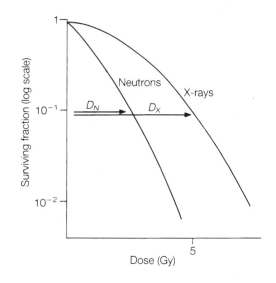

Figure 11.7. *Comparative survival curves for X-rays and neutrons. If 220 kVp X-rays have been used* RBE = D_x/D_N *for a given biological end point (10% survival in this example).*

If 220 kVp X-rays have been used, then as shown on the curve, the RBE is defined as

$$\text{RBE} = \frac{\text{dose of 220 kVp X-rays}}{\text{dose of radiation under test to cause the same biological end point}}.$$

The RBE of neutrons when determined in this way is frequently between 2 and 3.

Except at very high LET values (see section 1.14), the RBE of a radiation increases steadily with LET. However, it is an incomplete answer simply to state that 'neutrons cause more damage than X-rays because they are a higher LET radiation and therefore produce a higher density of ionization'. For equal absorbed doses measured in grays, the number of ion pairs created by each type of radiation in a macroscopic volume is the same. Therefore, for reasons not yet fully understood, and beyond the scope of this book, it is differences in the spatial distribution of ion pairs in the nucleus at the submicroscopic level that cause the difference in biological effect. This is illustrated in figure 11.8. For further discussion see, for example, Alper (1979).

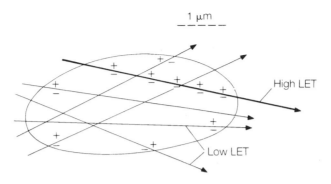

Figure 11.8. *Illustration of the difference in spatial distribution of ion pairs across a cell nucleus for low LET and high LET radiations. In each case five ion pairs are formed (same dose) but whereas these are likely to result from the same high LET particle, and thus be quite close together, they are more likely to result from five different low LET photons and hence be much more widely separated.*

11.4. RADIATION WEIGHTING FACTORS, EQUIVALENT DOSE AND THE SIEVERT

RBE-type experiments demonstrate clearly that, when attempting to predict the possible harmful effects of radiation, the purely physical concept of absorbed dose, as measured by the gray, is inadequate. However, in numerical terms RBE is a very difficult concept since it varies with dose, dose rate and fractionation, physiochemical conditions such as the presence or absence of oxygen, the biological end point chosen, the biological species and the time after irradiation at which the measurement is made. Furthermore, since the shapes of the survival curves are different, inspection of figure 11.7 shows that, at the very low doses that are important in radiological protection, the RBE may be somewhat higher than the value of 2 or 3 quoted for higher doses.

Largely for these reasons, the International Commission on Radiation Units introduced a new term, 'quality factor' (see e.g. ICRU 1980). This is a dimensionless, invariant quantity for a given type of radiation, determined solely by the type of radiation. However, this quality factor was applied to absorbed dose at a point and in radiological protection it is the absorbed dose averaged over a tissue or organ and weighted for the radiation quality that is of interest.

Thus in 1991 the International Commission on Radiological Protection (ICRP 1991) introduced a new concept, the radiation weighting factor w_R. This should now be used to calculate the **equivalent dose** to a tissue T, H_T, according to the equation

$$H_T = w_R D_T$$

where D_T is the dose averaged over tissue T from the radiation type R.

This is now considered to be an equivalent dose because for equal values of H_T the damage to tissue T will be the same for different types of radiation.

w_R is dimensionless so H_T still has the units J kg^{-1} but is now given the special name sievert (Sv).

Values of w_R are representative of the range of RBE values for that radiation in inducing stochastic effects (see next section) at low doses. A simplified version of the ICRP recommendations is given in table 11.2.

Note the following.

(i) An important exception to this table may be the radiation weighting factor for Auger electrons, especially if they are emitted from a radionuclide that is closely bound to DNA. There is evidence that they may be as harmful on an equi-absorbed dose basis as α-particles.

Table 11.2. *Radiation weighting factors*

Radiation type and energy range	Radiation weighting factor w_R
Photons (all energies)	1
Electrons	1
Protons	5
Neutrons (10 keV–100 keV)	10
(100 keV–2 MeV)	20
Alpha particles, fission fragments	20

(ii)   The concepts of radiation weighting factor and equivalent dose should only be applied in the context of radiological protection.

11.5.   RADIATION EFFECTS ON HUMANS

11.5.1.   *Deterministic and stochastic effects*

The biological effects of ionizing radiation on humans can be divided into two general categories. For some effects there appears to be a definite threshold dose below which no damage, in terms of measurable biological response, can be detected. Effects of ionizing radiation such as skin erythema, epilation and opacification of the lens of the eye and death resulting from acute exposure are in this category.

Such effects are described as deterministic (or sometimes non-stochastic) and are characterized by a dose–response curve of the type shown in figure 11.9(*a*). Deterministic effects can be thought of as acting at the tissue level, especially where the cells are mitotically active, and have the following features:

(a)   150 mGy or more are required to observe an effect;
(b)   the dose threshold varies from one tissue to another;
(c)   it is a somatic effect affecting the exposed individual;
(d)   repair and recovery can occur;
(e)   the severity of effect depends on dose/dose rate/number of exposures;
(f)   the effect usually occurs early, i.e. in days or weeks, and may be repaired quickly afterwards;
(g)   mechanisms are relatively well understood, e.g. in radiotherapy.

Threshold doses for the most sensitive deterministic effects are summarized in table 11.3.

These values are all well above the doses received by patients in conventional radiology. However, there have been isolated reports of radiation-induced skin injuries to patients resulting from prolonged fluoroscopically guided invasive procedures.

Absorbed dose rates to the skin from the direct beam of a fluorescence X-ray system are typically in the range 0.01–0.05 Gy min^{-1} but may reach 0.2 Gy min^{-1} or even higher. A few minutes screening at this higher dose rate could cause early transient erythema and any additional fluorographic dose required for film or digital image recording must not be overlooked. For a review of possible hazards from interventional procedures see Wagner *et al* (1994).

Stochastic effects, literally those governed by the laws of chance, are thought to occur primarily at the cellular level. Since a single ionizing event may be capable of causing radiation injury, e.g. to the DNA, and even the lowest dose diagnostic examinations cause millions of ionizations in each gram of irradiated tissue, it is normal to assume that there is no threshold dose for stochastic effects of ionizing radiation. Thus the curve relating probability of effect to dose has the form shown in figure 11.9(*b*).

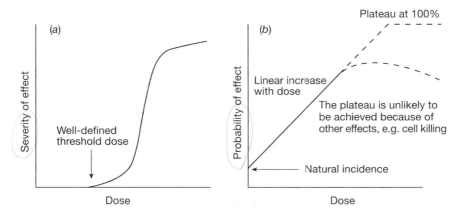

Figure 11.9. *Typical dose–response curves for (a) deterministic effects, (b) stochastic effects.*

Table 11.3. *Typical threshold doses for the most sensitive deterministic effects.*

Tissue and effect	Absorbed dose for a brief exposure (Gy)
Testes	
Temporary sterility	0.15
Permanent sterility	>3.5
Ovaries	
Sterility	>2.5
Lens	
Detectable opacities	>0.5
Visual impairment (cataract)	>2.0
Bone marrow	
Depression of haematopoiesis	>0.5
Skin	
Early transient erythema	2
Temporary epilation	3

The two most important long term effects of radiation, namely carcinogenesis and mutagenesis, are thought to be stochastic in nature. The former is presumably the result of damage to a somatic cell, which either initiates or promotes a malignant change, the latter the result of damage to a germ cell. A further important feature of stochastic effects is that whereas their frequency will increase with increasing dose, their severity does not. Thus for example the degree of malignancy of a radiation-induced cancer is not related to the dose. A further reason for believing that radiation-induced carcinogenesis is a stochastic effect is that there is no evidence for a threshold dose for radiation-induced cancer in the Japanese survivors. Finally a truly stochastic mechanism would exclude the possibility of recovery. There was little or no evidence for recovery effects in the multiple fluoroscopy work (see section 11.5.2) where patients received many small exposures over a period of time.

Carcinogenesis and mutagenesis may be contrasted in that the former is somatic, i.e. the effect is observed in the exposed individual, whereas the latter is hereditary, the effect being detected in the descendants.

11.5.2. Carcinogenesis

There is ample evidence from a wide range of sources that ionizing radiation can cause malignant disease in humans. For example, occupational exposure results in a greatly increased incidence of lung cancer among uranium miners, and in the period 1929–1949 American radiologists exhibited nine times as many leukaemias as other medical specialists. A frequently quoted example of industrial radiation-induced carcinogenesis is the 'radium dial painters'. They were mainly young women employed during and after the First World War to paint dials on clocks and watches with luminous paint. It was their custom to draw the brush into a fine point by licking or 'tipping' it. In so doing, the workers ingested appreciable quantities of radium-226 which passed via the blood stream to the skeleton. Years later a number of tumours, especially relatively rare osteogenic sarcomas, were reported.

A limited amount of evidence comes from approved medical procedures. For example, between 1939 and 1954, radiotherapy treatment was given to the whole of the spine for more than 14 000 patients suffering from ankylosing spondylitis. Statistically significant excess deaths due to malignant disease were subsequently observed, especially for leukaemia and carcinoma of the colon. Other data come from the use as a contrast agent in radiology of thorotrast which contains the alpha emitter thorium-232, X-ray pelvimetry, multiple fluoroscopies and radiation treatment for enlargement of the thymus gland. Unfortunately, radiation exposure in medical procedures is associated with a particular clinical condition. Therefore it is difficult to establish suitable controls for the purpose of quantifying the effect.

Finally, there is information gathered from the survivors of the Japanese atomic bombs. This has been fully reported in a series of articles published over many years in *Radiation Research* (Shimizu *et al* 1990).

Five key points, which all have a bearing on radiation protection procedures, can be made from this follow-up:

(a) the risk of cancer is not the same for all parts of the body; many parts are affected, but not all to the same degree;

(b) there is a long latent period before the disease develops—leukaemia peaked at about 5–14 years since when the relative risk (i.e. expressed relative to the natural incidence) has decreased; for solid tumours the relative risk has continued to rise (see figure 11.10);

(c) there is no evidence of a threshold dose;

(d) in terms of relative risk, cancer was highest in those who were under 10 years of age at the time of exposure;

(e) evidence is in reasonably good agreement with that from other sources and permits a risk estimate.

Sources of evidence that ionizing radiation causes cancer in humans are summarized in table 11.4.

Table 11.4. *Sources of evidence that ionizing radiation causes cancer in humans.*

Occupational	*Medical diagnosis*
uranium miners	prenatal irradiation
radium ingestion (dial painters)	thorotrast injections
American radiologists	multiple fluoroscopies (breast)
Atomic bombs	*Medical therapy*
Japanese survivors	cervix radiotherapy
Marshall islanders	breast radiotherapy
	radium treatment
	ankylosing spondylitis and others

11.5.3. Mutagenesis

The circumstantial evidence that ionizing radiation causes mutations in humans is overwhelming. For example, mutations have been observed in a wide variety of other species including plants, bacteria, fruit flies and mice, and radiation is known to impair the learning ability of mice and rats.

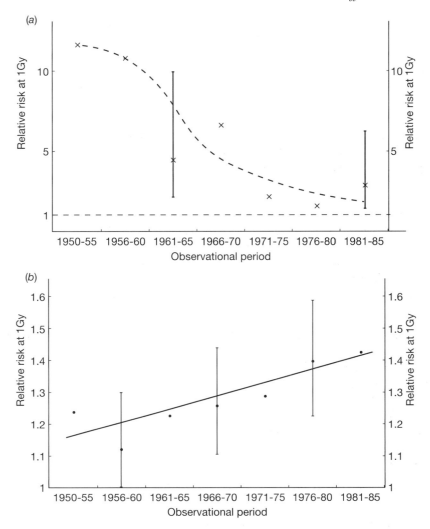

Figure 11.10. *(a) Follow-up of relative risk of radiation-induced cancer (leukaemia) in the Japanese survivors (adapted from Shimizu et al 1990). (b) Follow-up of relative risk of radiation-induced cancer (all cancers except leukaemia) in the Japanese survivors (adapted from Shimizu et al 1990). Some error bars are shown to indicate the accuracy of the data.*

Furthermore, radiation is known to cause extensive and long lasting chromosomal aberrations. These may occur either pre-replication (chromosome aberrations) or post-DNA-replication (chromatid aberrations). One mechanism by which these aberrations can arise from breaks and faulty rejoining is shown in figure 11.11. An important method of retrospective dosimetry is to score such aberrations in cells circulating in the peripheral blood. For a dose of 50 mSv whole body radiation, one or two dicentric chromosomes would be scored for every 1000 mitotic cells examined (Brill 1985). The normal incidence is negligible.

Notwithstanding, researchers have so far failed to demonstrate convincing evidence of hereditary or genetic changes in humans as a result of radiation, even in the offspring of Japanese survivors. For the latter group four measures of genetic effects, ranging from untoward pregnancy outcomes to a mutation resulting in an electrophoretic variant in blood proteins, have been studied. The difference between proximally and

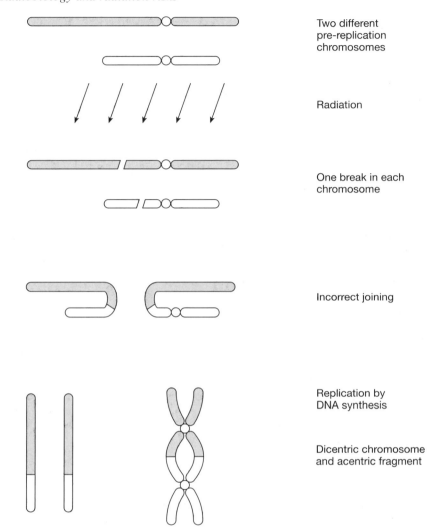

Two different
pre-replication
chromosomes

Radiation

One break in each
chromosome

Incorrect joining

Replication by
DNA synthesis

Dicentric chromosome
and acentric fragment

Figure 11.11. *Steps in the formation of a dicentric chromosome and acentric fragment as a result of radiation breaks and faulty rejoining.*

distally exposed survivors is in the direction expected if a genetic effect had resulted from the radiation, but none of the findings is statistically significant.

This failure to record a significant effect is presumably caused by the statistical difficulty of showing a significant increase in the presence of a high and variable natural incidence of both physical and mental genetically related anomalies. For severe disability the natural incidence is between 4 and 6%.

There are additional problems in assessing genetic risk arising from the diagnostic use of radiation. For example, only radiation exposure to the gonads is important and even this component can be discounted after child-bearing age. Second, many radiation-induced mutations will be recessive, so their chance of 'appearing' may depend on the overall level of radiation exposure in the population. Finally, the risk to subsequent generations will depend on the stability of the mutation once formed.

Based on the measured mutation rate per locus in the mouse, adjusted for the estimated comparable

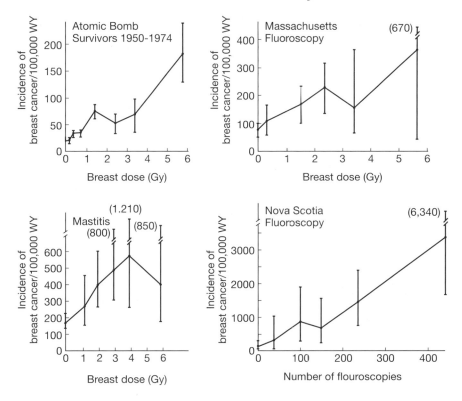

Figure 11.12. *Incidence of cancer of the female breast as a function of dose in atomic bomb survivors, women treated with X-rays for acute post-partum mastitis and women subjected to multiple fluoroscopic examinations of the chest during treatment of pulmonary tuberculosis with artificial pneumothorax (redrawn from Boice* et al *1979).*

number of loci in the human, the effective dose required to double the mutation rate in humans is estimated to be at least 1 Sv. Hence 10 mSv per generation parental exposure might increase the spontaneous mutation incidence by about 1%. An alternative way to express the mutation risk will be discussed in the next section.

11.6.   RISK FACTORS AND COLLECTIVE DOSES

Various international bodies have examined the data on cancer induction and genetic damage caused by radiation and used it to make an assessment of the risks of exposure to low doses of radiation. For example, as shown in figure 11.12 the incidence of breast cancer in atomic bomb survivors has been compared with three other sources of data (Boice *et al* 1979, Upton 1987). All give reasonably good agreement that the probability of breast cancer in females is about 1% Sv^{-1}.[1]

Not surprisingly, in view of the large statistical errors on available data and the extrapolation required from high doses, these estimates often vary considerably. It is important to keep this in mind in what follows. Some of the figures will be very approximate.

---

[1] This notation will be used throughout to express risk factors. It means, literally, that if a large population of females were each exposed to 1 Sv, on average 1 in 100 would contract a radiation-induced breast cancer. Note however the remarks on collective doses later in this section, which provide a more realistic interpretation.

Increasingly the figures quoted by the International Commission on Radiological Protection (ICRP) are being used for the purposes of risk estimation and they will be quoted here. ICRP has relatively recently (1991) made a reassessment of risk factors in the light of information that has become available since its last review (ICRP 1977). It is important to keep this in mind because current legislation (see chapter 12) is based on 1977 figures.

Most of the recent information has led to a higher estimate of risk. For example, dosimetry at Hiroshima and Nagasaki has been recalculated. This has shown that the previous estimates of neutron doses were too high, especially at Hiroshima. Because neutrons have a high radiation weighting factor, they make a relatively high contribution to the equivalent dose. Reducing the calculated neutron contribution increases the cancer risk per sievert.

A second point is that prolonged follow-up has continued to show an excess of solid tumours in Japanese survivors exposed to radiation. Follow-up to 1985 has given better information on the increased risk to those who were under 10 years of age at the time of irradiation. Different risk factors are now quoted for the work force (18–65 years) and the whole population.

Third, the model for cancer induction has been reviewed. In the past it has been assumed to be additive. Now there is evidence that, for some cancers, e.g. breast, a relative risk model is more appropriate. This assumes there will be a proportional increase in cancer at all ages and, as the population gets older and the natural incidence of cancer increases, the extra cancers will also increase on this model.

Finally, it is now considered appropriate to make some allowance for the detrimental effect on quality of life of non-fatal cancers.

For all these reasons the assessed overall risk has been increased substantially as shown in table 11.5.

Table 11.5. *A summary of the main risk factors comparing the 1991 and 1977 ICRP recommendations. Note, in extrapolating from high doses ICRP considers that risks may be reduced by a factor of about 2 owing to dose and dose rate effects. This correction has been made to the figures shown below which apply to low doses of radiation (see model A in figure 11.13).*

	1991 % Sv^{-1}		1977 Sv^{-1}
	Work force	Whole population	
fatal cancer	4.0	5.0	1.25
non-fatal cancer	1.0	0.8	
hereditary effects	0.8	1.3	0.8
total (rounded)	5.6	7.3	2.0

There is a body of opinion that considers the potentially harmful effects of low doses of radiation are overstated. For example the point is made that there is no direct evidence that the very low doses used in, say, dental radiology have caused any cancers at all. This is of course true. Observed excess cases have all been with much higher doses, e.g. the dose to breast tissue in fluoroscopy was estimated at 0.04–0.2 Gy and the number of examinations on one individual frequently exceeded 100.

Thus any extrapolation from observed data to lower doses is bound to be a model. Three possible models, which have all been widely discussed in the literature, are illustrated in figure 11.13. However we have no direct means of checking which of the three curves, or indeed any other shape of curve, is correct. For further details see the figure legend.

Resolution of this problem would require a much better understanding of the mechanism of carcinogenesis. However, if cancer arises as the final result of a series of successive changes, it is at least plausible that radiation-induced damage could be one key step in the chain of events. Since even a few micrograys of radiation will cause in excess of 10^8 ion pairs per gram of tissue and a single ion pair is capable of breaking a

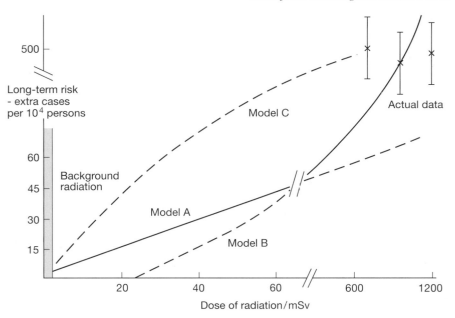

Figure 11.13. *Three curves showing the possible variation in long term risk with radiation dose. The solid line (model A) represents current thinking—a linear increase at low doses with no threshold. At higher doses (well above protection and diagnostic levels) the curve turns upwards. The dotted curves represent a non-linear increase, possibly with a threshold (model B), and a supra-linear model in which low dose effects would be higher than predicted (model C). Doses from background radiation and typical radiological examinations are shown shaded.*

chromosome and hence causing a translocation, we must also accept that there is, effectively, no lower limit to the dose of radiation that will cause cancer. Clearly the vast majority of changes at the molecular level do not manifest themselves in disease!

This principle is extremely important in the diagnostic use of X-rays. If a linear response with no threshold is assumed then doubling the dose of radiation will double the number of fatal cancers, however low the dose. Also it is reasoned that 10 mSv given to 10 000 persons or 100 μSv given to a million persons are going to cause the same number of fatal cancers. A physical explanation for this reasoning would be that the same number of ion pairs has been created in each case. Therefore in situations where large numbers of persons are exposed to low doses, e.g. in diagnostic radiology, it is the **collective dose** (100 person Sv in this example) that will determine the detriment. Hence with a risk of fatal cancer of 5% Sv^{-1}, a collective dose of 100 person Sv to the whole population will cause five extra fatal cancers.

For further discussion on the biological basis of risk assessment following low dose irradiation see Baverstock and Slatter (1989).

11.7.   RISKS FROM RADIOLOGICAL EXAMINATIONS

11.7.1.   *Tissue weighting factors and effective dose*

One of the main conclusions from the Japanese work was that many tissues are affected but not all to the same degree. Therefore, in situations where only part of the body is irradiated, some mechanism is required that will allow the risks from different types of exposure to be compared.

This is done by assigning to each tissue a **tissue weighting factor**, w_T. Current recommended values for w_T are given in table 11.6.

Table 11.6. *Tissue weighting factors (ICRP 1991).*

Tissue or organ	w_T
gonads	0.2
bone marrow (red)	0.12
colon	0.12
lung	0.12
stomach	0.12
bladder	0.05
breast (averaged for population)	0.05
liver	0.05
oesophagus	0.05
thyroid	0.05
skin	0.01
bone surface	0.01
remainder	0.05

remainder = adrenals, brain, upper large intestine, small intestine, kidney, muscle, pancreas, spleen, thymus, uterus.

The effective dose, i.e. that whole body dose carrying the same risk, is then the weighted equivalent dose for all the organs and tissues in the body. Hence

$$\text{effective dose} = \sum H_T w_T \text{ Sv}$$

where H_T is the equivalent dose in tissue T.

11.7.2.   *Effective doses and risks in radiology*

Starting with information on patient entrance doses, see section 6.8, it is now possible to use tissue weighting factors to calculate the effective dose for any specified radiological examination. However, to do this it is first necessary to compute the H_T values for all relevant organs. These may be divided into organs and tissues in the direct beam, where body attenuation will be the main factor to consider, and organs/tissues outside the beam where scatter will also be important. NRPB have tabulated data for a number of examinations and these lead, typically, to the figures shown in table 11.7.

11.7.3.   *Radiation risks in context*

Risks from medical exposures to X-rays are justified in terms of the medical benefit. This is not readily quantifiable, certainly not by members of the public, therefore it is helpful to put radiation risks in context, i.e. in terms that the public can understand.

Both the Royal Society and the Health and Safety Executive in the UK have made studies of the public perception of risk. The Royal Society approach was to gauge public opinion on the risk associated with certain activities and then to work out the actual risk involved. Risks as low as 1 in 10^6 a year are commonly regarded as trivial but an imposed risk at a level of 1 in 10^4 is likely to be challenged. This is comparable with the risk from some of the higher dose examinations in table 11.7. For further discussion see NRPB (1993). Some other risks taken from Sumner *et al* (1989) are shown in table 11.8.

Table 11.7. *Typical effective doses and risks for radiological examinations.*

Examination	Effective dose (mSv)	Risk (7.3% Sv^{-1})
chest X-ray (PA)	0.04	4×10^{-6}
skull (lat)	0.15	1×10^{-5}
abdomen (AP)	1.4	1×10^{-4}
head CT	2.3	1.6×10^{-4}
chest CT	8.0	6×10^{-4}
abdomen CT	8.0	6×10^{-4}
barium meal	3.8	3×10^{-4}
barium enema	7.7	5×10^{-4}

Table 11.8. *Typical risks from well known activities.*

Activity	Risk of death per year
travelling 300 miles by car	10^{-5}
work accidents	2×10^{-5}
home accidents	10^{-4}
smoking 10 cigarettes a day	1.5×10^{-4}
coal mining	1.5×10^{-4}
deep sea fishing	2×10^{-3}

11.8.   SPECIAL HIGH-RISK SITUATIONS—IRRADIATION OF CHILDREN OR *IN UTERO*

Although there are conflicting reports in the literature, the balance of evidence suggests that exposure *in utero* or during early childhood carries an enhanced radiation risk. For irradiation of children in the age range 0–10 years most of the available data come from Hiroshima and Nagasaki and relate primarily to leukaemia. It is likely that the probability of induction is about twice as high as in adults.

For irradiation *in utero* the position is more complex and there are several possible harmful effects to be considered. To understand this, consider figure 11.14 which shows a typical result of delivering a dose of 2 Gy of X-rays to mice at different stages during pregnancy. Initially there will be a small number of rapidly dividing, highly radio-sensitive cells and damage is likely to result in prenatal death. In the intermediate stages of pregnancy there may be sufficient cells overall for the embryo to survive but only small numbers are performing any one specialized function so radiation damage causes abnormalities and neonatal death. In the later stages, when the foetus has fully formed, the effects are likely to be qualitatively similar to those in the adult.

Each of the effects will be considered briefly.

11.8.1.   Lethal effects

At the pre-implantation stage this effect will be indistinguishable from a spontaneous abortion. As the foetus develops it will increasingly become a deterministic effect and work with animals suggests that at the later stages of gestation the threshold dose could approach 1 Gy.

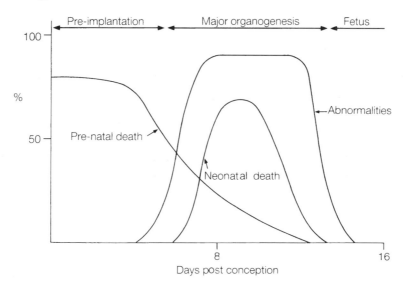

Figure 11.14. *Incidence of pre-natal death, abnormalities and neonatal death for mouse embryos irradiated with 2 Gy at different stages during pregnancy (redrawn from Russell and Russell 1954).*

11.8.2.   Malformations and other developmental effects

In utero irradiation during the period of early organogenesis has been shown to cause a range of malformations, depending on the timing of exposure, in every animal model tested. Confirmatory evidence for malformations in the Japanese survivors is lacking but there is some evidence for a high incidence of miscarriages, still births, neonatal and infant deaths, as well as a decrease in the mean height and head circumference in offspring that had been irradiated *in utero* with fairly high doses.

The threshold for malformation is probably not less than 250 mGy even for acute exposures in the most sensitive first month of pregnancy. In radiological examinations the risk must be negligible compared with the natural, spontaneously arising risk.

11.8.3.   Radiation effects on the developing brain

Most of the information on the effects of pre-natal irradiation on the nervous system comes from follow-up of the Japanese atomic bomb survivors. This is a very complex subject but basically the results demonstrate clearly that the period of maximum organogenesis is also the period of maximum radiation sensitivity.

Otake and Schull (1984) have shown that severe mental retardation is greatest during the period from 8 to 15 weeks when there is a rapid increase in the number of neurones before they migrate to their ultimate development sites in the cerebral cortex and lose their capacity to divide. Severe mental retardation is less for radiation exposures before and after this critical period. More recent work has shown that intelligence test scores mirror the observations on severe mental retardation (Schull 1991).

The best current estimate is that irradiation during the 8–15 weeks of gestation will reduce intelligence at the rate of about 30 IQ points per Gy (the standard deviation about a mean of 100 in the control population is about 15; a person with an IQ less than 70 is considered to be retarded). The data are consistent with no threshold, but the shift in IQ that might be caused by a diagnostic exposure is most unlikely to be detectable.

11.8.4.   Cancer induction

The most detailed data come from the Oxford Survey of Childhood Cancers which was started in 1955 and now includes deaths during the period 1953–1979. This work has not been without controversy but recent analyses, see for example Muirhead and Kneale (1989) and Bithell (1989), seem to be in reasonable agreement that the risk of cancer up to the age of 15 following *in utero* irradiation will be about 6% Sv^{-1} (6×10^{-4} for 10 mSv), half of which will be fatal.

One reason for stating the risk to 15 years is that direct comparison can be made with the natural cumulative risk of cancer in the UK up to that age (1 in 1300 or 7.7×10^{-4}). About half of these cancers will be fatal. The lifetime risk will of course be higher and ICRP 60 suggests 'at most a few times that for the population as a whole. If 'a few' were taken as 3, the lifetime risk of fatal cancer would be 15% Sv^{-1} or 15×10^{-4} for 10 mSv'.

The risk will be lower in the very earliest stages of conception (up to three weeks) but may not be zero.

11.8.5.   Pre-conception risk (leukaemia clusters)

There does appear to be a tendency for the incidence of childhood leukaemias to cluster and ionizing radiation has been a prime suspect for many years.

In 1983 a television documentary on the nuclear waste reprocessing plant at Sellafield, UK included the allegation that the incidence of leukaemia in the surrounding villages was about ten times higher than would be expected for the size of population. The increased incidence was confirmed by an independent advisory committee (Black 1984) and this has led to a number of intensive studies, probably the most widely publicized of which was by Gardner *et al* (1990). They carried out a case-control study which basically assumed that the extra cancers had not arisen by statistical chance, and looked for possible correlation with a wide range of 'suspect' causes using locally matched controls.

Many possible causes of leukaemia were examined including antenatal X-ray examinations, viral infections, habit factors (e.g. eating fish, eating shellfish, playing on the beach, growing own vegetables, using seaweed as a fertilizer), proximity to Sellafield and employment of parents. When relative risks were calculated on the basis of father's occupation, a link with Sellafield began to emerge but it was when comparisons were made on the basis of doses of external gamma radiation that some striking results were found.

Of the five Seascale cases of leukaemia, the fathers of four of them had received significant doses of external gamma radiation prior to conception and none of the area or local controls had received greater doses of radiation. Overall, results showed an approximately eightfold increase in relative risk if fathers had received doses in excess of 100 mSv (total) or 10 mSv in the 6 months prior to conception.

There were many unexplained features of this work and a number of subsequent epidemiological studies failed to substantiate the main findings. In a key test case the judge concluded that the balance of probabilities was decisively against paternal preconception irradiation being the cause of the excess leukaemias. This is important because 10 mSv is less than a factor of two greater than the effective dose associated with a number of standard diagnostic examinations. Further studies into a possible link between parental exposure to radiation and childhood cancer are continuing.

11.8.6.   Hereditary effects

There is no evidence to indicate that the foetal gonads are any more radiation sensitive than the adult gonads. Therefore the risk factor is assumed to be the same at 2.4% Sv^{-1} before and after birth. Note that this figure is higher than that in table 11.5 where the risk has been averaged over the whole population, a significant percentage of whom will be past the reproductive age.

11.8.7. *Summary—current thinking on risks of foetal irradiation (NRPB 1993b)*

(1)   Deterministic effects—gross malformations, uncertain at days 1–5, thereafter a threshold of at least 0.2 Sv.

—mental retardation, period of maximum sensitivity 8–15 weeks with a detriment of 30 IQ points Sv^{-1}

(ii)   Cancer induction—a risk of 6% Sv^{-1} to the age of 15 years, half of which will be lethal.

(iii)   Preconception risk—inappropriate at the present time to withold the use of X-rays clinically on the basis of preconception exposure risk.

(iv)   Hereditary effects—risk set at 2.4% Sv^{-1}, the same level as after birth.

11.9.   WORKED EXAMPLE ON RISK ASSOCIATED WITH AN ABDOMINAL EXAMINATION DURING UNDECLARED PREGNANCY

All practising radiologists are likely to be asked as some stage in their career to advise on a patient who has discovered she is pregnant after X-ray procedures involving the pelvis or lower abdomen have been performed. The worked example given below illustrates how the risk may be estimated. It also serves as an important indicator of *all* the information (italicized) that must be on record in relation to the examination if the assessment is to be made reasonably accurately. The availability or otherwise of these data would be one measure of good radiographic practice.

11.9.1.   *Basic data on good radiographic set-up*

Slim to average build woman estimated *7 weeks pregnant*

   $2\times$ *AP lumbar spine views* at *70* kVp, *80* mAs, *100* cm FFD (*70* cm FSD estimated)
   $1\times$ *lat lumbar spine view* at *80* kVp, *140* mAs, *100* cm FFD (*50* cm FSD estimated) *(L1–4)*
   $1\times$ *lat lumbar spine view* at *90* kVp, *150* mAs, *100* cm FFD (*50* cm FSD estimated) *(L5-S1)*
   These data are entered in columns 1–3 and 5 of table 11.9.
   X-ray tube output data are entered in column 4

kV		70	80	90
air kerma rate (μGy $(\text{mA s})^{-1}$) at 75 cm		*28.7*	*39.1*	*50.5.*

Table 11.9. *Foetal dose assessment from recorded exposure factors.*

View	kVp	mAs	μGy $(\text{mA s})^{-1}$	FSD (cm)	Air kerma (mGy)	BSF from NRPB 186	Skin dose (mGy)	Uterus dose/ unit skin dose	From NRPB 1986 Uterus dose
AP($\times$2)	70	80	28.7	70	2.64	1.36	3.8	0.179	0.684
									$\times 2$
									1.37
Lat	80	140	39.1	50	12.3	1.28	16.7	0.017	0.28
LSJ	90	150	50.5	50	17.0	1.27	22.9	0.014	0.32
									1.97
									mGy

11.9.2.   *Calculation*

(1)   Multiply the output factor (column 4) by the mA s and correct for the inverse square law to find the air kerma at the skin (column 6).

(2) Use NRPB-R 186 (Jones and Wall 1985) to look up the back-scatter factor (BSF) for this particular examination. This corrects for the fact that the output measurements in column 4 have been made in air whereas a significant amount of radiation will be scattered back to the skin from the patient.

(3) The skin dose (column 8) is now equal to air KERMA at the skin $\times$ BSF $\times$ 1.06 (the ratio of mass absorption coefficients for soft tissue and air).

(4) Look up the uterus dose per unit skin dose for the appropriate examination in NRPB-R 186 (column 9).

(5) Calculate the dose to the uterus (column 10).

It is useful to check at this stage that the answer is consistent with information in NRPB 200 (Shrimpton *et al* 1986) for the appropriate examination.

11.9.3. Risk assessment

The natural risk of fatal cancer up to age 15 is 1 in 1300.

The risk of fatal cancer associated with the estimated dose of 1.97 mGy is 1 in 17 000.

The estimated effect on mental retardation (assuming maximum sensitivity at 8–15 weeks) is less than 0.01 IQ point.

11.9.4. Notes

(1) A period of screening would be dealt with similarly. A record of the screening time is essential. The kV and mA may have to be estimated if they are automatically adjusted by the equipment to maintain the correct image brightness. Ideally a dose–area product meter should be fitted.

(2) Software has recently been produced—for use in conjunction with Hart *et al* (1994)—to expedite these calculations but there is some merit in working through them from first principles at least once!

(3) If the necessary information is not available the risks have to be estimated from published data or typical doses to the uterus or foetus. The use of mean foetal effective doses is not recommended. Risks should be worked out from first principles wherever possible since it has been shown that actual doses for standard examinations range over a factor of 10 or more (see section 6.8).

Table 11.10. *Risk of cancer and mental retardation (exposure 8–15 weeks) following typical in utero medical exposure to radiation.*

Examination	Mean foetal effective dose (mSv)	Fatal cancer to age 15 years	Loss of IQ points
X-ray			
Barium enema	16	5×10^{-4}	0.5
IVU	3.2	1×10^{-4}	0.1
Lumbar spine	3.2	1×10^{-4}	0.1
Abdomen	2.6	8×10^{-5}	0.1
CT			
Pelvis	25	8×10^{-4}	0.8
Abdomen	8.0	2.4×10^{-4}	0.2
Lumbar spine	2.4	7×10^{-5}	0.1
Nuclear medicine			
99m-Tc brain scan	5.1	1.5×10^{-4}	0.2
99m-Tc bone scan	4.2	1.3×10^{-4}	0.1

Some examinations carrying the highest risks are shown in table 11.10. Most other frequently performed examinations carry a lesser risk (an important exception is when radionuclides other than 99m-Tc are used in nuclear medicine—see next section).

On the basis of such risk estimates NRPB is now recommending that mean foetal doses of 5 mSv or less can generally be considered acceptable. However above this figure counselling on an individual basis may be necessary. A dose of 25 mSv will double the natural risk of fatal cancer before the age of 15 years.

11.10.   RISK FROM INGESTED OR INJECTED ACTIVITY

When dealing with the dosimetry of radionuclides, we need a relationship between the activity of the source and the exposure rate at some point near the source. This depends on both the number and energy of gamma ray photons from the source in unit time.

The factor relating exposure rate and activity is variously known as the exposure rate constant, the specific gamma ray constant or 'k' factor. A suitable unit in which to define 'k' would be coulomb kg^{-1} h^{-1} at 1 metre from a 1 MBq point source. The corresponding absorbed dose rate in air or in tissue can then be calculated.

Note that some radionuclides have complex decay schemes. This is undesirable if some of the decays contribute no useful gamma rays to the image forming process but contribute locally to the dose to the patient. Thus, somewhat paradoxically, a high k factor is a desirable property for an imaging radionuclide.

Calculation of the effective dose due to radioactivity within the body follows the same principles as for external beams, i.e. the dose to each organ is computed and multiplied by the appropriate tissue weighting factor, but is rather more complicated. Some of the reasons for this are

(a)   the total activity in the body will decrease as a result of a combination of physical and biological half-lives (see section 1.8),

(b)   the activity will redistribute throughout the different organs and tissues during its residence in the body,

(c)   activity in a particular organ will irradiate not only that organ, but also, if it is a gamma emitter, a number of other organs and

(d)   doses to other organs will depend on activity in the source organ, the inverse square law, attenuation in intervening tissues and the mass absorption coefficient of the target organ.

Methods of calculation and relevant data are contained in a series of publications from the Medical Internal Radiation Dose Committee (MIRD) of the Society of Nuclear Medicine (see Loevinger and Berman 1968, Snyder *et al* 1975). Johansson *et al* (1992) have recently recalculated effective doses for a wide range of radiopharmaceuticals using the latest ICRP tissue weighting factors. Values for some of the more common examinations, together with the associated risk, are shown in table 11.11. For further detail see section 12.6.

Note that when administering radiopharmaceuticals it is important that, in addition to the effective dose being acceptably low, the doses to individual organs must also be checked to ensure that deterministic effects cannot occur.

11.11.   SUMMARY

The potentially harmful effects of ionizing radiation must be recognized and understood. Furthermore, it is important to appreciate that increasingly sophisticated experiments have failed to provide evidence of a safe level of radiation for the two most important long term effects, namely carcinogenesis and mutagenesis. Therefore it is important that radiologists should have a good appreciation of the risks associated with the examinations they carry out. Indeed, the public is increasingly expecting to be kept informed of the risks involved.

Table 11.11. *Effective doses and risks for some common nuclear medicine examinations.*

Radiopharmaceutical	Study	Max usual activity (MBq)	Effective dose (mSv)	Risk ($7.3\%\ \mathrm{Sv}^{-1}$)
51-Cr EDTA	GFR	3	0.006	4×10^{-7}
99m-Tc MAG3	renography	100	0.7	5×10^{-5}
99m-Tc microspheres	lung perfusion	100	1.0	7×10^{-5}
99m-Tc DTPA	renography	300	1.6	1.1×10^{-4}
99m-Tc white blood cells	abscess	200	2.8	2×10^{-4}
99m-Tc phosphates	bone	600	3.5	2.5×10^{-4}
111-In white blood cells	abscess	20	9.4	7×10^{-4}
99m-Tc pertechnetate	first pass blood flow	800	9.5	7×10^{-4}
67-Ga citrate	tumour/abscess	150	16.5	1.2×10^{-3}
201-Tl ion	myocardium	80	18.4	1.3×10^{-3}

Some situations carry a higher risk and the importance of minimizing exposures to children and during pregnancy cannot be over-emphasized.

However, it is necessary to keep a sense of perspective and calculations show that, when diagnostic X-rays are used correctly and doses are carefully controlled, risks are acceptable within the broader context of both the clinical value of diagnostic information gained and the risks associated with daily living.

REFERENCES

Alper T 1979 *Cellular Radiobiology* (Cambridge: Cambridge University Press)

Baverstock K F and Slatter J W (ed) 1989 *Low Dose Radiation: Biological Basis of Risk Assessment* (London: Taylor and Francis)

Bergonié J and Tribondeau L 1906 De quelques resultats de la radiotherapie et essai de fixation d'une technique rationnelle *C. R. Seances Acad. Sci.* **143** 983 (Engl. transl. Fletcher G H 1954 Interpretation of some results of radiotherapy and an attempt at determining a logical treatment technique *Radiat. Res.* **11** 587)

Bithell J F 1989 Epidemiological studies of children irradiated in utero *Low Dose Radiation—Biological Basis of Risk Assessment* ed K F Baverstock and J W Stather (London: Taylor and Francis)

Black D (chairman) 1984 Investigation of the possible increased incidence of cancer in West Cumbria *Report of the Independent Advisory Group* (London: HMSO)

Boice J R Jr, Land C E, Shore R E, Norman J E and Tokunaga M 1979 Risk of breast cancer following low dose exposure *Radiology* **131** 589–97

Brill A B (ed) 1985 *Low Level Radiation Effects: A Fact Book* (Society of Nuclear Medicine)

Coggle J E 1983 *Biological Effects of Radiation* (London: Taylor and Francis)

Gardner M J, Snee M P, Hall A J, Powell C, Dounes S and Terrell J D 1990 Results of a case control study of leukaemia and lymphoma among young people near Sellafield nuclear plant in W Cumbria *Br. Med. J.* **300** 423–9

Hall E J 1988 *Radiobiology for the Radiologist* 3rd edn (Philadelphia, PA: Lippincott)

Hart D, Jones D G and Wall B F 1994 *Estimation of Effective Dose in Diagnostic Radiology from Entrance Surface Dose and Dose Area–Product Measurement* NRPB R262 National Radiological Protection Board.

International Commission on Radiological Protection (ICRP) 1977 Recommendations of the International Commission on Radiological Protection, ICRP 26 *Ann. ICRP* **1** 3

International Commission on Radiological Protection (ICRP) 1991 1990 Recommendations of the International Commission on Radiological Protection, ICRP 60 *Ann. ICRP* **21** 1–3

International Commission on Radiation Units and Measurements (ICRU) 1980 radiation quantities and units *(ICRU Publication 33)*

Johansson L, Mattsson S, Nosslin B and Leide-Svegborn S 1992 Effective dose from radiopharmaceuticals *Eur. J. Nucl. Med.* **19** 933–8

Jones D G and Wall B F 1985 *Organ Doses from Medical X-ray Exposures Calculated using Monte Carlo Techniques* NRPB-R186 (National Radiological Protection Board)

Loevinger R and Berman M 1968 A schema for absorbed dose calculations for biologically distributed radionuclides *Medical Internal Radiation Dose Committee of the Society for Nuclear Medicine (Pamphlet 1)*

Muirhead C R and Kneale G W 1989 Prenatal irradiation and childhood cancer *J. Radiol. Prot.* 209–12

National Radiological Protection Board (NRPB) 1993a Board statement on the 1990 recommendations of ICRP *Documents of the NRPB* vol 4, No 1

National Radiological Protection Board (NRPB) 1993b Board statement on diagnostic medical exposures to ionising radiation during pregnancy *Documents of the NRPB* vol 4, No 4, pp 1–14

National Radiological Protection Board (NRPB) 1994 *Radiation Doses—Maps and Magnitudes* 2nd edn (NRPB)

Otake M and Schull V J 1984 *In utero* exposure to A-bomb radiation and mental retardation—a reassessment *Br. J. Radiol.* **57** 409

Russell L B and Russell W L 1954 An analysis of the changing radiation response of the developing mouse embryo *J. Cell Comp. Physiol.* **43** (Supplement 1) 103

Schull W J 1991 Ionising radiation and the developing human brain *Risks Associated with Ionising Radiations (Ann. ICRP* **22***)* pp 195–218

Shimizu Y, Kato and Schull W J 1990 Studies of the mortality of A bomb survivors. 9 Mortality 1950–1985 Part 2: Cancer mortality based on the recently revised doses DS86 *Radiat. Res.* **121** 120–41

Shrimpton P C, Wall, B F, Jones D G, Fisher E S, Hillier M C and Kendall G M 1986 *A National Survey of Doses to Patients Undergoing a Selection of Routine X-ray Examinations in English Hospitals* NRPB-R 200 (National Radiological Protection Board)

Snyder W S, Ford M R, Warner G G and Watson S B 1975 'S' absorbed dose per unit cumulated activity for selected radionuclides and organs *Medical Internal Radiation Dose Committee of the Society for Nuclear Medicine (Pamphlet 11)*

Sumner D, Wheldon T and Watson W 1991 *Radiation Risks* 3rd edn (Glasgow: Tarragon)

Upton A C 1987 Cancer induction and non-stochastic effects *Biological Basis of Radiological Protection and its Application to Risk Management, Br. J. Radiol.* **60** 1–16

Wagner L K, Eifel P J and Geise R A 1994 Potential biological effects following high X-ray dose interventional procedures *J. Vasc. Interven. Radiol.* **5** 71–84

EXERCISES

1   State the 'law' of Bergonié and Tribondeau on cellular radiosensitivity. Discuss the extent to which cell types follow the 'law'.

2   Explain why it is difficult to predict quantitatively the effects of very low doses of radiation. How and why might the prediction depend on the linear energy transfer (LET) of the radiation.

3   What is the difference between stochastic and deterministic radiation effects? What is the evidence for believing that radiation-induced carcinogenesis is a stochastic effect?

4   Discuss the precautions that should be taken to minimize the risk of deterministic effects of radiation during invasive procedures and the actions that should be taken to ensure that they are identified if they occur.

5   Sketch the most common forms of chromosome defect detectable after whole body irradiation. Discuss the feasibility of chromosome aberration analysis as a method of retrospective whole body monitoring.

6   What is the most sensitive period for the production of abnormalities in humans by irradiation *in utero*? Why does irradiation at other times, both later and earlier, effectively produce few abnormalities?

7   Review the evidence that ionizing radiation can cause harmful genetic effects.

8   List the organs and tissues of the body identified by ICRP publication 60 (1991) as the most sensitive with regard to causing long term detriment by ionizing radiation. Give an approximate risk of fatal radiation-induced cancer for a whole body dose of 5 mSv.

9   List the variables that must be known in order to estimate the dose to the foetus when a patient who is pregnant is examined with X-rays. Outline the method of calculation.

10 Discuss the factors that determine the effective dose to a patient when a radiopharmaceutical is administered.

CHAPTER 12

PRACTICAL RADIATION PROTECTION AND LEGISLATION

12.1. INTRODUCTION

Although the emphasis in this chapter is on radiation protection and safety it should be remembered that radiation safety is only a part of total safety and should not be over-emphasized to the exclusion of other aspects of safety.

Radiation exposures can be split into three broad categories; occupational, medical and public exposures. In radiology departments all three categories can be encountered. Any activity which increases overall exposure is referred to as a 'practice' and any activity to decrease an existing exposure as an 'intervention'. The terms radiation or radiological safety and radiation or radiological protection are often used interchangeably. Strictly speaking radiation protection is concerned with the limitation of radiation dose whereas radiation safety is concerned with reducing the potential for accidents. Since the distinction is largely academic, both are described in this chapter without any attempt to distinguish between them.

12.2. ROLE OF RADIATION PROTECTION IN DIAGNOSTIC RADIOLOGY

12.2.1. Principles of protection

The principles of radiation protection are common to all uses of ionizing radiation. They are the following.

Justification

Since there is no safe threshold dose for stochastic effects (section 11.5.1) ionizing radiation should not be used in any 'practice' unless the net benefit from the exposure for an individual or society is greater than the radiation detriment. Although justification of a practice should be generic and broad in nature and should not necessarily be carried out each time a practice is undertaken, some aspects of it can require consideration each time the practice is undertaken. A patient must receive more benefit than detriment from an exposure. The benefit to the radiologist and radiographer from any dose they receive in the course of this investigation can only be quantified in terms of employment. The detriment to them must also be balanced against the benefits the patient and society gain from the exposure. Likewise the radiation doses members of the public receive as small adventitious doses through being in the proximity of radiographic exposures or nuclear medicine patients are justified by the benefit to society. The much larger doses which can be received by family or friends who are caring for or comforting patients who are emitting ionizing radiation must also be justified by a benefit to society.

Optimization

The magnitude of individual doses, the probability that exposures will occur and the number of persons that will be exposed should all be kept as low as reasonably achievable (the ALARA principle[1]) taking into account economic and social factors. The process of optimization is aided by setting 'constraint doses'. Constraint doses are planning doses which should not be exceeded by a practice. They are not legal limits but are set by local or national bodies as levels of exposure which good working practice should achieve. If measurements show they are being exceeded an assessment should be made to ascertain why and remedial action, if deemed necessary, should be taken.

It is also recommended that under some circumstances a risk constraint should be set in the case of potential exposures to limit the risk to any one individual particularly in the event of an accident.

Justification is applied to the practice as whole whereas optimization can be applied to individual components of practice.

Limitation

After following through the processes of justification and optimization there is, for occupational and public exposure, a legal limit on the radiation dose that can be received. There are, currently, no limits (only constraints) on the doses that can be given for medical purposes. These are still at the discretion of the doctor involved who has to be prepared to clinically justify them if necessary.

The legal limits are set to restrict stochastic effects (section 11.5.1) to acceptable levels and to be well below thresholds for deterministic effects.

12.2.2. *Patient protection*

The most efficient way to reduce patient dose is not to carry out the radiographic examination! Non-essential radiographs such as reassurance radiographs or radiographs for insurance purposes should not be taken. Past films should be utilized where possible and films should be moved with the patient between hospitals. National protocols such as those recommended by the Royal College of Radiologists (1998) should be used so that the minimum number of films required for diagnosis are taken. Where possible, techniques using non-ionizing radiation such as ultrasound or MRI should be used.

Reduction of patient dose can be achieved by a wide variety of means but it must always be remembered that to obtain a radiograph a dose must be given to the patient. If the dose is reduced too far the radiograph does not contain the information required and in fact the dose given is wasted, as the benefit to the patient does not compensate for the risk from it.

(i)   *Constraints or 'investigation levels'.* Levels of dose have been set in terms of skin dose or dose–area product (see section 6.13) which should not be exceeded for common radiographic and screening examinations. These constraint doses are all some way above those required by adopting best practices and the circumstances under which they are not achieved should be quite rare. If circumstances are identified where they are being exceeded the techniques and equipment used should be investigated and remedial action taken as necessary.

(ii) *Use of high kVp techniques.* As pointed out in section 6.14 the use of higher kVp should reduce patient dose.

(iii) *Collimation.* Reducing the volume of the patient irradiated not only improves contrast because of the reduction in scatter but can significantly reduce the effective dose. Ensuring that even a small border exists round the edge of a radiograph can significantly reduce the area exposed (coning the field to 1 cm inside an 18 cm × 24 cm cassette reduces the area irradiated by 18.5%). Whilst modern units

---

[1] Sometimes presented as ALARP—as low as reasonably practicable.

which automatically adjust the field size to the cassette size ensure no unrecorded area of the patient is exposed, they do not necessarily produce the optimum field size for the anatomy being radiographed. The optimum size could sometimes be smaller.

(iv)   *Optimization of imaging system.* This generally refers to the use of fast film–screen combinations as described in section 4.5 and efficient image intensifier–television systems as described in section 4.11.

(v)   *Reduction of screening dose.* This can be achieved by keeping screening time to a minimum. For older units this is essentially at the discretion of the radiologists but modern units allow a variety of methods to be used. The use of last frame hold can significantly reduce dose particularly during processes involving the insertion of catheters, orthopaedics or cardiology. Some units now allow the collimators to be repositioned using a last frame hold which further reduces doses. Pulsing the output of the tube during screening also reduces the dose. Obviously the more pulses per second that are required to visualize the process involved, the larger the patient dose per unit time. To see movements of the heart requires many more pulses per second than to see movement in the large bowel. It should be noted that the use of a digital unit, as described in section 9.4, can lead to an increased dose being given for a screening examination but offers many more image manipulation options.

The method of automatic control of kVp and mA to maintain the dose rate of an image intensifier input can also be optimized on some units. Initially the kVp should be raised to the maximum value that still provides acceptable contrast with further output, if necessary, obtained by raising the mA. This will ensure that the dose to the patient is minimized whilst diagnostic information is maintained.

(vi)   *Filtration.* The use of aluminium filtration (above 2.5 mm if the operating voltage exceeds 100 kVp) is mandatory in X-ray units. As described in section 3.8 this removes the lower photon energies from the X-ray beam. Special K-edge filters can be used in paediatric radiography (section 9.8) and in mammography (section 9.16) to remove unwanted high and low energy photons. Some sophisticated screening units now automatically insert between 0.1 mm and 0.2 mm of additional copper filtration to further harden the beam. These units effectively calculate the attenuation of the patient and choose the thickness, if any, of copper to insert. Depending on kVp, the first 0.1 mm of copper reduces the patient skin dose by 30–40% and the next 0.1 mm by approximately 10%. (The effective dose reduction will vary but will be no more than approximately half of these figures.)

(vii)   *Output wave form.* The more closely the output of the generator approaches a steady (DC) voltage supply the higher the quality of the beam (section 2.3.4) and the fewer low energy photons are present. Even quite simple X-ray units are now using medium to high frequency generators, which produce essentially a DC supply.

(viii)   *Use of low attenuation materials.* After the X-ray beam leaves the patient it may travel through several structures before reaching the imaging medium. Each of these can absorb some of the beam. It has been estimated by Hufton and Russell (1986) that by replacing the covers on cassettes, table tops and the spacing material in grids by carbon fibre based materials a dose reduction of up to nearly 60% can be achieved, depending on tube kVp.

(ix)   *Choice of grid.* As described in section 5.8 the higher the grid ratio the greater the dose required to produce the required film density.

(x)   *CT slit width.* The dose to the patient from contiguous slices should not vary with slice width except at very narrow slices. On some units it is found that, at the very narrowest of slices, the first collimator does in fact stay wider than the second collimator, resulting in the patient receiving a higher dose (as much as twice) that for a normal set of contiguous slices. Most units will also increase the mA s at narrower slice widths in order to maintain the signal-to-noise ratio. The dose from CT scans is high, so the number of slices should be kept to a minimum and repeating the scan with contrast medium should only be undertaken if definitely required.

(xi)   *Gonad shields.* These must be used whenever possible.

12.2.3. Staff protection

Radiography and radiological staff have a vested interest in keeping the dose to the patient as low as possible because in principle their own doses are proportional to the patient dose. As described in section 12.4.1, despite the very high dose rates produced by diagnostic X-ray equipment, by taking relatively simple precautions the doses received by staff are kept very low.

The dose received is equal to the dose rate times the period of exposure so a reduction in either dose rate or period of exposure reduces the dose received.

The design of modern units ensures that it is virtually impossible to receive an inadvertent whole body exposure during normal operations. Some radiologists still occasionally image their own fingers because they move them from just outside the main beam to just inside during the investigation. (Some orthopaedic pinning operations are virtually impossible to do without 'seeing' the surgeon's fingers.)

Scattered radiation is present whenever an X-ray set is operated so protection against scatter is important. The amount of scattered radiation produced and its distribution varies with the kVp (see section 3.4.3). The scattered radiation dose in the vicinity of the patient when taking a large film could be reduced by 50–65% by increasing the kVp from 50 kVp to 90 kVp.

Reduction of dose rate

Distance. As described in section 1.12 the dose rate in a radiation field from a point source falls as the square of the distance. For X-ray units this rule only strictly applies to the main beam where the focal spot can be assumed to be a point source. The scatter from the patient and other equipment is not from a point source and the rate of fall-off with distance is more complicated. However the general rule still applies that the further away from the X-ray tube, patient and table, the lower the dose rate in the radiation field. During an exposure personnel should stand at least two metres from the source of the scattered radiation whenever possible.

Shielding. Section 3.2 showed how materials could reduce the intensity of a radiation field. At diagnostic energies lead is one of the most efficient materials because it has a large linear attenuation coefficient. As described in section 12.5 it is incorporated into some fixed shielding in cubicles and doors and it is also used in personal protective clothing. Lead–rubber aprons are used for body protection. These can have a lead equivalence of up to 0.5 mm. However these aprons are heavy and uncomfortable to wear and most X-ray rooms would have aprons of no more than 0.35 mm lead equivalence. A thickness of 0.35 mm is laid down in the *Guidance Notes* (1988) as the minimum allowed when X-ray photons are generated above 100 kVp. (Below 100 kVp only 0.25 mm lead equivalence is required.) Lead–rubber gloves should have a lead equivalence of at least 0.25 mm and thyroid collars a similar lead equivalence to aprons. If using an over-couch screening unit, it is also worth considering the use of lead glass spectacles if eye measurements show high doses are being received.

Protective clothing is not designed to shield against the main beam. It will only protect against scattered radiation or the beam after it has passed through a patient. Protective clothing must be properly stored so that it does not become cracked and must be periodically checked to ensure that it is in good condition. Suffice to say it must also be worn when provided. If neither protective clothing nor a permanent screen is available, the person should leave the room. 'Hiding' behind someone wearing a lead–rubber apron is not acceptable.

12.2.4. Public protection

Members of the public are distinguished from patients if they receive an exposure but gain no clinical benefit. This can happen in waiting rooms, changing cubicles or as a consequence of being in the corridor adjacent to the X-ray department. These doses are kept to a minimum by permanent shielding incorporated into the walls of the X-ray rooms and doors. When mobile units are being used in wards the situation is rather different (and more difficult). Here the radiographer must ensure that the primary beam is not directed towards a patient in an adjacent bed. Any patient in a nearby bed who cannot be moved should be protected from scattered radiation by laying a lead apron over them.

Members of the public who hold or support patients (such as parents holding children) knowingly and willingly after the risks are pointed out to them are not subject to the 'member of the public' dose limit set out in section 12.3. They must however be subject to a dose constraint. Compliance with the dose constraint must be demonstrated by suitable monitoring.

Members of the public who become involved in medical research are also not subject to the normal dose limits. A Local Ethical Committee must accept all medical research projects. If the radiation exposure received is part of the normal treatment protocol then it is of direct benefit to the patient and is considered as a normal medical exposure. If it is additional to the normal treatment protocol or is to be delivered to a healthy volunteer, then the Ethical Committee considering the research project must take any risk into account.

Although the Euratom Directive 97/43 (1997) recommends that member states set dose constraints for such work, none have yet been set in the UK. A useful document is ICRP 62 (1992). This has a section dealing with the use of ionizing radiation in medical research and identifies, in broad terms, the levels of benefit to science and society that must be gained from the exposure of volunteers to different radiation doses.

A prerequisite of research work is that the informed consent of the volunteer must be obtained. Presentation of the possible risks of an exposure, extrapolated from the information given in chapter 11, in a form that is informative but not frightening is difficult. It is however essential.

12.3.   EUROPEAN AND UK LEGISLATION

12.3.1.   Introduction

Increasing recognition of the harmful effects of ionizing radiation led to the establishment, in 1928, of the International Commission on Radiological Protection (ICRP). This commission is a non-governmental body and through an expert committee structure uses the knowledge of independent advisers, from any country in the world, to give guidance on all aspects of radiation protection. This guidance is issued in the form of recommendations and can range from broad concepts to detailed review of scientific research. Because of this structure ICRP recommendations are not mandatory. However they are very influential internationally and few countries adopt regulations dealing with radiation protection which differ to any extent from them. On a worldwide basis the legislation controlling the use of ionizing radiation is generally very similar. The major differences are found in the methods of enforcement.

The ICRP updated their general recommendations following an appraisal of the current epidemiological data in 1991. These recommendations, generally known as ICRP 60 (1991), set out new maximum permissible doses and developed further the conceptual framework of justification, optimization and limitation, first laid down in ICRP 26 (1977).

12.3.2.   European legislation

ICRP recommendations are promulgated within the European Community through a series of Directives. Member states must respond to a Directive, within a laid down time scale, by introducing national legislation. This legislation must, as a minimum, comply with the Directive but can be more restrictive if nationally deemed necessary.

12.3.3.   Basic Safety Standards Directive 96/29/Euratom (1996)

This Directive replaces Directive 80/836/Euratom (1980) (amended by Directive 84/467/Euratom 1984). It implements the recommendations of ICRP 60 and applies to all practices which involve a risk from ionizing radiation arising from natural or man-made sources. The basic requirements are summarized as follows:

(a)   All practices above the minimum levels below which the Directive does not apply must be reported and some must be specifically authorized.

(b)   Occupational and public exposures are distinguished from medical exposures and those of carers and comforters (other than those carrying out the tasks as part of their work). Whilst all medical exposures are, and always have been, excluded from dose limitation within the limits outlined below, those of carers and comforters were not considered under the previous Directive 80/836/Euratom. This new provision states that if the dose is willingly and knowingly (i.e. the risks have been explained) received then the 'member of the public' dose limits do not apply. Constraint doses for carers and comforters should be set however.

(c)   The new dose limits for exposed workers and students or apprentices above the age of 18 are shown in table 12.1. In addition there is a requirement that during breast feeding there is no risk of bodily contamination.

(d)   For students and apprentices below the age of 18 years who are obliged to use ionizing radiation, the effective dose limit is 6 mSv/year and the equivalent dose limits are 50 mSv/year for the lens of the eye and 150 mSv/year for any 1 cm^2 of the skin.

(e)   The dose limits for members of the public are an effective dose of 1 mSv/year (in special circumstances this can be exceeded but the dose must not exceed an average of 1 mSv/year over any 5 year period), an equivalent dose to the lens of the eye of 15 mSv/year and an equivalent dose to any 1 cm^2 of skin of 50 mSv/year.

(f)   Work places where there is a possibility that the exposure to ionizing radiation could exceed the limits set for members of the public must be classified as *controlled* or *supervised*. Controlled areas must be suitably demarcated and delineated with controlled access. Appropriate work instructions (local rules) must be issued for operations in these areas. Radiological monitoring is required.

For a supervised area there is a requirement to monitor the work force, but the other requirements of a controlled area are discretionary.

(g)   Exposed workers are divided into two groups.

Category A—those who are liable to receive an effective dose greater than 6 mSv/year or an equivalent dose greater than 3/10 of the limit for the lens of the eye, skin or extremities.

Category B—exposed workers who are not category A workers.

(h)   Category A workers must be monitored. Category B workers should have sufficient personal monitoring to demonstrate they do not need to be in category A.

(i)   All exposed workers, apprentices and students who, in the course of their work, are dealing with sources of ionizing radiation, must be given adequate training. Women must be informed of the need to declare if pregnant or breast feeding so as to avoid any hazard to the foetus or neonate.

(j)   A suitably qualified expert must be appointed to advise undertakings on radiation risks and assess radiation protection arrangements (this person is frequently called the Radiation Protection Adviser).

(k)   All work places using ionizing radiation must have suitable monitoring equipment available.

(l)   Records of monitoring must be kept and made available to individuals. (For category A workers this is for at least 30 years following termination of work involving exposure or until the age of 70 years.)

(m)   Category A workers must receive suitable medical surveillance. This must include a medical prior to starting work and a periodic review of health. Medical records must also be kept until the person is 75 years of age or for 30 years after work involving exposure to ionizing radiation ceased.

(n)   A system of inspection to enforce the requirements of the Directive must be established.

12.3.4.   *Medical Exposures Directive 97/43/Euratom (1997)*

This Directive acknowledges the benefits that the use of ionizing radiation bring to medicine but identifies that it is the major source of exposure to artificial sources of ionizing radiation. Therefore medical practices

Table 12.1. *Maximum permissible dose limits as specified in 96/29/Euratom.*

Effective dose in any consecutive 5 year period	100 mSv
Effective dose in any year	50 mSv
(the option for a yearly effective dose of 20 mSv is maintained)	
Equivalent dose to the lens of the eye in any year	150 mSv
Equivalent dose to any 1 cm^2 of the skin in any year	500 mSv
Equivalent dose to extremities in any year	500 mSv
Foetal dose once a pregnancy has been declared	1 mSv

need to be undertaken in optimized radiation protection conditions.

The Directive applies to all medical exposures including those for medico/legal reasons and the exposure of carers and comforters. It places considerable emphasis on the justification of medical exposures, requiring that they show a net benefit for the individual or society, identifying in particular

(a)   consideration of new types of practice,

(b)   consideration of the specific objectives of the exposure and the characteristics of the individual involved,

(c)   a requirement that exposures for research purposes must be examined by an ethics committee and

(d)   a requirement that exposures for medico-legal reasons or other reasons with no direct health benefits to the individual should not be undertaken.

The Directive requires all doses to be optimized. To this end it recommends that diagnostic reference levels for radiodiagnostic examinations should be established and used for members of the public. Whilst establishing that carers and comforters are not subject to dose limits (see section 12.3.3) it requires that national constraint doses be established for these persons. It also requires that nuclear medicine patients or their guardians are issued with written instructions with regard to the steps to be taken to minimize the exposure of anyone coming in contact with that patient.

The responsibilities of both the person prescribing and clinically responsible for the exposure and the person carrying it out are identified. Both groups must have suitable training. This applies to qualification under the regulations and to on-going education after qualification. Some emphasis is placed on having written protocols for standard radiological procedures and recommendations for referral criteria. Medical physics experts must be available to give advice on all aspects of exposure from ionizing radiation in medicine.

The equipment used for medical exposures must be included in an inventory kept by the competent authority, must be subject to acceptance testing before it is used for medical exposures, must be kept under radiation protection surveillance and must be subject to regular quality control. The Directive prohibits the use of direct fluoroscopy without an image intensifier and recommends that all fluoroscopic dose rates be controlled. It also recommends that, where practicable, all new radiodiagnostic equipment should have in-built dose measuring devices.

The Directive draws particular attention to the requirements to keep doses to a minimum for paediatric exposures, health surveillance programmes, potential high dose techniques such as CT scanning and interventional radiography, pregnant women and, with regard to nuclear medicine investigations, breast feeding females.

12.3.5.   Outside Workers Directive 90/641/Euratom (1990)

This Directive was designed to avoid the situation where itinerant workers could receive more than the yearly maximum dose by working for several different employers in several different countries. It requires that a category A employee of one employer working in a controlled area of another employer must carry a passbook in which his/her previous doses are recorded. On leaving that controlled area the controller of that area must enter into the passbook an estimate of the dose received.

12.3.6. UK legislation

UK legislation complies with the Euratom Directives in various ways.

12.3.7. Radioactive Substances Act 1993

This act essentially controls the holding and disposal of radioactive materials and has been formulated primarily to protect the environment. It essentially lays down three conditions:

(1) Before an undertaking can hold radioactive sources, the premises must be registered under section 7 of the Act with the Environmental Protection Agency. This registration certificate is very specific. It specifies the radionuclides, the activity of the nuclides and, for sealed sources, the number of sources that can be held. Occasionally it may allow a general beta/gamma emitting radionuclide holding for a small amount of activity to accommodate new projects. Records must be kept to show that the registration is not exceeded and inspectors visit the premises periodically.

Occasionally a sealed source is used in more than one location. An example would be an americium-241 source used to check the lead equivalence of the walls of X-ray rooms. In these situations a mobile registration under section 10 of the Act must be obtained. One normal condition of a section 10 registration is that a clear record must be kept of the location of the source at any time.

(2) Radioactive waste can only be accumulated if the user is authorized under section 15 of the Act and only disposed of if the user is authorized under section 13 of the Act. Both authorizations are generally placed in the same certificate. The authorization is very specific. It will state where the waste can be accumulated, the route of disposal, the radionuclides that can be disposed of and the activity that can be disposed. Again records must be kept to show compliance.

Currently there are various exemptions to the Act. (Implementation of Euratom Basic Safety Standards will probably require some of these to be revoked or amended.) The two exemptions probably most relevant to the Health Service are the following.

(a) The Radioactive Substances (Testing Instruments) Exemption Order 1985. This order allows small sealed sources used in measuring instruments or for testing instruments to be exempt from registration. Examples are the external standard source in a liquid scintillation counter and the 3 MBq Co-57 sources used as markers in nuclear medicine.

(b) The Radioactive Substances (Hospitals) Exemption Order 1990. This order allows the occasional patient from a peripheral hospital to be investigated in a central main nuclear medicine department without having to transfer the patient to the main hospital and without having to obtain a registration and authorization for the peripheral hospital. It lays down conditions about the storage of any sources, procedures for emergencies and the maximum amounts that can be held on the premises. The maximum amount of most relevance is the 1 GBq limit for Tc-99m. There are similar conditions with regard to any waste produced; the limit of most relevance is 1 GBq for technetium-99m discharged to the sewage system of the premises in a month. If the amounts exceed these limits then a full registration or authorization must be obtained.

12.3.8. The Medicines (Administration of Radioactive Substances) Regulations 1978

These regulations are operated through a committee called the Administration of Radioactive Substances Advisory Committee (ARSAC) who advise Health Ministers on the issuing of certificates under the regulations. Because of this Committee these are commonly called the ARSAC regulations and the certificates issued under the regulations ARSAC certificates. These regulations are not primarily concerned with worker doses but are more concerned with the protection of patients or volunteers during clinical or research use of radioactive substances. They cover all administrations of radioactive materials no matter how small the amount being administered. Only a doctor or dentist can hold a certificate to administer radioactive materials but it is acceptable that other persons acting on their behalf and under their guidance may actually carry out the administration. This transfer of authorization to administer should be a formal written transfer even though the responsibility still rests with the certificate holder. Certificates are normally only issued to senior staff of consultant status and are time limited.

Notes for Guidance (ARSAC 1998) on these regulations and application forms are available from the ARSAC. These guidance notes are particularly useful for routine tests as they give descriptions of classes of radioactive materials that are considered acceptable for given purposes. As long as the maximum recommended activity is not to be exceeded no justification need be given for the request to use the radioactive material for the purpose described in the *Notes for Guidance*. Application for an ARSAC certificate must include a description of the applicant's experience and training, a description of the equipment available to carry out the tests, the agreement and description of the scientist(s) associated with the work and the agreement of the Radiation Protection Adviser. If a research proposal is being considered or a routine test not listed in the *Guidance Notes* then, in addition to the above, the ARSAC also require a full justification and description of the use including effective dose calculations.

These certificates are specific to the appointment held at the time of application and the hospital in which the investigation or treatment will be done. If either of these change permanently the ARSAC should be informed, as the certificate will require amendment. If, in a specific case, for clinical reasons, a change in hospital premises must be made in order to administer the radiopharmaceutical this is permitted with the agreement of the local Radiation Protection Adviser.

A certificate from the ARSAC to carry out a research project that involves the administration of a radioactive substance does not remove the requirement for ethical committee approval. Most ethical committees would require an ARSAC certificate to accompany the application or assurance that an application has been made to the ARSAC.

12.3.9. The Ionising Radiations (Protection of Persons Undergoing Medical Examination or Treatment) Regulations 1988

The regulations are generally referred to as the POPUMET regulations. They were introduced in response to the Euratom Patient Protection Directive, which has been replaced by the Medical Exposures Directive, described in section 12.3.4. As a result these regulations are being changed. One change is the title of the persons involved with exposures. The new names are given in brackets below.

They specifically apply to diagnostic and therapeutic exposure but do not include research exposures. Under the regulations all examinations should be carried out using accepted radiological practice with as low a dose as practicable to achieve the required clinical result. The regulations distinguish between those clinically directing (the practitioner) an exposure and those physically directing (the operator) an exposure. Both are required to receive adequate training to ensure that the patient dose is kept to a minimum and that patients are not unnecessarily exposed. A syllabus is suggested in the regulations and is often referred to as the 'core of knowledge'. A certificate of training must be issued.

Most radiological examinations are under the clinical direction of a radiologist. This includes normal requests (by the referrer) for radiographic examination or examinations following a strict protocol in, say, an accident and emergency unit. Where the clinician makes a conscious decision to expose a patient without radiologist scrutiny or not following an established protocol, a POPUMET certificate is required. Some orthopaedic surgeons and cardiologists quite clearly do require certificates. These 'core of knowledge' courses are neither designed nor intended to give clinicians or surgeons the same status radiologically as qualified radiologists nor do they provide practical training in the technique.

The persons physically directing an exposure are those actually effecting the exposure. Sometimes the same person clinically and physically directs an exposure as in fluoroscopy. The training is common to both groups. Radiologists and radiographers are deemed to have received adequate instruction in the course of their training. Everyone else should attend a POPUMET course.

An employer must be able to demonstrate that all members of staff who are physically or clinically directing an exposure hold the required training certificate. The employer must also have a full inventory of all equipment used to deliver radiation doses to patients.

12.3.10.   Radioactive Material (Road Transport) Great Britain Regulations 1996

These Regulations will only concern a few readers. They differ from all other regulations in that they must comply with international standards as laid down in International Atomic Energy Agency Regulations (IAEA 1990) and the *European Agreement* (1995).

Any substance with an activity greater than 70 Bq g^{-1} comes within the terms of the regulations unless the radioactive material is contained in the body of a person undergoing medical treatment or is a luminous device worn by a person. (Military applications, domestic smoke detectors and gaseous tritium devices also have a limited exemption.) Because of the exemption applied to patients, many hospitals will never be directly concerned with these regulations because they do not send radionuclides to other hospitals.

There are three types of radioactive package.

(a)   *Excepted packages.* An excepted package can only contain a limited amount of activity and the dose rate on the surface must not exceed 5 μSv h^{-1}. It has limited labelling requirements but requires documentation. The activity allowed is set out for each radionuclide in the regulations.
(b)   *Type A package.* These packages must go through limited type testing and will be labelled with white I, yellow I or yellow II labels. The type of label depends on the package surface dose rate and the transport index. The transport index is the maximum dose rate in μSv h^{-1} at 1 metre from the package divided by 10. There is a limit on the activity that can be in the package.
(c)   *Type B packages.* These packages go through stringent testing and require competent authority approval. They are subject to the same labelling requirements as type A packages. They are used to carry large quantities of radioactive material.

A hospital that sends out radionuclides will often only send sufficient activity for a single patient, which should come within the excepted package category most of the time. Sufficient activity for, say, several bone scan injections would have to travel as a type A package. The drivers of vehicles carrying these packages require formal training and should understand the hazard of the material they are carrying.

12.3.11.   Ionising Radiations Regulations 1985

The system of regulation adopted for worker protection in the UK for all aspects of safety is based on three tiers of documentation. The first tier is legislative where a Statutory Instrument is passed by Parliament. The regulations in such an instrument are mandatory and can only be changed by a new Statutory Instrument in Parliament. The second tier is in the form of Approved Codes of Practice (ACOPs). These are prepared by the Health and Safety Commission and support the relevant statutory instrument by expanding each regulation with an interpretation as to what is required to comply. In legal terms an ACOP is evidentiary and one does have the option of non-compliance with it. However if this option of non-compliance is chosen it could be necessary to prove in a court that the action taken was at least as good as the one recommended in the ACOP. The advantage of an ACOP is that it can be changed by the Commission without recourse to Parliament. The third tier is called 'notes for guidance'. The 'notes for guidance' are generally prepared by a committee of experts drawn from operators, advisers and legislators and are normally specific to an area of use. These notes for guidance give a practical guide on how the legislation and ACOP can be complied with in that area of use. The notes for guidance do not have a direct legal standing. The balance between the material in the ACOP and notes for guidance is variable. In the current set of documents dealing with ionizing radiations the ACOP is quite extensive (Guidance Notes 1988). In the new legislation referred to in section 12.3.13 a new philosophy is proposed with a small ACOP and most of the material in new notes for guidance.

The Ionising Radiations Regulations apply to all users of ionizing radiation and medical users must comply fully. The implementation of the regulations is the responsibility of the employer who will be prosecuted in the event of a breach. There are some 41 regulations and 10 schedules in the regulations. The following are identified as the ones of most interest in radiological practice.

Regulations 6 and 7 with schedule 1

These identify dose limits and the need for employers to keep doses as low as reasonably practicable by engineering controls and suitable systems of work. In regulation 7 the requirement not to exceed the dose limits specified in schedule 1 of the act is laid down.

Regulations 8 and 9 with schedules 6 and 2

These describe the criteria adopted for designation of *controlled* and *supervised* areas. A controlled area is one where an employee is likely to exceed 3/10 of the dose limit. A supervised area is any area between 1/10 and the above 3/10 of a dose limit. In practice a controlled area is any area where the dose rate is in excess of 7.5 μSv h^{-1} or contamination levels as identified in schedule 2 are exceeded. This means that all X-ray rooms are classed as controlled areas. A controlled area must have an engineered barrier and access must be under the control of the employer. Access is limited to classified workers (persons who are likely to exceed 3/10 of the dose limit) or persons working to a written system of work and patients. This system of work controls working practices such that it ensures that the 3/10 dose cannot be exceeded. (In medical circumstances virtually all employees work to systems of work and are not classified.) In order to demonstrate that the systems of work are successful, it is common practice to issue everyone regularly exposed to ionizing radiation is issued with a personal dosimeter.

A classified area must be physically demarcated, generally in practice by the walls of the X-ray room, and have clear warning signs.

Supervised areas can be identified by warning signs but there is no restriction on access for employees. Supervised and controlled areas must be identified in the local rules (see below).

Regulation 10

If an employer has a controlled area, a Radiation Protection Adviser (RPA) must be employed. This appointment must be in writing and the RPA must be given the facilities to carry out the necessary work for the employer. The RPA is strictly an adviser. The responsibility for ensuring compliance stays with the employer.

Regulation 11. Local rules

Every employer who undertakes work with ionizing radiations must have, in writing, a set of *local rules*. Local rules ensure both compliance with the requirements of the regulations in practice and that exposures are kept to a minimum. The local rules are specific to an area, not general to an establishment. A newcomer to a department should be able to identify quickly the procedures that must be followed. Local rules should be clear and concise. They should also include a section on how to deal with foreseeable emergencies.

The essential contents of local rules would be the following.

(a) A clear description of the area to which they apply, identifying which areas are controlled and which are supervised.
(b) The name of the radiation protection supervisor of the area (see below), the name of the radiation protection adviser for the area and the method of contacting them.
(c) The responsibilities of the radiation protection supervisor.
(d) The working instructions for the area including a clear system of work if non-classified persons are entering controlled areas.
(e) The procedures to deal with any emergencies which may be anticipated.

Additionally it may be useful to add to the rules the local methods of ensuring compliance with other requirements of the regulations. For possible inclusion would be a description of the management and supervision structure for the area, the method and the frequency of testing safety controls and warning devices, the method of measuring dose rate and, particularly, contamination, the testing of instruments and the arrangements for personal dosimetry.

Regulation 11. Radiation Protection Supervisor (RPS)

The RPS has local responsibility on behalf of the employer for ensuring compliance with the working procedures identified in the local rules. The RPS must be appointed in writing and must have sufficient experience, knowledge and status in the management structure to be able to ensure the implementation of the local rules and to action the emergency procedures should an emergency arise.

Regulation 24. Monitoring

An employer must supply adequate area monitoring equipment, which must be calibrated at least every 14 months. Records of any monitoring undertaken must be kept. Suitable personal dosimeters should be worn. Employees must be told of the doses recorded on personal dosimeters.

Regulation 32. Duties of manufacturers

A manufacturer has a responsibility to ensure that any equipment they supply is both constructed and designed to keep all exposures to staff as low as reasonably practicable. To ensure that this is the case, a 'critical examination' must be undertaken by an RPA (employed by either the manufacturer or the employer buying the equipment). This examination will check the correct operation of the safety features and warning devices, ensure that there is sufficient protection for staff and that adequate operational information about use, testing and maintenance has been provided.

Regulation 33. Equipment used for medical exposures

Under this regulation employers must ensure that any equipment used for medical exposure is designed, installed and maintained such that the objectives of use, e.g. diagnosis, treatment or research, are achieved with the minimum practical exposure. The word 'equipment' is interpreted quite liberally and includes not only the ionizing radiation generating device but also ancillary equipment such as image receptors, calibrators and film processing units. In radiology departments this has clear implications for the choice of equipment. It also implies the need for a quality assurance/routine performance program. A document recommending suitable standards for such a program has been published by the Institute of Physics and Engineering in Medicine (1997).

12.3.12.   The Ionising Radiations Regulations (Outside Workers) 1993

These regulations implement the Euratom Outside Workers Directive (1990). Although these regulations were intended to control doses in certain industries, there are some instances where they may need to be considered in hospital practice. A classified engineer working on an X-ray unit in a radiology department or a classified radiologist from one trust working in the radiology department of another are two examples. The latter is probably rare but the former can occur. One way to avoid the regulation is formally to transfer control of the X-ray room to the engineer for the period of work. Alternatively, non-classified engineers can be requested.

12.3.13.   New UK legislation

The Euratom Directives of 1996 and 1997 will have to be implemented into amended UK legislation. Where the directives may result in changes, these have been indicated in the text. At the time of writing major changes are not envisaged, but the numbering of some regulations will change.

12.4.   DOSES TO STAFF AND PATIENTS

12.4.1.   Staff doses

In many countries the number of classified workers exceeding 20 mSv/y is very few (less than 20 in total in the UK excluding non-coal miners) and a reduced yearly limit of 20 mSv with the option of exceeding it under special circumstances should present no undue problems. In other countries this may be a problem. However the introduction of a 20 mSv/y dose limit should not restrict work in a radiology or nuclear medicine department in any country. The introduction of 6 mSv/y as the threshold dose for becoming a classified worker could have a greater impact. Very few hospitals will have any member of the radiology department receiving sufficient dose to be considered likely to exceed the current threshold for classification of 15 mSv/y. The systems of work (section 12.3.11) used to restrict radiation dose have worked very successfully. However many more hospitals will have a small number of persons who could potentially exceed 6 mSv/y. Likely candidates would be interventional radiologists, orthopaedic surgeons or cardiologists.

The limit on the dose to the foetus of 1 mSv once a pregnancy has been declared could also have some impact. It is generally accepted that limiting the surface dose to the abdomen of the mother to 2 mSv during the rest of the pregnancy will ensure that the foetal dose does not exceed 1 mSv. Film badge and TLD-based personal dosimeters are working to their detection limit if they are used to show that 2 mSv is not being exceeded over a 6 month period, particularly if just one positive dose is recorded (see section 12.7). If necessary it will be prudent to move staff away from work that may result in potential or unpredictable exposures. Examples would be using a mobile unit, working in a screening room or preparing radiopharmaceuticals.

Table 12.2 shows the distribution of staff doses from diagnostic radiology departments. Very few radiographers exceed 1 mSv/y, just less than 1% of the total reported. For radiologists the number exceeding 1 mSv/y is 5% but none exceed 5 mSv/y. Cardiologists however have some 12% exceeding 1 mSv/y and 2% exceeding 5 mSv/y. A few nurses exceed 1 mSv/y.

As can be seen from the above figures, it should be relatively easy for radiographers, following good working practices, to continue working when pregnant and to stay well within a limit of 2 mSv to the abdomen once the pregnancy has been declared. Radiologists and nurses may have to be a little more careful and avoid interventional investigations especially in a screening room which uses an over-couch X-ray tube[1]. For

---

[1] Note: All the above figures relate to total body doses. Doses to the extremities, especially the hands and eyes, may be much higher, especially during interventional procedures, and appropriate extremity monitoring may be required. Dose limits are, of course, higher than for the whole body (table 12.1) and there are no plans to change these as there is no evidence of a change in threshold doses for deterministic effects (see section 11.5).

Table 12.2. *Occupational exposures in some diagnostic radiology departments in the UK during 1991 (Hughes and O'Riordan 1993).*

Occupational group	Number of workers in dose range (mSv)					
	0–1	1–5	5–10	10–15	15–20	>20
Radiographers	5663	55	1	0	0	0
Radiologists	729	38	0	0	0	0
Cardiologists	171	22	2	0	1	0
Other clinicians	465	9	0	0	0	0
Dept nurses	1522	38	0	1	0	0
Science/tech. staff	1070	27	1	0	0	0
Others	937	5	2	0	0	0

all females working in radiology departments it is prudent to tell the Radiation Protection Supervisor or Superintendent as soon as a pregnancy is confirmed so that any changes to work activity can be made as soon as possible. As described in section 11.8 the risk to the foetus of a dose of 1 mSv is extremely small compared to the natural risk of a physical malformation or mental defect occurring. Depending on the classification adopted, between 1 and 5 in 100 neonates are considered to have a physical or mental defect.

12.4.2. *Patient doses*

The measurement of patient doses and the role of constraint doses in their reduction have been described in chapter 6. Occasionally however the dose received by a patient during an examination is greater than intended due to failure of equipment or software control. If the radiologist or radiographer reacts quickly enough, the dose is limited. Any incident of suspected over-exposure must be reported immediately to the departmental Radiation Protection Supervisor or Superintendent. No further exposure must be taken until an investigation has taken place. Any failure of equipment or software must be reported so that the manufacturers and the Department of Health can assess the fault and, if necessary, warnings can be issued to other users.

If the dose given is large enough, a contravention of the Ionising Radiations Regulation 33 may have occurred and the Health and Safety Executive (HSE) must be informed.

Guidance Note PM77 (HSE 1998) on the fitness of equipment used for medical exposure to ionizing radiation contains guidelines for such notification. These are in the form of multiplying factors of the normal doses above which notification is necessary (table 12.3).

Table 12.3. *Levels of over-exposure due to equipment malfunction at which HSE notification is necessary.*

Type of diagnostic examination	Multiplying factor
Barium meals and enemas, IVUs, angiography, Fluoroscopic examination (including digital) and CT	3
Extremities, skull, chest, dental and other simple examinations	20
Lumbar spine, pelvis, mammography and all other examinations not referred to above	10

In the event of a suspected over-exposure the dose must be estimated. If a dose–area product meter (see section 6.6) is fitted this is relatively easy by using the recorded DAP reading and the information in the NRPB SR-262 (Hart *et al* 1994) or, if a paediatric exposure, NRPB SR-279 (Hart *et al* 1996). If a DAP meter is not fitted, measurements using the same exposure factors (if possible) in free air must be made (and then adjusted for patient backscatter) to estimate the skin dose or alternatively a measurement can be made on a suitable phantom. The NRPB data can then be used. Sometimes the exposure factors cannot even be estimated and under these circumstances one must resort to calibrating X-ray film against known doses and stripping the emulsion off one side of the over-exposed film to reduce its density.

12.5.   X-RAY ROOMS

12.5.1.   *Introduction*

X-ray rooms must be designed to accommodate the equipment being used in them. This design requires the co-operation of the radiologist, the radiographer and the physicist. Generally the walls and the door constitute the barrier of the controlled area produced by the X-ray unit and access is controlled through the doors of the room. The shielding required in each wall, the ceiling and floor depends on the fraction of the total work load for which the X-ray beam is directed towards them when calculating the primary beam shield. Only the total work load need be considered when calculating the shielding required from the scattered and leakage radiation as these are considered isotropic. This work load can be averaged over a suitable period (eight hours in the UK) but many countries do not allow any adjustment for occupation of the area being shielded and always assume 100% occupation. This is a very conservative assumption in all but a very few cases.

The maximum permissible dose rates outside the walls after shielding depend on whether a member of the public or a hospital employee could be present. They would normally be 2.5 and 7.5 μSv h^{-1} respectively.

12.5.2.   *Points of note on room design*

(i)   The shielding can be achieved by a suitable thickness of concrete blocks, barium plaster on bricks or plaster lath or lead sheet. Lead sheet is not common because of cost and difficulty in hanging it.

(ii)   Doors must have an equivalent shielding to the walls. All cracks in the door frame must be covered by lead sheet. Sliding doors are attractive for saving space but can be difficult and tiring to move. This can result in them not being closed properly in a busy room.

(iii)   The radiographers' cubicle in the room should have sufficient shielding to ensure that the dose rate does not exceed 1 μSv h^{-1}. It should be positioned and have sufficient shielded windows in it to ensure the radiographer can see the patient at all times. It should also be large enough to accommodate every person who will require to go behind a shield unless supplementary mobile lead shields are also present in the room.

(iv)   A warning light should be positioned at the entrance to each room. Ideally this would be a two stage device with a 'controlled area' warning illuminating when the electricity is switched on at the unit and a 'do not enter' warning illuminating on 'prepping' the X-ray tube. Rooms relying on a single warning light must also have a notice indicating the action to be taken when the light is illuminated.

(v)   If there are two X-ray units in the room, warning lights must be placed on the head of each unit to indicate which one has been selected.

(vi)   Hanging facilities for lead–rubber aprons must be provided and storage for lead–rubber gloves and gonad shields.

(vii)   Emergency 'cut-off' and aid buttons must be positioned in suitable locations.

(viii) On completion of building or modification of a room, a Radiation Protection Adviser should carry out a 'critical examination' to ensure that all safety devices are working properly and that the shielding is as designed. The leakage dose rate from the unit in the room must not exceed 1 mGy h^{-1} at one metre from the X-ray head.

12.6.   NUCLEAR MEDICINE

12.6.1.   Introduction

The radiation hazards in a nuclear medicine department are the external hazard from the radiation field around the radiopharmaceutical container (bottle or patient) and the internal hazard that would result from ingestion of some of the radiopharmaceutical. Ingestion of the radiopharmaceutical could take place in one of three ways; through the mouth, through the nose or through the skin if a cut or abrasion is present. Some radioactive elements that can be found in a nuclear medicine department, in particular the iodine isotopes, can actually diffuse through unbroken skin. Once in the body the radionuclide is accumulated in the organ or tissue which would normally accumulate the stable element or, for some radionuclides, a close chemical analogue. Once incorporated, it is virtually impossible to remove the radionuclide medically. It will only leave by natural biological processes. As the radionuclide decays, all of the energy from any beta particles emitted will be absorbed in the body together with a portion of the gamma photon energy. Small ingested amounts of a radionuclide can result in significant radiation doses.

12.6.2.   Annual limit on intake

The dose received following ingestion of a radioactive material cannot be measured in the same way as external radiation doses. In order to ensure that the annual dose limit is not exceeded following ingestion, the ICRP have set an annual limit on intake (ALI) for all common radionuclides. This intake can be taken every year and the effective dose limit will not be exceeded in any year.

   ICRP recommends that the committed effective dose to a tissue T over an assumed working lifetime of 50 years ($H_{50,T}$) following intake of a radionuclide should be calculated. By summing over all tissues, allocating all of this dose to the year of intake and ensuring the dose limit for that year is not exceeded, the ALI will ensure that, even for radionuclides with a very long effective half-life, the limit for any year is not exceeded.

   For example assume

(1)   all exposure arises from internal sources,
(2)   the committed effective dose to tissue T over 50 years for unit intake (say 1 MBq) is $h_{50,T}$ and
(3)   the intake is I (measured in MBq). Then the total committed effective dose is $I w_T h_{50,T}$ summed over all tissues. Hence the annual limit on intake (ALI) is that value of I which satisfies the equation (for 20 mSv/y effective dose limit)

$$I \sum_{\text{all T}} w_T h_{50,T} = 20 \text{ mSv}.$$

To find the value of I, $h_{50,T}$ must be calculated. To do this, the factors to consider are:

(1)   the total energy radiated per unit time by the radionuclide, perhaps separated into that from charged particles and that from gamma rays;
(2)   absorption effects from gamma rays, originating both in that organ and elsewhere, and taking into consideration geometrical factors, the inverse square law and the mass absorption coefficient;
(3)   the distribution of radionuclide within the body;
(4)   the time scale of exposure. If the radionuclide localizes quickly, this may be expressed in terms of the effective half-life, but note that after leaving an organ the activity may localize elsewhere—e.g. the kidney excretes via the bladder.

   Methods of calculation and relevant data are contained in a series of publications from the Medical Internal Radiation Dose Committee (MIRD) of the Society of Nuclear Medicine (e.g. 1968, 1975). Exact calculation of the absorbed dose to an organ can be extremely difficult.

12.6.3.   *Derived limits for airborne and surface contamination*

When the ALI is known, the permissible derived air concentration (DAC can be calculated if simplifying assumptions are made. Assuming a working year of 2000 h (50 weeks at 40 h), a rate of breathing of 0.2 m^3 per min and that inhalation is the only route of intake

$$DAC = \frac{ALI}{2000 \times 60 \times 0.02} = \frac{ALI}{2.4 \times 10^3} \, Bq \, m^{-3}.$$

For Tc-99m the ALI is 9.1×10^8 Bq, hence the maximum permissible DAC, rounded to the nearest whole number, is 4×10^5 Bq m^{-3}. For I-131 the ALI is 9.1×10^5 Bq and the DAC 4×10^2 Bq m^{-3} (ICRP 1991, 1995).

Note that the concept should be used with care since it only applies to the ICRP reference man working under conditions of light activity and makes a number of other assumptions about the metabolic breathing pattern.

The derived working limit for surface contamination that will ensure the maximum permissible DAC is not exceeded varies from one radionuclide to another. For radionuclides used in nuclear medicine, it may be as low as 370 Bq spread over an area of 10^{-2} m^2. This causes two problems.

(1)   The activity 'seen' by a small detector may be no more than 20–30 Bq. A Geiger–Müller counter is insufficiently sensitive to detect this level of activity and a scintillation crystal monitor must be used.
(2)   It may be impossible to decontaminate to maximum permissible levels by washing and cleaning after even quite a small spill (say 1 MBq). If the radionuclide has a long half-life, contaminated equipment will then have to be removed from service. Fortunately for Tc-99m (6 h half-life) the contamination may decay to an acceptable level overnight.

12.6.4.   *Special precautions in nuclear medicine*

With unsealed radionuclides the two hazards identified in the introduction to this section must be controlled. Although small doses from the external hazard are most frequent, the internal hazard is potentially far more serious.

The external hazard can be greatly reduced by following relatively simple precautions.

(1)   Always keep bottles containing radionuclides in lead pots in a shielded area—never leave the vial unshielded.
(2)   Never handle bottles containing radionuclides but always use tongs.
(3)   If possible stay at least 0.5 m away from a patient who has received radioactivity, and check that nurses do not remain unduly close to the patient for unnecessarily long periods. The dose rate 25 cm away from a patient who has received 500 MBq of Tc-99m for, say, a bone scan is about 33 μSv h^{-1}. At a metre the dose rate is only 9 μSv h^{-1}.
(4)   Whenever possible, use shielded syringes to give injections.
(5)   If a shield cannot be used, extra care should be taken when handling the syringe not to hold the end containing the radionuclide.

The internal hazard can be avoided by simple good house-keeping practices. Protective clothing, especially gloves, should always be worn. Syringes do back fire and contaminated skin is difficult to clean. All manipulations of the radionuclide from bottle to syringe must be performed over a tray so that any drops can be contained. Syringes must always be vented into a swab, never squirted generally over the room. All contaminated and potentially contaminated materials must be disposed of in a container that has been clearly labelled as suitable for the purpose. *Always* wash and monitor your hands before leaving the radioactive area.

12.7. PERSONAL DOSIMETRY

Two fundamentally different techniques are generally used for personal dosimetry in medical practice—the film badge and the thermoluminescent dosimeter (TLD). The physical principles of both these techniques of radiation measurement were discussed in chapter 6 so only those aspects of their performance that are relevant to personal dosimetry are discussed here. This will be done by considering the requirements of a ideal personal dosimeter and the extent to which each method satisfies this ideal.

12.7.1. Range of response

The monitor must be sensitive to very small exposures since they will be the norm, but it must also be capable of recording accurately a high exposure should this arise in an accident. Hence a wide range is required.

Because of the shape of its characteristic curve, photographic film is useful over only a limited range of doses. However, the range can be extended if a fast film is backed by a much slower film. If the fast film is over-blackened, it is carefully removed and readings are obtained from the slower film.

There are no such problems with a TLD which has a wide range of response from 0.1 mGy upwards.

12.7.2. Linearity of response

If the response is linear with dose, measurement at just two known doses will allow a calibration curve to be drawn. This is possible with TLDs although not recommended, but for film, because of the shape of the characteristic curve, calibration is necessary at a large number of doses so that the exact shape of the curve can be established. Furthermore, calibration is necessary each time a batch of film is developed because of variations in film blackening with development conditions.

12.7.3. Calibration against radiation standards

A measure in terms of a fundamental physical property—e.g. temperature rise—would be desirable. However, neither film blackening nor thermoluminescence is in the category. Therefore calibration against a standard radiation source is necessary on each occasion.

12.7.4. Variation of sensitivity with radiation energy

As discussed in section 6.11, because of the presence of silver and bromine in film, blackening per unit dose is much higher at low photon energies where the photoelectric effect dominates than at higher photon energies. Similar differences in sensitivity exist for electrons. Therefore the film badge holder contains a number of filters (figure 12.1(*a*)), which not only extend the range of radiation energies over which the blackening per unit dose is approximately constant (figure 12.1(*b*)), but also provide data which may be used to calculate the dose for low energy photons and electrons. The cadmium filter will capture neutrons with subsequent emission of gamma rays so additional blackening under the filter is evidence of neutrons.

The sensitivity of a lithium fluoride TLD is independent of radiation energy down to about 100 keV, then increases slightly due to the small difference in atomic number between LiF and soft tissue. At even lower energies this effect is counter-balanced by self-absorption in the lithium fluoride–disc chip and the overall change in sensitivity with energy is unlikely to exceed 20%.

12.7.5. Sensitivity to temperature and humidity

The fog level of film can increase markedly under conditions of elevated temperature or humidity and this may be misinterpreted as spurious radiation. TLDs show no such effects.

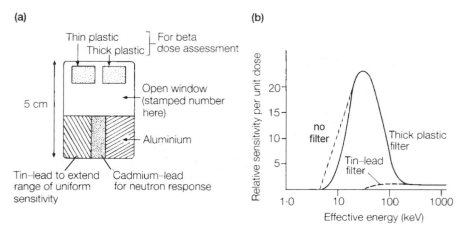

Figure 12.1. *(a) Details of a film badge holder. (b) Curves showing the variation of sensitivity (film blackening per unit dose) with radiation energy both without and with filters. Note that in the diagnostic range, correction for variation in sensitivity has to be made.*

12.7.6.   Uniformity of response within batches

Provided care is taken over storage and development, photographic films should nowadays be uniformly sensitive within a batch. The sensitivity of a TLD can vary within a batch so individual calibration may be necessary. Also careful annealing is required after use or the TLD tends to 'remember' its previous radiation history or change sensitivity.

12.7.7.   Maximum time of use

This is primarily governed by the risk of latent image fading when radiation is accumulated in small amounts over a long period of time. For both systems the effect is negligible provided that the time scale for calibration is comparable with the time scale for use. However, the effects of 12.7.5 restrict film to two months.

12.7.8.   Compactness

As shown in figure 12.1 a film badge holder is quite small. TLDs can be extremely small (figure 12.2) and are especially useful for monitoring exposure to the hands and fingers.

12.7.9.   Permanent visual record

When investigating the possible cause of a high reading (e.g. is it real or has it arisen because the dosimeter but not the individual has been exposed?), access to a permanent visual record can be useful. This is readily available with photographic film but with TLDs all the raw data are lost at read-out.

12.7.10.   Indication of type of radiation

Similarly, the pattern of film blackening can give useful information on the type of radiation as discussed above.

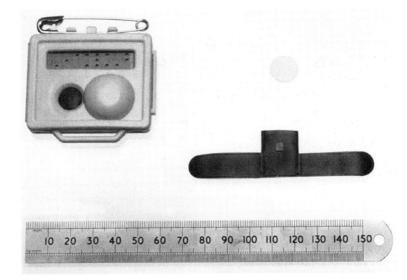

Figure 12.2. *Examples of TLD monitors. On the left is a badge holder for whole body monitoring. On the upper right is a LiF teflon disc. A very small LiF chip that would be useful for measuring extremity doses is shown on the lower right.*

12.7.11. Indication of pattern of radiation

Again, this information is occasionally useful and can be obtained from a film but not from TLD. For example, if a filter casts a sharp shadow, this suggests a single exposure from one direction but a diffuse shadow suggests several smaller exposures from different directions. The presence of small, intense black spots on the film suggests contamination with unsealed radioactive material.

Table 12.4. *Relative merits of film badges and thermoluminescent dosimeters as alternative methods of personal monitoring.*

	Film badge	TLD
1. Range of usefulness	0.2 mGy–6 Gy	0.1 mGy–10^4 Gy
2. Linearity of response	No	Yes
3. Calibration against radiation standards?	Yes	Yes
4. Response independent of radiation energy?	No	Yes (except at low kV)
5. Sensitive to temperature and humidity	Yes	No
6. Uniformity of response within batches	Yes	Yes (with care)
7. Maximum time of use	2 months	12 months
8. Compactness	Small	Very small
9. Permanent visual record?	Yes	No
10. Indication of type of radiation?	Yes	No
11. Indication of pattern of radiation exposure?	Sometimes	No
12. Cost to NHS	Expensive	Very expensive

12.7.12.   Cost

Both systems are relatively expensive. Photographic film has now become expensive and is a consumable. TLD dosimeters are re-usable but the capital cost for their purchase and for the read-out and annealing equipment is high. A typical cost in the UK at 1998 prices might be £2.00 per worker per month.

All these factors are summarized in table 12.4. There is no 'best buy'—each method is well suited to certain applications and both are used widely in the UK at the present time.

12.7.13.   Electronic personal dosimeters

For some time there have been available relatively small direct reading electronic personal dosimeters based on small Geiger–Müller tubes which can be clipped to a coat or carried in a pocket. Although reasonably accurate at high photon energies, because of the energy response of the Geiger–Müller tube, they very much underestimate the dose at low photon energies (below approximately 50 keV). They cannot therefore be used to show compliance with regulations.

An electronic personal dosimeter using simple photodiodes and microelectronics has however been developed which has an energy response as good as that of TLD and film. The photodetectors produce pulses of charge when exposed to ionizing radiation. These are integrated and the dose calculated by the device using an algorithm in the processor in each device. Each dosimeter has to be individully calibrated and adjusted. They can also record dose rate and activate in-built alarms if required to do so. Once recorded by the detector, no fading of the record takes place so the dosimeters can be worn for an indefinite period of time (subject only to battery life) and still give the accurate accumulated dose for the period. Although they can give no information as to how the dose was received, they are capable of measuring doses in the microsievert range which is lower than either film badges or TLD dosimeters can record. A special read-out instrument communicates with the detector using an infra-red beam. It is possible, by using linked computers through a modem, to read-out the dosimeter remotely thus removing the necessity to return the dosimeter to a central dosimetry service for read-out other than the periodic dose data check, battery change and function check.

Development work is still continuing on the device but at the time of writing the dosimeter cannot be used in all radiation environments. It can be affected by intense electromagnetic fields which sometimes produce spuriously high readings. Currently it is an expensive dosimeter to use in areas where there are no problems associated with operating a system using film badges or TLD dosimeters.

APPENDIX: PROTECTION ARRANGEMENTS IN THE USA

The legislation controlling use of ionizing radiation in medical practice in the USA is detailed and all embracing. As with European legislation it is based on the ICRP recommendations so the basic requirements and recommendations are the same as for the European legislation. Control is applied at the Federal, state and local level and as a consequence there is a great deal of variation in detail. Full implementation of the regulations produces a tighter control over the use of radioactive materials and, in particular, their disposal than in EU countries. However, the regulations controlling the use of X-rays are probably not as all embracing as those in force in Europe on the basis of EU Directives.

The main Federal agencies concerned are the Nuclear Regulatory Commission (NRC), the Food and Drug Administration and the Environment Protection Agency.

Nuclear Medicine

The issue of licences to persons allowing them to administer radionuclides to patients is a complex procedure. Several different types of licence exist and these are issued either directly by the Nuclear Regulatory Commission (NRC) or by State agencies acting on behalf of the NRC. Licences can be issued to individuals or to institutions which have a credible radiation safety committee. In the case of the latter, much greater flexibility exists as the use is at the discretion of this committee.

Individual licences will generally be for the holding and limited use of pre-packaged individual quantities of radiopharmaceuticals. These pharmaceuticals must have been themselves licensed by the Food and Drug Administration for use on humans. Individuals will have to demonstrate a level of expertise and knowledge and have access to the requisite equipment to carry out the investigation. Often these licences will not be all embracing but will be limited to certain procedures. Any additional work or modification of practice can only be undertaken if the licence is changed.

Diagnostic radiology

The qualifications required for performing general radiography vary widely between the various states and local jurisdictions. In principle any medical doctor can operate an X-ray facility for general radiography although in some states evidence of training is required. The level of training required is variable. Likewise the level of training required for the person operating the X-ray unit is not specified in detail. At the lowest level the practice secretary may carry out the exposures!

The exception to this is mammography. To carry out mammography the physician, radiographer (technician) and medical physicist must have appropriate qualifications, training and experience. This is a national requirement and applies in every state.

In hospitals the accreditation system ensures some degree of control. The Joint Commission for the Accreditation of Healthcare Organisations (JCAHO), the accreditation body, insists that personnel carrying out radiology and radiography are qualified. This would normally require an appropriate college degree, certification by the American Board of Radiologists (for radiologists) plus training and experience. The level required can vary from state to state.

The Nationwide Evaluation of X-ray Trends (NEXT) program carried out by the Food and Drug Administration's Center for Devices and Radiological Health has, since 1973, monitored the doses received by patients for specified radiological examinations using patient-equivalent phantoms. The use of phantoms in these surveys standardizes the measurement of skin-entrance air kerma and assessment of the image quality. Currently, one type of examination is investigated each year. As well as measuring the air kerma doses, comprehensive data with regard to the equipment used, the quality of film processing and the radiographic technique are also recorded. Mammography, in particular, has been surveyed on several occasions leading directly to the introduction of the mammography quality standards referred to above. This work has been complemented by the publication of 'Handbooks of Tissue Doses' by the Center and the Bureau of Radiological Health of the US Public Health Department which provide tables of the radiation doses received by a reference patient for a wide range of radiological procedures. The handbooks are distributed to medical physics and radiology departments. These data are used by state governments and by federal institutions as guidance when considering radiation doses to patients.

At the present time there is no formal program of inspection to see if the various guidance documents of the agencies are being implemented. Hospitals with medical physics staff would be expected to comply more or less completely. The situation where medical physics staff are not available is very variable. However, all radiological establishments could be subject to visits by field staff from the Food and Drugs Administration who would use the NEXT and other guidance in their surveys and reports.

REFERENCES

Administration of Radioactive Substances Advisory Committee (ARSAC) 1998 *Notes for Guidance on the Clinical Administration of Radiopharmaceuticals and Use of Sealed Radioactive Sources* (National Radiological Protection Board)

1980 Directive 80/836 Euratom (Basic Safety Standards) *Official J. Euro. Communities* No L246

1984 Directive 84/467 Euratom (Basic Safety Standards Amendment) *Official J. Euro. Communities* No L265

1990 Directive 90/641 Euratom (Outside Workers) *Official J. Euro. Communities* No L349

1996 Directive 96/29 Euratom (Basic Safety Standards Directive) *Official J. Euro. Communities* No L159

1997 Directive 97/43 Euratom (Medical Exposures Directive) *Official J. Euro. Communities* No L180/22

1995 *European Agreement concerning the International Carriage of Dangerous Goods by Road (ADR)*

Farr R F and Allisy-Roberts P J 1997 *Physics for Medical Imaging* (Philadelphia, PA: Saunders)

1988 *Guidance Notes for the Protection of Persons Against Ionising Radiations Arising From Medical and Dental Use* (London: HMSO)

1998 *Guidance Note PM77 (revised) Fitness of Equipment used for Medical Exposure to Ionising Radiation* (London: HSE)

Hart D, Jones D G and Wall B F 1994 *Estimation of Effective Dose in Diagnostic Radiology from Entrance Surface Dose and Dose–Area Product Measurements* NRPB-R262 (London: HMSO)

——1996 *Coefficients for Estimating Effective Doses from Paediatric X-Ray Examinations* NRPB R-279 (London: HMSO)

Hufton A P and Russell J G B 1986 The use of carbon fibre material in table tops. Cassette fronts and grid covers: magnitude of possible dose reduction *Br. J. Radiol.* **59** 157

Hughes J S and O'Riordan M C 1993 *Radiation Exposure of the UK Population—1993 Review* (London: HMSO) NRPB-R263

1990 *IAEA Safety Series 6—Regulations for the Safe Transport of Radioactive Materials* 1985 edn

Institute of Physics and Engineering in Medicine (IPEM) 1997 Recommended standards for the routine performance testing of diagnostic X-ray imaging systems *IPEM Report* **77** (York: IPEM)

International Commission on Radiological Protection (ICRP) 1977 Recommendations of the International Committee on Radiological Protection *ICRP Publication 26 Ann. ICRP* **1** (3)

——1991 Recommendations of the International Committee on Radiological Protection *ICRP Publication 60 Ann. ICRP* **21** (2/3)

——1992 Radiological Protection in Biomedical Research *ICRP Publication 62 Ann. ICRP* **23** (2)

——1995 Dose coefficients for intakes of radionuclides by workers *ICRP Publications 68 Ann. ICRP* **24** (4)

Ionising Radiation (Protection of Persons Undergoing Medical Examination or Treatment) Regulations 1988 *Statutory Instrument* **778** (London: HMSO)

Ionising Radiation (Outside Workers) Regulations 1993 *Statutory Instrument* **2379** (London: HMSO)

Ionising Radiations Regulations 1985 *Statutory Instrument* 1333 (London: HMSO)

1995 *Making the best use of a Department of Clinical Radiology* 4th edn (London: Royal College of Radiologists)

Medical Internal Radiation Dose Committee (MIRD) 1968 *Pamphlets* 1–3 (Maryville, TN: Society of Nuclear Medicine)

——1975 *Pamphlet* 11 (Maryville, TN: Society of Nuclear Medicine)

Medicines (Administration of Radioactive Substances) Regulations 1978 *Statutory Instrument* **1006** (London: HMSO)

1993 *Radioactive Substances Act* (London: HMSO)

Radioactive Substances (Hospitals) Exemption Order 1990 *Statutory Instruments* **2512** (London: HMSO)

Radioactive Substances (Testing Instruments) Exemption Order 1985 *Statutory Instruments* **1049** (London: HMSO)

Radioactive Material (Road Transport) (Great Britain) Regulations 1996 *Statutory Instrument* **1350** (London: HMSO)

EXERCISES

1   What do you understand by the ALARA principle? Give examples of its application in an X-ray department.

2   What is the purpose of local rules? What would you expect to find in the local rules for an X-ray department?

3   Summarize the radiological techniques that can reduce the dose to the patient.

4   Comment on the validity of the statment 'Any dose can be justified in diagnostic radiology'.

5   What precautions should be taken regarding radiation protection in a paediatric X-ray clinic?

6   As the consultant in charge of a department planning a new suite of rooms for neuroradiological investigations, what considerations would you have with regard to radiation protection when deciding on the structure and lay-out, and on the equipment installed?

7   List the precautions that must be taken when using a mobile X-ray image intensifier in an orthopaedic theatre.

8   What are the requirements of an ideal personal dosimeter?

9   What are the advantages and disadvantages of a film badge personal dosimetry system when compared with a thermoluminescent system?

CHAPTER 13

DIAGNOSTIC ULTRASOUND

T A Whittingham

13.1. INTRODUCTION

One of the principal attractions of ultrasonic imaging in its early days was its freedom from the harmful effects associated with ionizing radiations. Ultrasound continues to be one of the safest imaging modalities, but its growth to its present dominant role is also due to the unique nature of the diagnostic information it provides. This concerns the mechanical behaviour and properties of tissue, and is therefore fundamentally different to that provided by other imaging modes. The major limitations to ultrasonic investigations are the barriers presented by gas and bone, and its dependence on a high degree of operator skill and experience.

Ultrasound is sound with a frequency higher than the upper limit of human hearing, i.e. about 20 kHz. Medical diagnostic ultrasound mainly uses frequencies between 2 and 15 MHz, although up to 100 MHz may be used for very superficial structures, such as the cornea. In general, whereas the behaviour of X-rays and gamma rays can often be understood most easily by considering them as particles or photons, ultrasound is very much a wave phenomenon and visible light offers a more useful comparison. Thus, ultrasound travels at different speeds through different tissues, undergoes partial reflection at the boundary between two media and is subject to refraction, diffraction, scattering and absorption. Sound waves are, however, fundamentally different from light, X-rays, gamma rays and all other electromagnetic waves and particles in that they cannot travel through a vacuum. They exist only as vibrations of the particles of a medium and their propagation depends on the density and compressibility of the medium.

Wherever the density or compressibility of tissue changes in the path of an ultrasonic wave, echoes are sent back to the ultrasound probe. These may be weak reflections from the interfaces between different tissues, or even weaker scatter from the numerous small scale structures within tissue. In most applications, the diagnostic information from an ultrasound scan comes from scattered echoes rather than from echoes reflected from larger interfaces. Fortunately, ultrasound pulses travel at a fairly constant speed along narrow pencil beams, so that the direction and range of echo sources can be measured and plotted. However, distortions and artefacts do occur, and some understanding of the basic physics of sound propagation and of the techniques used in scanning equipment is necessary, if high quality scans are to be produced and their limitations appreciated.

13.2. ULTRASOUND WAVES AND THEIR PROPERTIES

13.2.1. Longitudinal and transverse waves

In gases, liquids and soft tissues, ultrasound takes the form of a **longitudinal wave**, consisting of high and low pressure disturbances which travel through a medium (figure 13.1). In describing the wave, the medium may be given a simplified representation as consisting of elastic 'particles'. These are small regions of identical material which can be considered to move as a unit. The particles temporarily acquire the wave energy as both kinetic energy and compressional energy (i.e. as a spring stores energy) and pass it on to their neighbours. They are alternately pushed forward and compressed as the pressure increases above normal, and then pulled back again and stretched as the pressure falls below normal.

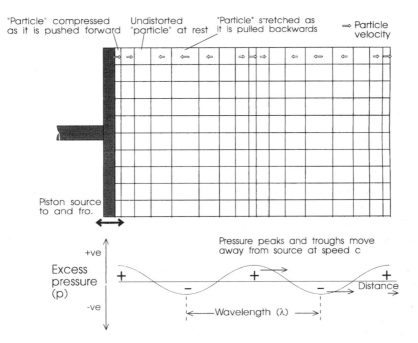

Figure 13.1. *Schematic representation of a longitudinal sound wave at one instant. 'Particles' move back and forth about fixed mean positions, being alternately compressed and pushed forward and stretched and pulled back. The pressure variations are passed from one layer of particles to the next at the speed of sound c.*

Another form of ultrasonic wave, known as a **transverse** or **shear wave**, can occur in solids. Here the local motion of the medium is perpendicular to the direction in which the wave is travelling, i.e. the particles are alternately forced from side to side rather than back and forth as in a compressional wave. Transverse waves do not propagate well in fluids or soft tissues because such media lack the very strong coupling between molecules (stiffness) needed for each layer of particles to exert a sufficient sideways force on the next. Where they do occur in tissues, such as bone or cartilage, transverse waves generally propagate only for short distances before all their energy is converted into heat. Henceforth in this chapter it should be assumed that longitudinal waves are being discussed.

13.2.2.   *Excess pressure and frequency*

The simplest type of wave to consider is a plane continuous wave (CW). In a continuous ultrasound wave the pressure variation follows a sinusoidal curve, repeating itself exactly every cycle. The frequency (f) of a continuous wave is the number of cycles occurring in one second. Frequency is expressed in units of hertz (Hz), or more typically megahertz (MHz) in the case of medical ultrasound, where one hertz is one cycle per second. The time duration of one cycle is known as the period ($=1/f$). When discussing sound waves, the term 'pressure' normally means the excess pressure or acoustic pressure (p). This is the contribution to the pressure at the chosen point, solely due to the passage of the sound wave. Note that, since it is the difference between the absolute pressure at any instant and that which there would have been in the absence of the sound wave, it can be either positive or negative. Figure 13.1 shows how the (excess) pressure at one instant in a medium varies with distance for a continuous sound wave.

13.2.3.   *Speed of sound*

The speed with which the pressure disturbances (both positive and negative) travel away from the source is known as the **speed of sound** (c). The speed of sound is a constant for any medium and is completely determined by the density (ρ) and compressibility of the medium. It does not, therefore, depend on the frequency of the wave. The relevant measure of compressibility is the **bulk modulus of elasticity** (κ), which is the ratio of the pressure applied to a fixed mass of medium to the fractional change in volume. It is high for relatively incompressible media such as solids, water or tissues, but low for compressible media such as gases. The speed of sound may be expressed in terms of κ and ρ by:

$$c = \sqrt{\frac{\kappa}{\rho}}.$$

In practice, tissues differ much more in compressibility than in density, so that bone (very incompressible, high κ) has a higher speed of sound than muscle, despite the fact that it is more dense. The mean value of the speed of sound in soft tissue is generally taken to be 1540 m s^{-1}. The speeds in individual tissues (table 13.1) vary from about 5% above this mean value (some samples of muscle) to about 10% below it (some samples of fat). The speed of sound in water is also in this range, being 1482 m s^{-1} at 20 °C, but increasing with temperature at the rate of about 3 m s^{-1} °C^{-1}.

13.2.4.   *Wavelength*

As shown in figure 13.1 the wavelength (λ) is the length of one cycle of the wave, measured in the direction of its travel. As for electromagnetic waves (see section 1.10), since wavelength is the distance travelled by a wave in one period, it is related to speed (c) and frequency (f) by:

$$\lambda = \frac{c}{f}.$$

Although ultrasound equipment is normally characterized by the frequencies of the waves it uses, wavelength is more useful than frequency when considering a spatial dimension such as the width of a beam or the length of a pulse. Taking an approximate value of c in tissue to be 1500 m s^{-1}, it may be seen that wavelengths associated with medical applications of ultrasound in tissues are typically between about 0.1 mm (for 15 MHz waves) and 0.5 mm (for 3 MHz waves). Note that the frequency of a sound wave stays constant as the wave passes from one medium to another, whilst the speed and wavelength may change.

13.2.5. *Characteristic acoustic impedance*

As described in section 13.2.1, the pressure rises as particles are compressed and pushed forward, and falls as they are pulled back and stretched. Since it is the convention to describe the velocity of a particle as positive when it is moving away from the source, and negative when it is moving towards the source, this means that particle velocities are positive when the excess pressure is positive, and negative when the (excess) pressure is negative. In fact, at any instant, the particle velocity (v) is directly proportional to the excess pressure p.

$$p = \text{constant} \times v.$$

The constant of proportionality between p and v in a particular medium is known as the **characteristic acoustic impedance** (z) of that medium. Thus the definition of z is:

$$z = p/v = \text{constant (characteristic acoustic impedance)}.$$

The value of this constant can be found by multiplying the density ρ by the speed of sound c in the medium concerned, i.e.:

$$z = \rho c.$$

The characteristic acoustic impedance does not depend on frequency. Nor does it vary by more that a few per cent between soft tissues, since neither the speed of sound nor density vary much between soft tissues (table 13.1). Typical values of z for water and soft tissue are around 1.5 kg m^{-2} s^{-1}. Note that this rather unwieldy formal SI unit is often referred to simply as a rayl, in honour of Lord Rayleigh who did much early work on the theory of sound (see also section 13.3.6).

The importance of z lies primarily in the fact that the strength of the echo produced at a reflecting interface is determined by the relative values of z in the two media on each side of the interface (see section 13.3.4). It is also important because it is needed in order to calculate the intensity of a wave from measurements of its pressure amplitude, as discussed next.

13.2.6. *Energy, power and intensity*

The source does work and gives energy to the first layer of particles of the medium as it pushes and pulls them. This energy is passed from particle to particle as the wave propagates, eventually being absorbed as heat. A single ultrasound pulse from a diagnostic scanner leaving the probe might typically carry with it a few microjoules (μJ) of energy. Over any specified time period, any source of ultrasound will transmit a certain amount of energy. Power is the rate at which energy is transferred, its unit being the watt (W), where 1 watt equals 1 joule per second. The rate of working by the source, and hence the transmitted acoustic power, varies from instant to instant. The instantaneous acoustic power is zero when the source momentarily stops and changes direction, and reaches its peak when it is pushing the adjacent medium forwards, or pulling it backwards, at maximum speed. The temporal average acoustic output power of the source is the total energy transferred from the source to the medium in every second. This may be up to a few hundred mW for a medical diagnostic scanner, which might typically transmit a few thousand pulses per second.

A quantity that is often of more interest than power is **acoustic intensity**. This is a measure of the local concentration of power, and is defined, as for X-rays (see section 1.10), as the energy flow per unit area per second, or the power per unit area (assuming the area considered is perpendicular to the direction of travel of the wave). Although defined in terms of an area, intensity describes the situation at a point. It equals the power that would be measured passing through a tiny area centred on the point, divided by that area. Strictly the SI unit of intensity is a watt per square metre (W m^{-2}), but ultrasound intensities are usually quoted in W cm^{-2} or mW cm^{-2}.

Just as with power, the instantaneous intensity (I_{int}) at any point varies from instant to instant, being zero whenever the particles are momentarily at rest and maximum whenever they are moving with maximum speed forwards or backwards. Considering it as the rate of working of a layer of unit area in pushing and pulling the next, it can be shown to be equal to the product of the pressure p and the particle velocity v. However, we have seen above that p and v are always proportional to each other, being related by the characteristic acoustic impedance z. It is thus possible to substitute p/z for v and obtain the more practically useful result:

$$I_{int} = p^2/z.$$

The temporal peak intensity (I_{TP}) at a particular point is the maximum value of I_{int} produced by the wave at that point. If the peak pressure within a continuous wave or pulse is p_{pk}, then $I_{TP} = p_{pk}^2/z$.

The temporal average intensity (I_{TA}) at a particular point is important for determining the temperature rise that might be produced in a sound absorbing medium. It equals $\frac{1}{2}p_{pk}^2/z$ for a continuous wave, but for diagnostic scanners the waves reaching a particular point will be a series of pulses, each probably with a different energy, so I_{TA} must be found by measurement of p^2 over the time period of the scanning sweep. The largest value of I_{TA} to be found at any point in the scanned area is known as the **spatial peak temporal average intensity** (I_{SPTA}). The value of I_{SPTA} that an ultrasound scanner might produce depends on the settings of the machine controls and the choice of scanning mode. If machines are set to give their greatest I_{SPTA} values, these are typically 100 mW cm^{-2} for B-mode and M-mode, 300 mW cm^{-2} for Doppler imaging and 1000 mW cm^{-2} for pulsed Doppler.

13.2.7. Decibels

The decibel is a special unit reserved for expressing the ratio of two energy-related quantities, such as two intensities, two powers or two energies. Consider the example of two intensities I_1 and I_2 that have a ratio of 2, i.e. $I_1/I_2 = 2$. To express this ratio in units of bels, one simply gives the logarithm (base 10) of the ratio. Thus, since $\log_{10} 2 = 0.3$, I_1 can be said to be 0.3 bels greater than I_2. In fact bels are rarely used directly, as it is more convenient to use the smaller decibel (dB) unit, where 10 decibels equals 1 bel. This means a two stage process of first finding the logarithm of the ratio and then multiplying by ten.

$$\text{No of dB} = 10\log_{10}\frac{I_1}{I_2}.$$

Thus, in the present example, it would be more usual to say I_1 is 3 dB greater than I_2. If we had chosen to compare I_2 to I_1 instead of the other way round, i.e. $I_2/I_1 = 0.5$, we would first find $\log_{10} 0.5 = -0.3$, and then multiply by 10 to give -3 dB.

The benefits of using decibels rather than ratios directly are twofold. Firstly the use of logarithms means that very large ratios can be expressed compactly. Thus, in scanning situations where one echo might have an intensity that is 10^{10} times that of another, it may simply be said that its intensity is 100 dB greater (since $\log_{10} 10^{10} = 10$). The second benefit derives from the fact that, where numbers are multiplied or divided, their logarithms are simply added or subtracted. Thus, suppose the energy of a pulse is reduced by a factor of $\frac{1}{2}$ due to one process and then by a further factor of 10 by a second process. Overall, the energy has been reduced by a factor of $\frac{1}{2} \times \frac{1}{10} = \frac{1}{200}$. Since dB are based on logarithms, an alternative approach would be to express the two factors in dB and simply add them together. Thus the first factor $(\frac{1}{2})$ is a reduction of -3 dB, and the second factor $(\frac{1}{10})$ is a reduction of -10 dB, giving a total reduction of -13 dB. This result could be converted back into the form of a ratio (using antilog$_{10}1.3 = 200$), but in those applications where decibels are used, results are normally left in dB form.

In practice, sufficient accuracy in converting between ratios and decibels can often be achieved just by remembering the two examples given above (a ratio of 2 is equivalent to 3 dB, and a ratio of 10 is equivalent to 10 dB). Any ratio that can be expressed in factors of 2 and 10 can then be converted to dB by adding 3 dB for every factor of 2, and 10 dB for every factor of 10. For example:

Ratio
4	$= 2 \times 2$	$= 3 \text{ dB} + 3 \text{ dB}$	$= 6 \text{ dB}$
80	$= 2 \times 2 \times 2 \times 10$	$= 3 \text{ dB} + 3 \text{ dB} + 3 \text{ dB} + 10 \text{ dB}$	$= 19 \text{ dB}$
1/100	$= 1/10 \times 1/10$	$= -10 \text{ dB} - 10 \text{ dB}$	$= -20 \text{ dB}$
1 000 000	$= 10^6$	$= 6 \times 10 \text{ dB}$	$= 60 \text{ dB}.$

One important use of decibels is for calculating the reduction in intensity due to a specified thickness of a particular tissue or other medium (see attenuation coefficient, section 13.3.3). Another is in calibrating the output power control of a scanner (e.g. -3 dB means the power is half the maximum).

Decibels are also used to express the ratio of the amplitudes of two waves or two electronic signals. As stated above, decibels are for use with energy or power quantities, so it is actually the ratio of the powers associated with the two amplitudes that is given in decibels. It is therefore necessary to square the two amplitudes before taking the logarithm, since the power associated with an ultrasonic wave is proportional to the square of the pressure (section 13.2.7), and the electrical power associated with an electrical voltage is proportional to the square of the voltage. A mathematically equivalent alternative to squaring the amplitudes is to use 20 instead of 10 in the dB formula. Thus, if A_1 and A_2 represent the two amplitudes

$$\text{No of dB} = 10 \log_{10} \left[\frac{A_1}{A_2} \right]^2 = 20 \log_{10} \frac{A_1}{A_2}.$$

Thus a doubling of amplitude is equivalent to a factor of four in power, or a change of 6 dB.

13.2.8. *Pulse waves, energy spectra and bandwidth*

The pulses used in medical ultrasound generally have a length of only about 2 cycles (figure 13.2(*a*)). Typically, peak positive and negative pressures are up to about a megapascal (MPa) or so (note that 1 MPa corresponds to about 10 times atmospheric pressure). Using $v = p/z$, with z approximately 1.5 Mrayl, shows that the corresponding peak particle velocities are typically less than 1 m s^{-1}. The peak back and forth displacements of particles are inversely proportional to frequency, but are also small, being typically around 0.01 μm.

Strictly, only a continuous wave can be characterized by a single frequency. However, as discussed in section 8.8.1, any waveform can be formed by adding together an infinite number of sinusoids with appropriate amplitudes and frequencies, the relative amplitude at each frequency being found using Fourier analysis. Although an infinite range of frequencies is needed to produce an exact equivalence, only sinusoids in a restricted frequency range will have a substantial amplitude. A plot of amplitude versus frequency is known as the amplitude spectrum of the pulse. Since energy is proportional to the square of amplitude, the energy spectrum (figure 13.2(*b*)) of the pulse, showing the relative energy at each frequency, is given by squaring each amplitude in the amplitude spectrum.

Two useful characteristics of the energy spectrum are the **centre frequency** (f_c), at which the spectrum has its maximum height, and the **pulse bandwidth**, which is defined as the width of the energy spectrum at half its maximum height. An important rule is that pulse bandwidth increases as pulse length decreases. In fact:

$$\text{pulse bandwidth (MHz)} = \frac{1}{\text{pulse length } (\mu s)}.$$

Thus a continuous wave, which might be considered to be a pulse of constant amplitude and infinite length, has an infinitely narrow bandwidth (i.e. it has a spectrum consisting of a single line). For a typical two-cycle imaging pulse, the bandwidth is about 50% of f_c. Thus, a '3 MHz' imaging pulse really means a pulse with a centre frequency of 3 MHz, but containing substantial energy at frequencies between about 2.2 MHz and 3.8 MHz.

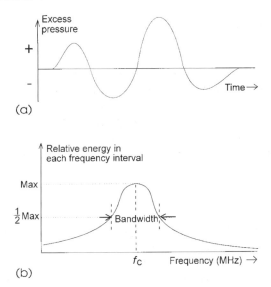

Figure 13.2. *Pressure–time waveform of a typical ultrasonic pulse and its energy spectrum. The pulse centre frequency (f_c) and the pulse bandwidth are indicated.*

13.3. THE PROPAGATION OF ULTRASOUND WAVES IN TISSUE

13.3.1. *Speed of sound*

Some typical values for c and z are shown in table 13.1. Note that since the speed of sound varies very little with frequency, it is not necessary to specify the frequency at which measurements are made. Note also that published data for a given soft tissue can vary greatly between studies, due to both variations between samples from different sites, and differences in measurement techniques.

13.3.2. *Attenuation*

Wave attenuation is the reduction of intensity with distance from the source. For a wave travelling through the body the causes of attenuation include divergence of the beam, partial reflection and refraction at tissue interfaces, and **absorption** and **scattering** within individual tissues.

  Absorption is the name given to the conversion of wave energy to heat. If a medium has a relatively simple molecular structure, like water, it reacts to the pressure changes of a sound wave in an elastic fashion, passing on the pressure stimuli without loss. (Strictly this is only true of waves of low amplitude; large amplitude waves in water become distorted due to non-linear propagation and lose intensity). However, even at low amplitudes, the large and complex molecules in tissue respond to changes in pressure in more complex ways, for example by changing their structure or vibrational behaviour. Such changes are reversible and are of no consequence to the wave if they take negligible time to occur. However, if the time occupied by each cycle is similar to that needed for these changes to occur, the wave energy 'borrowed' to implement them during the time of increasing pressure can be given back as an increase in pressure just as the pressure due to the original wave is falling again. This partially cancels out the amplitude of the ordered pressure swings, resulting in attenuation of the wave, and an increase in random motion, i.e. heat energy. Moreover, the fraction of wave energy converted into heat is the same in each cycle, so the more cycles there are per second, the greater will be the absorption. This means that absorption increases in proportion to frequency.

Scattering occurs whenever a wave meets a small obstacle about a wavelength or less in size. A useful analogue might be the scattering of ripples on the surface of a pond by a reed, where the scattered waves take the form of circular ripples spreading out from the reed. In tissue, ultrasound is scattered at all the innumerable tiny microstructures, wherever there is a change in compressibility or density. The energy of the scattered waves is transferred from the original wave, which is therefore attenuated. Scattering has both welcome and unwelcome consequences for ultrasonic imaging. Echoes produced by scattering are vital for differentiating between different tissue types and pathologies. Scattering is therefore discussed again in section 13.3.6 as a major cause of echo production, when it will be shown that the fraction of the incident power scattered by a small structure increases rapidly with frequency and structure size. From the point of view of attenuation, this means that the contribution of scattering, as well as that of absorption, increases with frequency.

13.3.3. Attenuation coefficient

The attenuation coefficient (α) of any particular homogenous tissue is the rate of decrease of intensity with distance of a plane wave (i.e. a wave that neither diverges nor converges) in that tissue. It takes account of absorption and scattering, but, by specifying a plane wave, it specifically excludes the effect of the beam shape (section 13.5), which would depend on the probe. Furthermore, it does not take account of any reflection or refraction effects (see below), such a might occur at tissue boundaries or at the boundaries of larger blood vessels or ducts within a block of tissue. Attenuation coefficient is normally expressed in units of dB cm^{-1}.

Unlike the speed of sound, the attenuation coefficient of tissues is very dependent upon the frequency of the ultrasound wave and varies considerably between tissues. For most soft tissues, it is found that, as the frequency f increases, α increases such that α/f remains approximately constant. This is consistent with the discussion on absorption given above. Approximate values for α/f are given in table 13.1. Note that for water and air, α is proportional to f^2.

The half-depth or thickness of tissue necessary to reduce the intensity of a wave by a factor of a half (-3 dB) is equal to $3/\alpha$ cm and values for this are also tabulated for a frequency of 3 MHz in table 13.1.

Table 13.1. *Ultrasound properties (approximate) of some human tissues at 37 °C, and other media (Duck 1990, [a]Kaye and Laby 1995, [b]Bass et al 1972).*

Medium	Speed of sound (m s^{-1})	Characteristic impedance (kg m^{-2} s^{-1})	α/f (dB cm^{-1} MHz^{-1})	Half-depth at 3 MHz (cm)
Blood (whole)	1585	1.68×10^6	0.2	1.2
Liver	1580	1.66×10^6	0.6	1.7
Muscle (skeletal)	1575	1.64×10^6	1	1
Fat (average)	1430	1.31×10^6	0.4–1.4	2.5–0.7
Bone (skull, outer)	2800	5.60×10^6	22	0.04
Water (20 °C, 1 atm)	1482	1.48×10^6	(0.0022 dB cm^{-1} MHz^{-2})	150
Air (20 °C, 10% humidity)	331[a]	392	(1.6 dB cm^{-1} MHz^{-2})[b]	0.2

13.3.4. Reflection

Wherever an ultrasound wave meets an interface where the characteristic acoustic impedance changes, a reflected wave is produced which carries with it a fraction of the power of the original wave. If the interface is smooth (on the scale of a wavelength) it is said to be a *specular reflector*, and behaves in the same way that a mirror (or partial mirror) reflects light waves. In particular, the angle of reflection equals the angle of incidence (figure 13.3(a)). This has important practical consequences since it means that where the source of the ultrasound is also the receiver, as in medical ultrasonic scanning, the wave reflected from a smooth surface can only be detected if the incident wave is perpendicular to the surface (figure 13.3(b)). This gives rise to incomplete visualization of some tissue boundaries, as described in section 13.7.10.

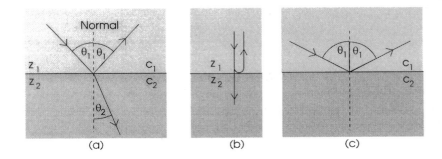

Figure 13.3. *(a) Partial reflection occurs when an ultrasound beam meets the boundary between two media of different characteristic impedances. If the speed of sound is different in the two media, as assumed here, the transmitted beam is refracted ($\theta_2 \neq \theta_1$). (b) Perpendicular incidence is assumed in the definition of reflection coefficient. (c) Total internal reflection occurs if $\sin \theta_1 \times c_2/c_1 \geq 1$.*

If the interface is large, smooth and flat, and the incident wave is plane and perpendicular to the interface, an intensity reflection coefficient R can be defined, equal to the ratio I_r/I_i, where I_r and I_i are the intensities of the reflected and incident waves respectively. Its value depends only on the ratio z_2/z_1 where z_1 and z_2 are the characteristic acoustic impedances of the media on the incident and transmission sides of the interface respectively.

$$R = \frac{I_r}{I_i} = \left[\frac{z_2/z_1 - 1}{z_2/z_1 + 1} \right]^2 = \left[\frac{z_2 - z_1}{z_2 + z_1} \right]^2.$$

Since all energy is either reflected or transmitted by the interface, the transmitted intensity I_t is equal to $I_i - I_r$. An intensity transmission coefficient (T) can therefore also be defined:

$$T = \frac{I_t}{I_i} = 1 - R.$$

If two media have the same characteristic acoustic impedance, i.e. $z_2 = z_1$, there will be no reflection from the interface between them. Furthermore, substitution of a few numbers quickly demonstrates that when z_2 and z_1 are very different, the fraction of energy reflected is high, irrespective of whether z_2 or z_1 is the larger. Therefore big differences in impedance must be avoided if one wishes to transmit appreciable intensities of ultrasound across an interface. This has great importance where one of the media concerned is a gas and the other is a solid or liquid, since gases have very low impedances. The very high reflection coefficient at such an interface (0.999 or greater) is the reason why ultrasound will not penetrate beyond the first gas-filled bowel or air-filled lung surface encountered, and why even the thinnest layer of air between the probe and the patient's skin must be excluded by using a coupling gel.

On the other hand, since the acoustic impedances of all soft tissues are rather similar, only a small fraction of the energy is reflected at a boundary between soft tissues. Even if two soft tissues at opposite ends of the impedance 'league table', such as fat and muscle (table 13.1), are considered, a boundary between them would have a reflection coefficient of around only 0.01. In other words the reflected wave would be a hundred times weaker (-20 dB) than the incident wave. Some soft tissue pairings, such as blood and blood clot, can produce reflection coefficients as low as 10^{-6} (-60 dB).

For many tissue boundaries, small surface irregularities produce weak scattered waves over a very wide range of angles. Such boundaries are described as diffuse reflectors by analogy with the way that a matt surface or ground glass plate produces diffuse reflection of light. The echoes they produce on an ultrasound image are weaker than those from a specular reflector, but they are much more likely to be registered, as they do not require that the interface is perpendicular to the incident wave direction.

13.3.5. Refraction

If an ultrasound wave meets, at an oblique angle, a boundary between two media having different speeds of sound, the transmitted wave will be deflected. This is known as refraction, and is illustrated in figure 13.3(*a*). The effect is analogous to that of a light beam meeting a glass or water interface. In common with optics, Snell's law applies:

$$\frac{\text{the sine of the angle of incidence}}{\text{the sine of the angle of transmission}} = \frac{c_1}{c_2}.$$

Here c_1 and c_2 are the speeds of sound in the first and second media respectively, and angles are measured from a line (*normal*) perpendicular to the boundary. The law shows that the transmitted beam is deflected further away from the normal when $c_2 > c_1$ or towards the normal (as in figure 13.3(*a*)) when $c_2 < c_1$. If $c_2 = c_1$ or if the beam strikes the boundary at right angles (regardless of the values of c_2 and c_1) then no refraction takes place. In soft tissues, because variations in the speed of sound are small, beam deviations are generally only slight, but they are often sufficient to degrade the image quality and produce image artefacts (section 13.7.10).

Where the speed changes from a lower to a higher value at an interface, and the angle of incidence is large, it is possible for the sine of the angle of transmission, as calculated from Snell's law, to be greater than 1. Since the sine of a real angle cannot be more than 1, this means there can be no transmission. The surface then acts as a complete reflector and the beam undergoes **total internal reflection** back into the first medium (figure 13.3(*c*)). This effect is one cause of edge effect artefacts, as described in section 13.7.10.

13.3.6. Scattering

As previously mentioned under attenuation, an ultrasound wave encountering a target that is smaller than a few wavelengths generates a much weaker scattered wave that radiates away in all directions. When the scattering structure is much smaller than a wavelength, the process is called Rayleigh scattering and the power (W) of the scattered wave depends strongly on frequency (f) and scatterer size (a), according to:

$$W = \text{constant} \times a^6 f^4.$$

This relationship shows that the contribution of scattering to attenuation increases with frequency, as mentioned previously. The strong dependence between the power (and hence amplitude) of a scattered echo and the size of the scatterer means that scattered echoes should offer a good way of differentiating between tissues. Although representing real tissue as a collection of discrete scatterers of a particular size is a gross over-simplification, differences between different tissues and between different pathologies should, and do, lead to different strengths and textures in the complex speckle patterns (section 13.7.10) produced by scattering from within them.

13.4.   THE ULTRASOUND TRANSDUCER

An ultrasound transducer is a device that converts electrical signals into ultrasound waves, and *vice versa*. Transducers used in medical ultrasound are based on the piezoelectric effect. Piezoelectric materials change in size when a voltage is applied across them and, conversely, generate a voltage in response to an applied pressure. The mineral quartz is a common piezoelectric material and was used in early ultrasound transducers. Many other naturally occurring substances are piezoelectric to some extent, including bone. However, modern medical ultrasound transducers usually consist of thin plates of lead zirconate titanate (PZT), a synthetic piezoelectric ceramic that can be cast or machined into discs, rectangular plates, bowls or any other desired shape. One important disadvantage of PZT is its high characteristic acoustic impedance. The large reflection coefficient (section 13.3.4) this produces at its faces poses substantial problems, as will become evident below.

Figure 13.4 shows a cross sectional diagram of a transducer, as might be used in the probe of a stand-alone A-mode or M-mode system, or in the probe of a single element mechanical sector scanner. The same basic features are associated with each transducer element of a transducer array probe (sections 13.7.4–13.7.6). The opposite faces of the thin PZT transducer plate are coated with a conducting layer of silver to form electrodes, to which electrical leads are attached. By applying an oscillating voltage between these leads the plate can be made to expand and contract (typically by only a few μm) at the same frequency. Both faces of the plate therefore act as piston sources of ultrasound waves. In reception, the pressure variations produced by returning echo waves cause the plate to expand and contract accordingly, and thus generate proportional (analogue) voltage variations between the two electrodes. These form the electronic echo signals that are further amplified and processed by the receiver.

As with all vibrating mechanical structures, the amplitude of the thickness vibration of the transducer is greatest at certain 'resonance' frequencies. Resonances occur when a wave returns from a round trip across the transducer and back (after partial reflections at each face), in phase with itself. Waves that have completed one or more round trips will then combine with each other, and the original wave, to produce particularly large amplitude vibrations. The particular resonance that gives the greatest amplitude of vibration for a thin disc is called the **half wave resonance**, since it occurs when the round trip distance equals one wavelength, i.e. the disc thickness is half a wavelength. Maximum transmission and reception efficiency can be obtained by using a PZT disc which has its half wave resonance occurring at a frequency equal to the required ultrasound frequency. Thus, for half wave resonance at 3 MHz, a PZT disc with a thickness of 0.67 mm would be used (the wavelength in PZT at 3 MHz would be 1.33 mm, using $\lambda = c/f$ and taking c in PZT as 4000 m s^{-1}).

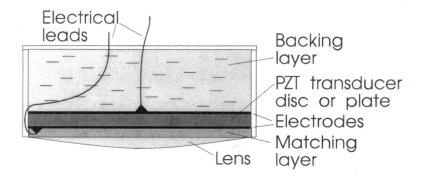

Figure 13.4. *Cross section of a typical ultrasound transducer.*

Short pulses are necessary for good axial resolution (section 13.7.1). However, when the electrical excitation stops, the ultrasound waves within the PZT continue to travel back and forth for some time. If the transducer were to vibrate in air, the reflection coefficient of each face would be so large that the vibrations would continue for thousands of cycles. The continuation of the vibration after the excitation has finished is called **ringing**, since it is precisely the same phenomenon that occurs within the metal of a bell when it is struck.

Even with one face placed against tissue and the other backed by air, the ringing would continue for tens of cycles. It is therefore necessary to reduce this ringing time by placing an absorbing backing layer, with a high acoustic impedance, close to that of PZT, immediately next to the back face of the transducer. The matching of the two impedances on each side of the back face reduces its reflection coefficient, and so allows some of the 'trapped' wave energy to pass more easily into the backing, where it is absorbed as heat. Placing a hand against a vibrating bell has the same damping effect.

If the front face of the PZT transducer were to transmit directly into tissue, the large reflection coefficient of the front face would severely impair the transducer's efficiency as a transmitter and a receiver. Early scanners had such poor sensitivity due to this problem that they had to employ frequencies as low as 1 MHz for abdominal imaging. The high attenuation associated with the use of 3 MHz, which is a more typical 'abdominal' frequency today, would have resulted in echoes that were too weak to be detected. In modern probes, the efficiency of transmission between the transducer and the patient's tissue is improved by having a thin impedance matching layer immediately next to the front face of the PZT.

The simplest form of matching layer (quarter wave matching layer) is made from a $\frac{1}{4}\lambda_{PZT}$ thick slab of low attenuation material having a characteristic impedance of $(z_{PZT} \times z_{TISSUE})^{1/2}$, i.e. intermediate between the PZT and tissue. Reverberations (repeated crossings due to partial reflections at each face) within this layer lead to the transmission of a diminishing sequence of substantially overlapping pulses into the patient. These add together to form a large resultant pulse in the patient, and so improve the transmission efficiency of the transducer. In reception, the transmission of tissue echoes back into the PZT is similarly enhanced by the same process acting in the opposite direction.

A single matching layer can produce 100% transmission into the patient for continuous waves, since such waves have a very well defined frequency and hence the layer thickness can be made precisely $\lambda/4$. However, as discussed earlier (section 13.2.8), short pulses are characterized by a frequency spectrum rather than a single frequency. This means that a quarter wave matching layer can provide perfect matching only for one frequency, usually chosen to be the centre frequency of the pulse spectrum. The efficiency of a single matching layer is therefore limited to a narrow range of frequencies. Multiple matching layers, consisting of several layers with various thicknesses and impedances, can give efficient matching over a wider range of frequencies, and therefore far superior results for short pulses.

In practice, a lens is usually incorporated between the matching layer and the patient, but this can be made from a material with a characteristic impedance close to that of tissue so that it does not affect impedance matching considerations.

13.5. ULTRASOUND BEAMS

The shape and size of an ultrasound beam, and the intensity variations within it, are determined by the size and shape of the transducer, and by the wavelength. The process is one of diffraction, whereby the instantaneous pressure at a given point in the beam is equal to the sum of contributions from every point on the transducer radiating face. Depending on the wavelength and the differences in distance from each point on the transducer face to a given beam point, the sum will be large or small. It will be large if most contributions arrive in phase (e.g. all producing positive pressures at the same instant), but will be small if, at each instant, as many contributions arrive producing positive excess pressures as those producing negative excess pressures.

Ultrasound of wavelength λ emitted from a flat disc transducer of radius a will travel as a slightly converging beam for a distance a^2/λ (figure 13.5), before diverging. At the range a^2/λ the width of the beam, as measured between points where the intensity is half that on the axis (-3 dB width), achieves its narrowest value of about a. Since the total power in any beam cross section should be the same at all ranges (assuming no absorption or scatter), this is therefore the range at which the on-axis intensity reaches its maximum. The region before this 'last axial maximum' is called the **near field** or **Fresnel zone**, and that beyond it is called the **far field** or **Fraunhofer zone**. The near field is characterized by a complex pattern of peaks and troughs

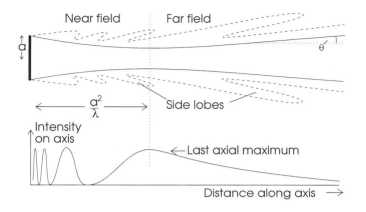

Figure 13.5. *Unfocused beam shapes produced by a flat disc transducer of radius a.*

of intensity. In the far field, the beam diverges at an angle θ to the axis, where $\sin \theta = 0.6\lambda/a$. It may be seen from these formulae that the higher the frequency (smaller λ) the longer the near field length and the less the beam divergence θ in the far field. High frequency unfocused beams are therefore better collimated than low frequency ones.

If the transducer is rectangular (sides $2a$ by $2b$), the beam shape has to be considered separately in the two planes that are respectively parallel to the two sides. In each plane the same basic features of near and far fields are produced, but in the plane parallel to the side of length $2a$, the near field length is a^2/λ and the far field divergence is given by θ_a where $\sin \theta_a = 0.5\lambda/a$. In the other plane these two formulae are the same, except that b replaces a. This transducer shape is relevant to the individual elements of linear and phased array probes (section 13.7.5 and 13.7.6).

A beam may be focused by using a bowl-shaped transducer, a concave mirror or, more usually, by inserting a lens in front of an unfocused transducer. This concentrates the wave into a narrower beam for a limited depth interval, known as the **focal zone** (figure 13.6). However, since the increased convergence of the beam before the focus is matched by an increased divergence beyond, the beam width at large range will be greater than that of the unfocused beam. The narrower beam width in the focal zone is accompanied by an increase in intensity, so that both lateral resolution and sensitivity (section 13.7.1) are improved in this region. The distance (F) between the transducer and the position of narrowest beam width (*focus*) is known as the focal length.

If the focal length is less than about $0.5a^2/\lambda$ (figure 13.6(a)), then the beam is said to be strongly focused, with a very narrow but relatively short focal zone. At the focus, the width (w_F) of a strongly focused beam from a transducer of radius a and focal length F is:

$$w_F = \frac{F\lambda}{a}.$$

Thus, as for unfocused beams, high frequencies (small λ) produce narrower beams. This formula also shows that a wider aperture would be required to achieve the same focal beam width for a deeper focus.

If the focal length is more than about $0.5a^2/\lambda$ (figure 13.6(b)), then the beam is said to be weakly focused, with a wider but longer focal zone. Where the focal length cannot be varied, weak focusing is preferred to strong focusing. This is because the region of greatest clinical interest might be anywhere from just beneath the probe to near the limit of penetration, and hence a long, moderately narrow focal zone is of more value than a very narrow but short one.

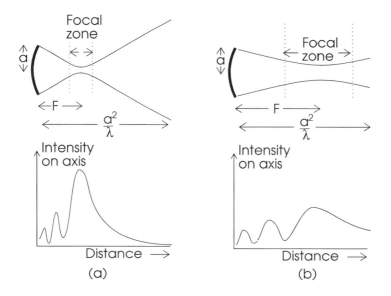

Figure 13.6. *(a) A strongly focused transducer. (b) A weakly focused transducer. The focal length F is measured from the transducer to the focus, at the narrowest part of the beam.*

The above simplified description of a beam actually refers only to its main lobe. At a more detailed level, the main lobe is flanked by weaker side lobes. These effectively widen the beam and manufacturers must take them into account when designing transducers for high performance imaging. Their relative amplitude can be reduced by a technique known as **apodization**, in which the radiating strength of the transducer is made non-uniform, generally being greatest at the centre and reducing towards to the edges. Unfortunately, this is a compromise process, since the decrease in side lobe amplitude is always accompanied by an increase in the width of the main lobe.

The discussion so far has considered a transmission beam. Transducers acting as receivers have receive beams, defined as the region in which a point source of sound would produce a detectable signal. If the same transducer acts for both transmission and reception (at different times, of course), the shape and size of the receive beam will be identical to that of the transmit beam. The difference is that a three dimensional plot of the receive beam would show the variation of receive sensitivity to a point source at different positions, whereas the transmit beam would show the peak pressure or intensity that might be measured at each point. In practice, the use of transducer array probes (sections 13.7.4–13.7.7) means that there are differences between the transmit and receive beams, because different apertures, apodization and focal lengths are used for transmission and reception. A further category of beam is the transmit–receive beam of a scanner. This depends on both the transmission beam and the receive beam, and a plot of this would show the variation of received signal amplitude produced by a point scatterer at different positions.

13.6. ULTRASONIC IMAGING MODES

13.6.1. *Pulse–echo range finding*

The pulse–echo method used in ultrasound imaging is based on the assumption that the speed of sound in all soft tissues is approximately constant. The system velocity assumed for most applications is 1540 m s^{-1}, corresponding to a 'go and return' time of 1.3 μs for every millimetre of range. Thus it is possible to deduce the range of a reflecting or scattering 'target' from the time elapsed between pulse transmission and the return of an echo from that target. A second assumption is that the ultrasound beam is straight and infinitely narrow, so that all echoes can be assumed to come from targets situated on the beam axis (scan line). In practice, neither assumption is totally justified, leading to some of the artefacts and limitations discussed later.

In all imaging modes the rate of transmission of pulses, called the **pulse repetition frequency (prf)**, must be sufficient to allow the scanner to keep up with any changes in the anatomical region of interest. However, an upper limit to the prf is set by the requirement that there should be time for the echoes from each pulse to return from the deepest reflecting or scattering structures before the transmission of the following pulse. Otherwise late echoes from one transmitted pulse will be confused with early echoes from the following pulse. Thus, for a 15 cm deep tissue region, the rule of 1.3 μs for every millimetre of range means the echo from the deepest target will return after approximately 200 μs. The highest prf that avoids range ambiguity is therefore equal to the number of times 200 μs goes into 1 second, i.e. about 5000 pulses per second. Higher prfs are clearly possible for scanning more superficial regions.

13.6.2.   A-mode scans

In an A-mode scan, the amplitudes of echoes along a single scan line are represented as vertical deflections on a display screen, with time elapsed since transmission (proportional to range) represented by the horizontal axis (timebase), as shown in figure 13.7. When an echo pulse is detected, a vertical deflection, proportional to the echo amplitude, is produced on the screen. The horizontal distance between two vertical deflections ('echoes') reveals the difference in range of the two reflecting interfaces in the tissue.

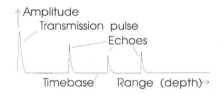

Figure 13.7. *A-mode scan display. The vertical co-ordinate represents echo amplitude; the horizontal co-ordinate indicates time elapsed since transmission (proportional to range of target). A-mode scans of a selected B-mode scan line are sometimes displayed with the time axis (timebase) running vertically alongside a B-mode image.*

A-mode scans provide the most accurate way of measuring the amplitudes of echoes and the distance between two targets. Dedicated A-mode equipment is rarely manufactured today, but A mode is sometimes provided as an additional feature on some B-scan equipment, for example in eye scanners.

13.6.3.   M-mode scans

This scan mode shows a two dimensional plot of the positions of reflecting surfaces or scatterers along a single scan line versus 'physiological time', i.e. as measured in seconds rather than the microseconds associated with ultrasound travel time (figure 13.8). It is used primarily in cardiac work, usually in conjunction with B mode (next section). The probe is manipulated so that a highlighted scan line on the B-mode image runs through the moving interfaces of interest (e.g. valve leaflets, chamber walls). On activating M mode, the B-mode scanning action stops and all further transmitted pulses are sent down the selected scan line. The echoes resulting from each transmission pulse are presented as brightness modulations along a corresponding line of the M-mode display, with the brightness (grey level) indicating echo amplitude. Each transmission–reception sequence results in a new M-mode line, displayed alongside the previous one, the data being continuously scrolled across the display. Some machines save up to say 10 seconds worth of the data that has been scrolled off the display, allowing it to be replayed later if required. There is often provision to show other physiological wave forms, such as an ECG or intra-cardiac pressure waveform, alongside the echo data on the display.

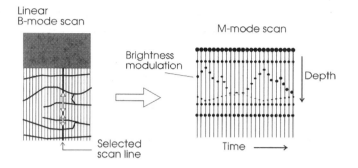

Figure 13.8. *M-mode scan display. Echo brightness (or grey level) indicates echo amplitude. The vertical co-ordinate represents range (depth) while the horizontal co-ordinate represents time.*

13.6.4.   B-mode scans

In a B-scan (figure 13.9), the beam axis (scan line) is automatically swept sideways through the patient's body to define a **scan plane**. Before the advent of real-time B-mode scanners, this beam sweep was achieved by moving a transducer by hand (such scanners are now referred to as 'static' B scanners). In modern scanners, the beam is swept automatically by a hand-held probe (section 13.7). The angle and position of the scan line, relative to the probe, are electronically monitored. Knowing the position and orientation of the scan line and the range of a target, the position of the target can be plotted on the display at the correct position. Echoes obtained from all scan lines build up on the display to form a two dimensional image of the scan plane. The amplitude of each echo determines the brightness (grey level) of the corresponding point on the display.

Note that the positions of all the targets on the B-mode image are displayed in relation to the principal axis of the probe, not true vertical. The usual convention is for targets situated on the principal axis of the probe to be shown on the central vertical scan line of the display, whatever the orientation of the probe in reality. Some machines, however, allow the operator to rotate the entire image through 90° or 180°. This may be helpful, for example, when viewing an image of a prostate, using an upward directed scan plane from a transrectal probe in a supine patient.

13.6.5.   Doppler imaging

These imaging modes include colour flow mapping and power Doppler imaging. They show the distribution of moving targets, particularly blood cells, generally as a colour overlay on the grey scale B-mode image. They are discussed further in section 13.10.

13.7.   B-MODE SCANNERS

13.7.1.   Factors that are important for the quality of a B-mode image

The width of the imaged region (*field of view*) is determined by the type of probe, linear probes offering the greatest width for superficial targets and sector probes giving greater widths for deep targets (figure 13.9). The

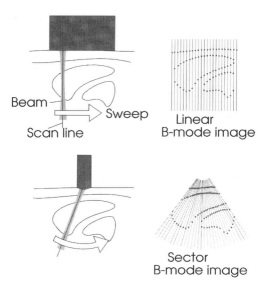

Figure 13.9. *Linear and sector B-mode scanning. The scan line (beam axis) is swept sideways to interrogate the scan plane. For each scan line, echoes are shown as brightness or grey level modulations (determined by echo amplitude) at the appropriate distance along a corresponding line on the display.*

maximum depth from which diagnostically useful echo information can be obtained is called the penetration. Use of a higher frequency means an increase in attenuation and hence a decrease in penetration. As transducer technology has improved, particularly in regard to impedance matching (section 13.4), the penetration to be expected for a particular frequency has increased. Sensitivity is a measure of the weakest scatterer or reflector that can be distinguished from noise. This varies with depth, generally being greatest close to the transmit focus. The sensitivity achievable at maximum depth determines the penetration. At any depth, sensitivity depends on the setting of the output power, the focus position and overall gain controls (section 13.7.9). A measure of the maximum sensitivity that a scanner can achieve is revealed by the quoted **dynamic range** of the scanner. This is the ratio of the intensity of the largest ultrasonic echo that can be handled by the receiver electronics without saturation (peak clipping) to that of the weakest echo that can be distinguished from electronic noise. Dynamic range can be deliberately reduced below its maximum by the dynamic range control (section 13.7.9).

The degree of spatial detail that can be seen within an image is described by the *spatial resolution*. There are two different forms of this for a B-mode image, both being better at higher frequencies. **Axial resolution** can be defined as the smallest distance between two point targets, lying one behind the other on the same scan line, that allows them to be resolved as separate targets. **Lateral resolution** may be defined in a similar way but applies to two targets lying side be side at the same range. A discussion of the spatial resolution limits for small targets in scattering media is deferred until *speckle* is considered in section 13.7.10, but the results are the same as those deduced below for simple targets in a liquid.

Axial resolution is approximately equal to half the length of the transmitted pulse, as may be seen by reference to figure 13.10. If one target is situated a distance d behind the other, the echo from the far target will be a distance $2d$ behind the echo from the nearer target as these echoes travel back to the probe. If d is reduced to half the pulse length, the distance between the leading edges of the two echoes will be equal to the pulse length, i.e. the beginning of the second echo will just coincide with the end of the first echo. Any further reduction in d would mean the two echoes would merge into one elongated echo, and the two targets

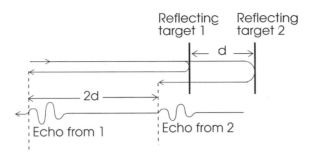

Figure 13.10. *Showing that axial resolution depends primarily on pulse length. If the distance d between the two interfaces is less than half the pulse length, the echo from the first interface will overlap with the reflected echo from the second interface.*

would be seen on the image as one long target. Pulse lengths are typically about two wavelengths, and since wavelength is inversely proportional to frequency, the use of higher frequencies means shorter pulse lengths, and hence better axial resolution.

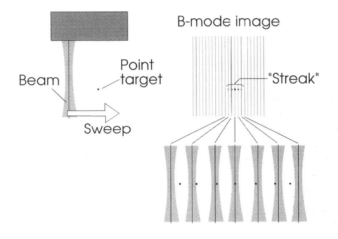

Figure 13.11. *Showing that lateral resolution depends on the width of the transmit–receive beam. As the beam passes over a point target it will be registered on several scan lines, producing a 'streak' with a length equal to the beam width.*

Lateral resolution is approximately equal to the width of the transmit–receive beam, as may be seen by reference to figure 13.11. If the beam is scanned across a small isolated target, a weak echo will be returned as soon as one side of the beam reaches it. This echo will register as a dim spot on the display scan line representing the beam axis, which is, as yet, half a beam width away from the target. As the beam passes over the target, successive transmitted pulses will produce stronger and stronger echoes until the axis of the beam passes through the target. Thereafter echoes of reducing strength will be returned until the beam has completely left the target behind. These will be registered as spots at the same range on a group of adjacent display scan lines, having maximum brightness on the central lines of the group. The point target is thus shown on the image as a line of spots (a 'streak') perpendicular to the scan lines, with a width equal to that of the transmit–receive beam. If two point targets lie side by side, they will both produce similar 'streaks', and these will merge into one if the targets are separated by less than the beam width. Beam widths become smaller, and hence lateral resolution becomes better, as frequency increases. Beam width varies with depth, being least, and hence lateral resolution being best, in the focal zone (section 13.5) of the transmit beam.

Contrast resolution is the name given to the ability of a scanning system to differentiate between two echoes, or two regions of a scan, on the basis of their echo amplitudes. It will be seen later (section 13.7.9) that this depends on the machine's ability to record echo amplitude with sufficient precision as a digital number, and on the grey scale transfer characteristic selected by the operator. It also depends on the width of the beam in the plane perpendicular to the scan plane, which defines the slice thickness (section 13.7.10)

Temporal resolution describes the ability to follow changes with time in the imaged tissue. It depends on the frame rate (which itself depends on the depth and width of the field of view and on the scan line density) and the degree of frame averaging that is selected (section 13.7.9).

13.7.2.   *Beam forming*

The first stage of all B-mode scanning systems is the beam former, concerned with generating, focusing and scanning the beam. In a single element mechanically scanned system, as described in section 13.7.3, this may be no more than a single transducer and a mechanical system to drive it and to monitor its position. In systems using array probes (sections 13.7.4–13.7.7) the beam forming stage includes delays, switches, transmission pulse generators, preamplifiers and their control logic. Many machines allow part of the field of view to be zoomed for improved resolution or frame rate (section 13.7.8), a procedure that is also essentially controlled by the beam former.

The stages after the beam former are common to all types of scanning system and so will be treated as a separate topic in section 13.7.9, after the beam forming techniques used by different types of scanner have been discussed.

13.7.3.   *Single element mechanical sector scanners*

This is the simplest form of scanner. It uses a single transducer with a weak fixed focus set at approximately half the intended maximum penetration depth. In the 'rocker' type of probe (figure 13.12), the transducer is driven by an electric motor and suitable coupling mechanism, so that the transducer, and hence the beam, oscillates through an angle of typically 45° on each side of the probe's principal axis. An angle sensing device keeps track of the direction of the transducer at all times so that echoes can be correctly positioned on the display. An alternative 'spinner' form of mechanical sector scanner uses a continuously rotating wheel with a number of transducers embedded around its rim. As the wheel spins, each transducer is swept in turn past an acoustically transparent window, thereby sweeping the beam through a sector. This arrangement is bulkier, but transducers with different centre frequencies and focal lengths can be accommodated within the one probe. With either type, as in all B-mode systems, each sweep of the transmit–receive beam generates a new frame, showing all echo-producing structures in their new positions.

The sector field of view of this type of probe is particularly appropriate to heart imaging and those abdominal situations where superficial acoustic barriers such as ribs, lungs or gas pockets restrict the acoustic window at the surface of the patient's body. It is also ideal for imaging the neonatal brain through the anterior fontanelle. These advantages are shared by other sector scanners, such as annular array scanners, considered next, and phased array scanners (section 13.7.6).

13.7.4.   *Annular array mechanical sector scanners*

These are very similar to the type just discussed, except that the transducer is constructed in the form of an annular array of concentric ring shaped transducer elements (figure 13.13). Use of an array transducer allows

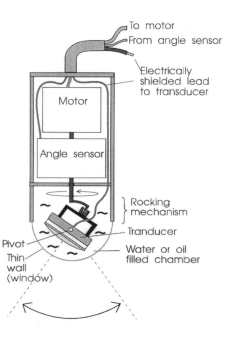

Figure 13.12. *Simplified representation of a 'rocker' type of mechanical sector scanner.*

the focus and other characteristics of the transmit and receive beams to be varied, thereby improving the lateral resolution and contrast resolution over the full depth range of the image. Each transducer element in the array operates independently of the others. It has its own pulse generator, which delivers an oscillating voltage to excite the element and thus produce an ultrasonic pulse, and its own pre-amplifier for giving initial amplification to the echoes.

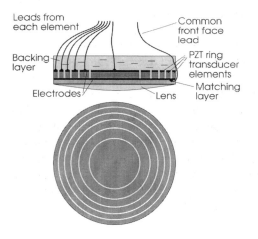

Figure 13.13. *Representation of an annular array transducer. Mechanical sector scanners that incorporate an annular array transducer, in place of a single element transducer, have excellent lateral resolution and slice thickness over a wide range of depths.*

In transmission, the operator can position the focus at the same depth as any region of interest, thereby improving the lateral and contrast resolution there (figure 13.14). A 'trigger' signal undergoes different time delays on the way to each pulse generator. These delays are chosen to compensate for the differences in path length (and hence travel time) between each element and the selected transmission focus. This ensures that the relatively weak individual pulses from all the elements arrive at the chosen focus at the same time, thereby reinforcing each other and producing a large resultant pulse. Appropriate time delays are pre-programmed for each possible setting of the transmission focus control. At any point away from the selected focal zone, the individual weak pulses from each element arrive at different times, producing no more than low amplitude background acoustic noise.

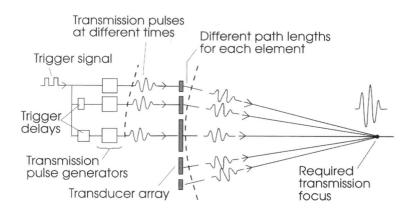

Figure 13.14. *The transmission focal length of an annular array transducer may be varied by the operator to minimize beam width (and hence optimize lateral resolution) at a particular depth. The principle is also used with linear and phased arrays but, for these probes, only the beam width in the scan plane is affected.*

In reception, a variable receive focus is achieved in a similar way by introducing time delays into the electrical signal paths of the returning echoes. These delays are pre-calculated to compensate for the small differences in travel time between the required receive focus and each element. When the delayed echo signals are summed together they produce a strong response only from a reflector at or near the focus. However, in contrast to the transmission situation, the operator is not required to select a receive focus. Instead, the machine automatically changes the receive delays every few μs over the period during which echoes return, advancing the range of the receive focus at an average rate of 1 mm every 1.3 μs (section 13.6.1). This ensures the receive focus always keeps up with the depth of origin of the echoes. This **dynamic focusing** technique produces an effective receive beam that consists of a string of narrow focal regions lying end to end, maintaining good lateral resolution from near the surface to the maximum depth of penetration (figure 13.15). Note that the outer elements are not used when focusing close to the probe, since they cannot contribute significantly to on-axis points so close to the probe.

The idea of stringing together the narrow focal zones of several beams to form a long narrow effective beam can also be applied to transmission. In a technique known as multiple zone focusing in transmission, several transmit–receive sequences are carried out along the same scan line. In each transmit–receive sequence the transmitted pulse is focused at a different range, and only those echoes arriving at times corresponding to ranges close to that focus are detected and stored. The operator is free to choose up to say four different transmit foci, positioned at whatever ranges he wishes. The more focal zones chosen, the greater depth range having improved lateral resolution. However, there is a frame rate penalty, since each additional transmission focus involves an additional transmit–receive sequence. This increases the time spent interrogating each scan line, and therefore each frame. In some advanced designs, the centre frequency of the transmitted pulse is

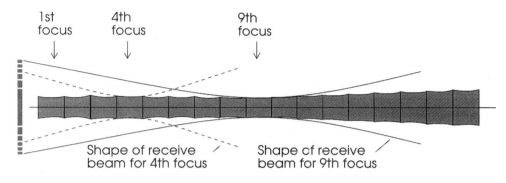

Figure 13.15. *The effective receive beam produced by dynamic focusing (and aperture) in reception. It remains narrow, and hence maintains good lateral resolution, from near the probe to the maximum scan depth. A beam with a fixed focus (say the ninth in this figure) would be wider than the 'effective beam' at all depths other than that of its focus.*

changed at the same time as the transmission focal length. This ensures the optimum resolution is achieved at all depths.

The independent electrical access to each element of the transducer array also means that the apodization (section 13.5) of the transducer array can be varied, in order to optimize the beam profile and side lobe level for different target depths. In transmission the voltage amplitude of the transmission pulses to individual elements may be varied. In reception the gain (amplification factor) of the individual pre-amplifiers may be varied.

Note: It is important to realize that the principles of variable focusing, aperture and apodization that have been introduced here as part of the description of an annular array transducer, also apply to linear array and phased array probes, as described next. In fact, from a historical perspective, they were first introduced for linear array scanners and only later applied to annular array mechanical scanners.

13.7.5. *Linear and curvilinear array scanners*

A simplified diagram of a linear array probe is shown in figure 13.16. A typical example might have 128 narrow rectangular transducer elements, each only about two wavelengths wide. The transducer array is usually constructed by taking a slab of polarized PZT with electrodes on both sides, and cutting channels through it to form separate elements. There is a separate electrical lead to the rear electrode of each element, but all elements share a common front electrode and lead. An 'active group' of adjacent elements is electronically selected to act as the beam forming transducer. This active group, and hence the beam, is stepped across the array by progressively dropping an element from one end of the group and adding a new one to the other end (figure 13.17).

As noted above, all the electronic beam forming techniques described in connection with the annular array scanner can be used in a linear array scanner, including operator-controlled transmission focus, automatic dynamic focusing in reception, variable active aperture and apodization. However, unlike annular arrays, which have two dimensional circular symmetry, linear arrays extend in one dimension only. Consequently electronic beam forming only controls the beam characteristics in the scan plane. However, in this plane, a narrow effective beam can be maintained to a much greater depth than is possible with other types of scanner. Recalling the formula $w_F = F\lambda/a$ (section 13.5), the ratio F/a, and hence the focal beam width (w_F), can be kept constant up to large values of F if the aperture ($2a$) is increased in proportion to F. Linear array scanners can achieve this by progressively adding extra elements to the active group as the receive focus is increased.

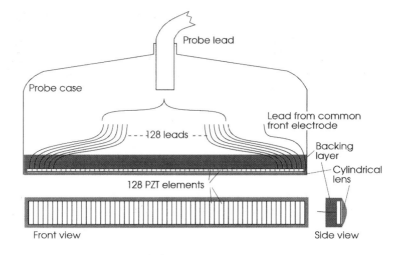

Figure 13.16. *Simplified cross sectional views of a linear array probe.*

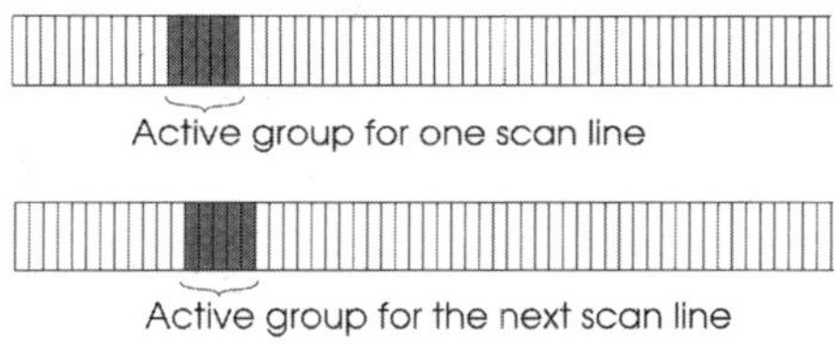

Figure 13.17. *The beam may be stepped along a linear array by dropping a transducer element from one end of the active group and adding one to the other.*

Most stop this aperture expansion when the active group contains around 20 elements, but some continue until it includes all the 128 elements (say) in the entire array. This requires more complex and expensive beam forming electronic hardware, but produces excellent lateral resolution for deep targets.

In the dimension perpendicular to the scan plane (elevation plane), the beam characteristics are fixed, determined by the width of the array in this dimension and a weakly focusing cylindrical lens running the length of the array. Some recent linear array probes provide a degree of variable beam forming in the elevation plane by subdividing the array of elements into several rows of smaller elements (figure 13.18). Such probes are sometimes referred to as '1.5D' arrays since they go part way to a full 2D transducer array. They allow additional control of the signals to and from all the elements in a given row, allowing the focus, aperture and apodization in the elevation plane to be varied, and thus improve slice thickness (section 13.7.10) and hence contrast resolution.

The rectangular field of view of linear arrays makes them popular for scanning superficial structures ('small parts') such as in the neck, breast, scrotum and limbs. Curvilinear array probes, which are basically linear array probes constructed with some degree of curvature (figure 13.19(*a*)), provide a wide field of view close to the probe and an even wider one at depth. They are widely used for obstetric and some abdominal applications, but cannot be used where the need to push the convex front face into full contact with the patient would cause unacceptable distortion of superficial structures. This problem has been overcome in the 'trapezoidal' scan format now being offered by some linear array probes (figure 13.19(*b*)). This involves a combination of linear array beam stepping with the beam steering technique described next.

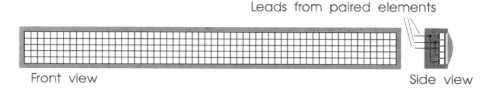

Leads from paired elements

Front view   Side view

Figure 13.18. *A '1.5D' linear array has several rows of transducer elements, allowing dynamic beam forming in the elevation plane, and hence producing an effective slice thickness that is narrow over a large depth range.*

A disadvantage of linear and curvilinear arrays is the presence of grating lobes. These are weak replicas of the beam, at angles of 30° or so from the main beam. They contribute unwanted echoes and so degrade contrast resolution

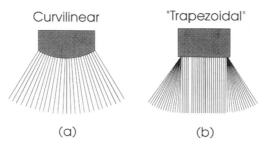

Curvilinear   "Trapezoidal"

(a)   (b)

Figure 13.19. *(a) Curvilinear arrays, and (b) linear arrays with peripheral beam steering, offer a wide field of view superficially, becoming even wider at depth.*

13.7.6.   *Phased array scanners*

A phased array probe also uses electronic means to form and scan the beam, but the beam is steered rather than stepped, producing a sector field of view. Again there may be about 128 elements, but this time they are less than half a wavelength wide, producing a much narrower probe. Unlike a linear array, all the elements are used to generate each beam in the sweep. Beam direction is controlled by delaying the pulses to and from each element to exactly compensate for the extra path length between each element and the desired focus. The principle is very similar to that already described for focus control with linear and annular arrays, except that the transmission focus and reception foci now lie on angled scan lines (figure 13.20). Again, as for linear array and annular array probes, variable aperture, apodization and focusing are possible in both transmission and reception. However, as in the case of a linear array, such electronic beam forming only affects the beam shape in the scan plane. In the dimension perpendicular to the scan plane, a cylindrical lens provides fixed, weak focusing.

An important advantage of phased arrays over mechanical sector scanners, particularly in heart scanning applications, is their ability to perform mixed mode scanning. Here, both a B-mode scan and an M-mode scan are displayed simultaneously. Since there is no moving transducer, there is no mechanical inertia associated with the scanning, and the beam can be made to jump virtually instantaneously from scan line to scan line in any sequence. This makes it possible for the beam to interrogate say three scan lines of the B-mode scan, then jump to a highlighted M-mode line and interrogate that, then jump back to continue the B-mode scan

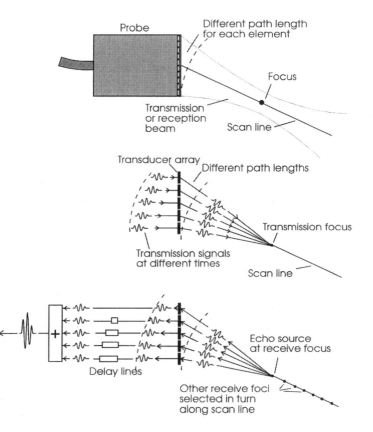

Figure 13.20. *Beam steering by a 'phased array' probe. The principle is similar to electronic focusing, except that delays are chosen to produce a focus on an angled scan line. Dynamic focusing in reception involves generating a sequence of receive foci, closely spaced along the same scan line.*

for three more lines, then back to the M-mode line, and so on. The two scans are seen to proceed together in real time, with the advantage that fine adjustments to the probe position, and hence the positioning of the M-mode line in the heart, can be monitored on the B-mode image. Similar mixed mode scanning is also possible with linear and curvilinear array probes, facilitating, for example, the M-mode recording of foetal heart movements.

A disadvantage compared to mechanical scanners is that the transmit–receive beam width increases, and hence the lateral resolution becomes poorer, towards the lateral edges of the sector scan. This is because the effective width of the transducer aperture is smaller when 'seen' from a scan line at an angle to the probe axis, and the beam width is inversely proportional to this aperture (section 13.5).

13.7.7. Intra-corporeal probes (endoprobes)

Intra-corporeal probes allow the transducer to be situated nearer to the tissue region of interest. The requirement for less penetration means that a higher frequency may be used, and consequently lateral and axial resolution are improved (section 13.7.1). Image quality is further improved by the reduction of beam distortion in the intervening tissue. As an example, an oesophageal probe for scanning the adult heart might have a centre frequency of 5 MHz, compared to the 3.5 MHz that would be more typical for a probe scanning through the chest wall.

In principle, any of the beam forming and scanning methods discussed above may be incorporated into an intra-corporeal probe. Thus the transducer could take the form of a suitably compact mechanical sector scanner (single element or annular array), a linear or curvilinear array or a phased array. A further choice is offered by radial scanning, in which the beam is scanned through 360° around the axis of the probe, in the manner of a circling lighthouse beam. Radial scanning probes commonly use a single element transducer, of fixed focal length, that is mechanically rotated within the probe (e.g. 5–7 MHz trans-oesophageal and trans-rectal probes), or on torque wire passed through a catheter (e.g. 30 MHz intra-arterial probes). However, modern developments in transducer array technology mean that electronic radial scanning of intra-catheter probes is now possible. This avoids problems due to the torque wire winding up, or the catheter tip vibrating excessively. It also permits electronic control of focal length. The technique uses a cylindrical array around the tip of the probe, and may be thought of as an extreme form of curvilinear array, where the curvature extends through 360°.

The particular scanning method chosen, the frequency and the overall shape and size of the endoprobe, depend on the anatomical constraints of the particular application. Thus a linear array, mounted on the side of a rigid cylindrical probe, is convenient for imaging the prostrate trans-rectally, whereas a forward looking curvilinear array mounted at the end of a trans-vaginal probe is better suited to imaging the ovaries or an early pregnancy. Sometimes, as in trans-oesophageal imaging of the heart, two compact phased arrays are mounted close together at right angles to each other, providing orthogonal B-mode sections of the target area.

13.7.8. *Write zoom*

This restricts the scanning action to a smaller region of interest, which the operator defines by positioning a box on the unzoomed image. The screen is then filled with a magnified real-time image of the selected area. The time saved by not scanning outside the box leads to an improvement in frame rate. In some applications, some of the potential increase in frame rate is traded for an increase in lateral resolution. For example, the time consuming technique of multiple zone focusing in transmission (section 13.7.4) may be used to produce a narrower effective transmission beam and hence improved lateral resolution.

13.7.9. *Amplifying, processing and displaying the echo signals*

Radio-frequency (RF) amplification and time gain compensation TGC

The weak electronic echo signals produced at the end of the beam forming stage must be amplified to a level at which they can be digitized and further processed. This is complicated by the fact that the desired echo signals are always accompanied by random voltage variations known as electronic noise. Contributions to this noise are generated at all frequencies, so the wider the range of frequencies that an electronic circuit can amplify, the greater the amplitude of the noise generated by that circuit. Thus, in order to keep the signal-to-noise ratio as high as possible, the amplifier should be able to handle just the range of frequencies that is in the spectrum of the echo pulses, and no more. This type of frequency selective amplifier is known as a radio-frequency (RF) amplifier, since it was first developed to amplify signals generated by radio waves.

The substantial and frequency dependent attenuation suffered by ultrasound in tissue has implications for the design of the RF amplifier. One effect of this attenuation is that the echo signals from targets deep within the body are much weaker than from identical targets closer to the surface. Thus, if 3 MHz ultrasound is used to image two identical targets in liver (attenuation coefficient of 0.6 dB cm^{-1} MHz^{-1}), one being 15 cm deeper than the other, the two echoes will differ in intensity by 54 dB (= 0.6 dB cm^{-1} MHz^{-1} × 15 cm × 3 MHz × 2, allowing for a two way trip). Because the strength of a reflected or scattered echo is used to indicate the nature of a target, this reduction in echo intensity with target depth would make interpretation very difficult. Simply amplifying all echoes more might make the smaller echo signals easier to see on the image, but the larger signals would become so strong that they would saturate the electronics and become indistinguishable from each other. To overcome this problem, **swept gain**, or **time gain compensation (TGC)**, is always used. The gain of an amplifier is the ratio of the amplitude of the output signal to that of the input signal. Swept gain means that the gain of the RF amplifier is increased with time so that the strong, early, echoes arising from targets close to the surface are amplified by a small gain factor, and the later echoes from more distant targets are amplified by a large gain factor (figure 13.21(*a*)). The gain versus time function is set by the operator, and may be a smooth progression (figure 13.21(*b*)) or more complex (figure 13.21(*c*)), according to the clinical application and the type of swept gain control provided. If the swept gain is correctly set, scattered echoes from within a particular type of tissue will be shown at a fairly even brightness at all depths, and echoes from larger interfaces with similar reflecting properties will be displayed at similar brightness levels.

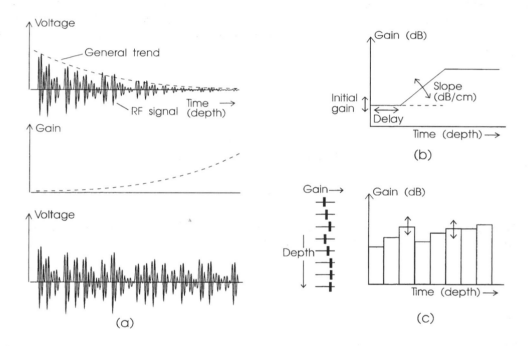

Figure 13.21. *(a) TGC attempts to compensate for the effect of attenuation by progressively increasing the amplification (gain) received by echoes as they return from greater and greater ranges. TGC controls may allow a simple increase of gain with depth (b), or independent control of gain in different range intervals (c).*

The dependence of echo strength on depth is more marked at higher frequencies because attenuation in tissue increases with frequency. Repeating the above calculation for a frequency of 6 MHz shows that the echo from the deeper target would be 108 dB weaker than from a similar target 15 cm nearer the probe in the liver. Hence, the spectrum of a deep echo will have relatively less high frequency energy than that of an echo from a closer target. The centre frequency and bandwidth (section 13.2.8) of echoes arriving at the transducer therefore become progressively lower as they return from increasingly greater depths. This means the beam width and pulse length become greater, and hence the lateral and axial resolutions become poorer, for deeper targets.

The variation of echo spectra with target depth is taken into account in high performance ultrasound scanners, where the centre frequency and bandwidth of the RF amplifier is continuously changed with time to match the expected changes in the echo spectra. The progressive reduction in amplifier bandwidth is accompanied by a reduction in electronic noise, leading to worthwhile improvements in sensitivity, contrast resolution and penetration for deeper targets.

An overall gain control is also provided to set the overall amplification given to all echoes, irrespective of depth. This, and the output power control, which determines the amplitude of the transmitted pulses, determine the sensitivity of the scanner.

Amplitude demodulation

Apart from the target position, the only echo information that is needed for the B-mode display (or A- or M-mode display) is the echo amplitude. It is therefore necessary to remove the RF oscillations within the echo signal to leave a simple pulse with the same amplitude as the original (figure 13.22). The receiver stage that achieves this is called the amplitude demodulator, and the simplified train of echo pulses it produces is called the demodulated signal.

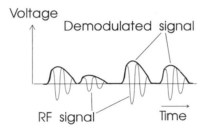

Figure 13.22. *Amplitude demodulation produces a simplified echo waveform with the same peak amplitude and a shape equal to the half envelope of the original.*

Digitization

The continuously varying voltage produced by the transducer is, in effect, an analogue of the pressure variation at the transducer produced by the returning echoes. Thus, the RF echo signal, the demodulated signal, or any signal in the form of a continuously changing voltage, may be described as an analogue signal. In order to take advantage of the power of digital signal processing technology, the analogue echo signal must first be converted into a digital signal (see sections 7.3.5 and 8.7). Increasingly, high performance scanners carry out this digitization process on the RF signal, allowing beam forming to be achieved digitally. Other scanners digitize the signal after amplitude demodulation.

The number of bits in the digital signal is crucial to contrast resolution and dynamic range (section 13.7.1). For example, an eight bit **analogue-to-digital converter** (ADC) can produce 256 different digital numbers (0–255), whilst a 12 bit ADC can produce 4096 (0–4095), allowing finer discrimination between signal amplitudes and hence better contrast resolution. Recalling that the dynamic range of a scanner is the ratio of the largest to the smallest of a range of signals that can be processed, the 12 bit system will have a dynamic range of 4095:1, while that of the eight bit system would be only 255:1. The dynamic range limit imposed by the ADC can be expressed in decibels, since every additional bit means a doubling of the largest number, and a doubling of amplitude is equivalent to an increase of almost exactly 6 dB (section 13.2.7). Thus the dynamic range following a 12 bit ADC is 72 dB, while that following an eight bit ADC is 48 dB.

Dynamic range compression

There appears to be relatively little benefit in displaying ultrasonic images with more than the 256 grey levels that can be stored in a typical eight bit deep image memory (see below). From what has just been said, this is a dynamic range of 48 dB. However, echoes from within the patient have a very much greater dynamic range. The echoes reflected from gas or bone may be well over 100 dB more intense than echoes scattered from microstructures within tissue or from groups of blood cells. The dynamic range of echoes at the transducer will have an even greater dynamic range, since attenuation will make echoes from deep targets much weaker than those from near targets. Ideally, the effect of attenuation will be removed by the TGC, so the dynamic range of echoes leaving the TGC stage should be only that due to differences in the reflection and scattering coefficients of soft tissue structures. Nevertheless, it is necessary to change (*compress*) this 100 dB or more dynamic range to match the 48 dB of a typical eight bit image memory. This may be achieved in two ways.

The first is with the dynamic range or reject control, which allows the operator to reject echoes below a certain threshold. Thus, selecting a dynamic range of say 40 dB means that only echoes with amplitudes within 40 dB of the maximum would be displayed. Reducing the dynamic range may be appropriate if an operator decides that very low level echoes are not of interest, although it should not be done without good reason as it represents a loss of information. A reduced dynamic range should lead to an improvement in contrast resolution for the remaining echoes, since the full range of display grey levels is available to the now smaller range of echo amplitudes.

The second means of compressing dynamic range is the transfer characteristic or pre-processing control. This determines the relationship between the amplitude of an echo (after the TGC and demodulation stages) and the number stored in the image memory. The most obvious relationship would be if the latter were linearly proportional to the former, giving a transfer curve or transfer characteristic in the form of a straight line (figure 13.23). However, many applications of diagnostic ultrasonic imaging are concerned with differentiating one tissue mass from another on the basis of the strengths of the low amplitude scattered echoes. One example is looking for tumours in tissue parenchyma. The contrast between such echoes is enhanced if most of the available range of grey levels is dedicated to the lower amplitude echoes, with larger amplitude echoes sharing only a few of the whiter grey levels. The operator can achieve this by selecting a low level enhancement transfer curve from the pre-processing options. In some applications, for example in the orbit or breast, the greatest contrast is required for distinguishing between echoes of medium amplitude, in which case an S curve, allocating most of the grey scale to mid-amplitude echoes, would be more appropriate. Note the similarity of this discussion and that in section 8.7 (figure 8.9).

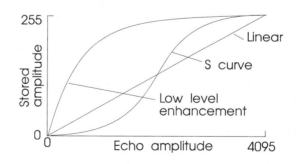

Figure 13.23. *Examples of pre-processing transfer curves. By selecting the curve most suited to the application, the operator may improve contrast resolution.*

Image memory

The image is presented as a rectangular matrix of small picture elements (*pixels*), each having a grey level determined by the amplitudes of the echoes from the corresponding part of the scan plane, following the TGC and compression stages described above. Each pixel has a horizontal and a vertical address, and for each pixel a number defining the grey level is stored at the same address in an image memory. Although one pixel represents only a tiny area of tissue, nevertheless there will be a short sequence of echoes from a scan line crossing that region, rather than a single echo. The operator may be offered a choice of sampling algorithms to determine how a single number is chosen to represent the amplitude of this short echo sequence. Choosing the echo amplitude at the instant corresponding to the first edge of the pixel is simple but, since the signal has a random electronic noise component as well as genuine echo information, this leads to a noisy image; averaging the digital echo samples across the pixel region is a more common method; storing the difference between average values for a pixel and its neighbour further along the scan line produces an image in which boundaries are emphasized.

A further choice to be made by the operator is how much the number stored for each pixel should change from frame to frame. It might be thought that this number should be simply replaced on a new for old basis in every frame. However, again electronic noise accompanying the genuine echo information may cause a particular pixel to be unduly bright in one frame and unduly dark in a later frame. This noise may be reduced by frame averaging, whereby the value stored for each pixel is the average of the previously stored value and that generated in the new frame. This leaves unchanging genuine echo information unaffected, but averages out the time-varying noise component. Thus, with a 50:50 averaging ratio, the number stored for each pixel would be 50% of the previous stored number plus 50% of the number representing the echo amplitude in the new frame. A higher averaging ratio, such as 80:20, would reduce noise even more effectively but would lead to a longer persistence effect. A lower averaging ratio, such as 20:80, would give a noisier image but a faster response to changes in the tissue, or movements of the probe from one site to another (better temporal resolution).

The process of storing echo amplitude information in the correct memory locations is known as writing. At the same time, the process of reading takes place, in which the same memory locations are interrogated row by row, in synchrony with the TV display scan raster. Because data are written into memory in the sequence defined by the ultrasonic scan format, but read out in a TV raster scan sequence, the process is called scan conversion. By selecting frame freeze, an operator can inhibit the writing action. The reading action continues, allowing the TV monitor to receive and display the now unchanging stored image data. Ideally, selecting frame freeze should also stop the probe from transmitting and thus prevent the patient receiving unnecessary ultrasound exposure, but this should not be assumed. Post-processing is a facility for modifying the relationship between the number stored in each memory location and the brightness or grey level of the corresponding pixel on the display. It contributes along with pre-processing to the overall transfer characteristic, relating the displayed grey level to the post-TGC echo amplitude. Post-processing has an advantage over pre-processing in that it can be done on a stored image but, since the memory usually has a smaller dynamic range than the ADC, it is less important than pre-processing as a means of adjusting image contrast.

A facility known as **read zoom** is sometimes provided. This allows the operator to select an area of interest within the frozen image and to magnify it to fill the TV screen. The read interrogation of the image memory is restricted to the reduced set of memory locations corresponding to the defined area. The data from this smaller area are thus displayed across the full extent of the screen. Interpolation between data from adjacent memory locations is necessary, otherwise the pixels of the zoomed image would be large and conspicuous. Read zoom does not have the benefits of improved frame rate and lateral resolution that are provided by write zoom (section 13.7.8).

13.7.10.   *Artefacts of B-mode images*

Image artefacts are misrepresentations of tissue structure in the ultrasonic image. This may be in the grey level, size, position or even the presence or absence of an echo. Experience and a good knowledge of the relevant anatomy allow most artefacts to be recognized. Some common artefacts and their causes are discussed below.

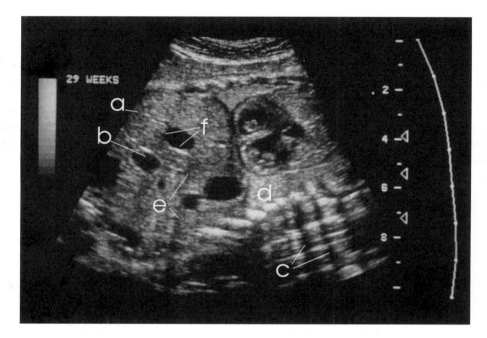

Figure 13.24. *Example of a B-mode image showing a number of artefacts: (a) speckle pattern, (b) reverberations, (c) shadowing, (d) post-cystic enhancement, (e) edge shadows, (f) incomplete visualization of specularly reflecting boundaries.*

Speckle pattern

Close examination of a B-mode image shows that scattering tissue parenchyma is represented by a complex mottled (speckled) pattern of bright and dark regions (figure 13.24). Although, for stationary tissue, this speckle pattern repeats itself exactly from frame to frame, it is random and there is no one-to-one correspondence between an individual bright 'speckle' and an individual scatterer in the tissue. The numerous individual scatterers are too close together, and their individual echoes are too weak, to be individually resolved on an image. At any instant, echoes will arrive from targets distributed over a volume defined by the time since transmission, the beam cross section and the pulse length. These will add together (*interfere*) to form a resultant signal that is either weak or strong depending on the precise spatial distribution of the scatterers within that volume. The resultant signal will therefore vary randomly from large (bright) to small (dark) along each scan line. Bright intervals on several adjacent scan lines make up a bright speckle.

Although each speckle is different in size and shape, the average speckle width (perpendicular to the scan lines) is equal to the beam width, and the average speckle height (along the scan lines) is equal to half the pulse length. A cyst in solid tissue is unlikely to be resolved if its axial and lateral dimensions are smaller than the corresponding average dimensions of the surrounding speckle pattern. The requirements for resolution of cysts are therefore consistent with those for resolution of point targets (section 13.7.1). Of course, a solid lesion generating its own internal speckle pattern will be more difficult to see due to the poorer contrast against the surrounding speckly pattern. Whether or not it is detected depends on its size in relation to the speckle pattern, the difference in scattering strength between its internal scatterers and those of its surroundings and the contrast resolution of the scanner.

Reverberations (multiple reflections)

When a transmitted pulse encounters a pair of relatively strongly reflecting parallel interfaces, the echo from the deeper interface can be partially reflected at the nearer interface to form a new, albeit weaker, 'transmitted pulse' travelling back into the body again. This will be partially reflected again at the deeper interface etc, leading to the production of a chain of such pulses at regular intervals and diminishing amplitudes (figure 13.25). These result in each target producing not one echo on the image, but a series of reverberation echoes at regular spacings, gradually weakening with range. Examples of suitably strong reflecting interfaces might be muscle fascia or fat–muscle interfaces. The commonest reverberation situation involves the probe face itself acting as the nearer reflector, since this is often large, fairly flat and relatively strongly reflecting.

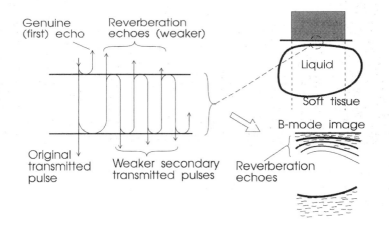

Figure 13.25. *Reverberations between strongly reflecting interfaces produce strings of uniformly spaced false echoes. Reverberation echoes produced by irregular tissue can appear more random and noiselike than this idealized representation. Reverberation echoes are also produced beyond the distal liquid boundary, but are usually lost in the speckle pattern produced by the soft tissue there.*

Reverberation echoes are usually only noticed as acoustic noise within the otherwise echo free images of liquid-filled structures. Classic examples are beyond the anterior wall of the bladder, pregnant uterus, gall bladders, the larger abdominal blood vessels. The regular spacing of the reverberation echoes helps to identify them as such (figure 13.24).

Beam width artefacts

Isolated targets surrounded by liquid, for example a gall stone, a foetal finger or limb, or a section of intestine in ascitic fluid, are often represented on the image by a wide streak. The explanation for this 'handlebar', 'umbrella', or 'Mexican hat' artefact has already been given in the discussion of lateral resolution in section 13.7.1. It may make two close targets at the same range appear as one elongated target, or it may cause a hole in a surface to be obscured due to the merging of the 'streaks' extending from tissue on each side.

Slice thickness artefact

The fact that the beam has a certain width perpendicular to the scan plane means that echoes will be detected from targets lying within half an elevation beam width (section 13.7.5) on each side of the scan plane. In effect, the beam is interrogating a slice rather than a plane, and the elevation beam width at any depth defines the slice thickness at that depth. The spurious echoes from targets within the slice, but not actually in the scan plane, are a form of acoustic noise, reducing the contrast resolution of the image. The effect is mainly noticeable when the scan plane passes through liquid, but solid (scattering) tissue lies within half a beam width of the scan plane (figure 13.26). Thus the image of a small spherical cyst, or a longitudinal section through a blood vessel that is narrower than the slice thickness, is likely to show weak acoustic noise within it.

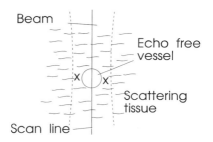

Figure 13.26. *Slice thickness artefact. The diagram represents the elevation plane, i.e. the plane perpendicular to the scan plane. Echoes from tissue (x) within the scanned slice, but not in the scan plane itself, appear as acoustic noise and degrade contrast resolution.*

Annular array scanners are least affected by this problem since, having beams of circular cross section, the beam width reductions achieved by array beam forming techniques apply equally in the elevation and scan planes.

Mirror images

A strongly reflecting boundary can act as a mirror, deflecting both the transmitted and reflected pulses. Echo producing targets then appear as virtual images on the far side of the boundary. A classic example of this is the air-backed diaphragm acting as a mirror to make a structure within the liver appear as a structure in the lung.

Missing interfaces

As mentioned in section 13.3.4, some smooth reflecting interfaces may be difficult to visualize unless the beam is almost perpendicular to the interface. For example, the smooth membrane around the kidney and the walls of arteries are usually incompletely visualized, since the ultrasound from the probe meets such interfaces at normal incidence only over a limited region (figure 13.24).

Registration errors

In a B-mode scan, the speed of sound through different tissues is assumed to be the same. In reality there can be an error in range, due to an echo returning at a time that is different to that predicted by the assumed speed (axial registration error), or an error in lateral position, due to lateral deflection of the beam by refraction at

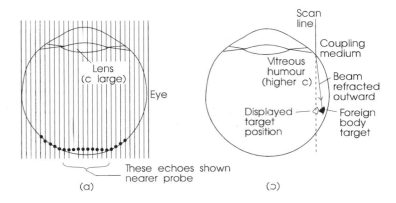

Figure 13.27. *Eye scanning examples of misregistration. (a) Axial misregistration of the retina image due to the relatively high speed of sound in the lens. (b) Lateral misregistration of a foreign body due to refraction at the surface of the eyeball.*

an intervening oblique interface. An example of axial misregistration is the forward bulge in the image of that part of the retina lying behind the lens, due to the higher speed of sound in the lens (figure 13.27(a)). An example of lateral misregistration, also from eye scanning, is the lateral shift of the image of an intra-ocular foreign body or tumour, caused by refraction at the oblique wall of the eyeball, having vitreous humour (higher speed) on one side and the soft tissue or a water bath on the other (figure 13.27(b)).

Acoustic shadows

A structure such as a bone or a calcified gall stone can reflect and/or absorb virtually all the ultrasound power incident upon it, creating a region of little or no echo strength (acoustic shadow) in the area on the image beyond it (figure 13.24). The problem can usually be solved by moving the transducer to another position from where the sound wave can avoid the obstruction. Even soft tissue masses can absorb the incident beam sufficiently strongly to create a shadow. This can often be of diagnostic value since it identifies the mass as being of a highly absorbing nature, like some tumours, or strongly reflecting, like a calcified gall stone.

Post-cystic enhancement

This is the opposite to shadowing in that the area behind a weakly attenuating structure is shown with enhanced brightness. It occurs where the TGC has been set up to compensate for absorption in solid tissue, but part of the field of view consists of a liquid region or some other mass of very low absorption. Echoes returning from beyond the liquid are given the benefit of extra gain, but have not actually suffered the attenuation that the extra gain was intended to compensate for. The artefact is useful in that bright echoes in a region beyond a relatively echo free area are a good indication that the echo free area represents a liquid (figure 13.24).

Edge shadows

These are shadows cast by strongly curved walls of rounded structures, such as a blood vessel, lesion or small organ such as the gall bladder (figure 13.24). The beam is strongly reflected by the convex surface so that targets further along the scan lines receive very little intensity and produce hardly any echoes. The reflection from the convex surface is strongly divergent (figure 13.28), and so loses intensity rapidly. Thus echoes from any targets in the reflected beam are too weak to be noticeable as mirror images beyond the surface. The resulting lack of echoes beyond the curved surface is known as an edge shadow. If the speed of sound changes from lower to higher at the curved surface, total internal reflection of the beam (section 13.3.5) can occur, leading to a more marked shadow than that caused by normal specular reflection.

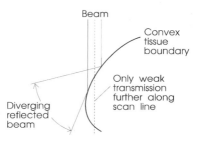

Figure 13.28. *Edge shadow artefacts can be produced by strong specular reflection or total internal reflection at a convex boundary.*

Beam distortion

This is due to refraction at interfaces within inhomogeneous subcutaneous tissue. The distorted beam is effectively widened and has an uneven intensity distribution, producing blurred images. This effect is rather similar to the distortion of visible images when light passes through mottled bathroom glass. This is more of a problem in some patients than in others, and is one of the reasons why some patients make poor scanning subjects.

13.8.   THE DOPPLER EFFECT

If a particle or tissue interface is moving towards the source of a wave, the wave reflected or scattered back to the source has a frequency that is higher than that of the incident wave. Conversely, if it is moving away from the source the frequency of the reflected or back-scattered wave is reduced. This is the Doppler effect. Its major application in diagnostic ultrasound is in the investigation of blood vessels, where the frequency changes of the echoes scattered by blood cells give detailed information about blood flow.

This may be understood by reference to figure 13.29 where successive wavefronts of an incident wave are scattered by a small particle into spherical wavefronts expanding away from the particle. (A wavefront is a surface joining all points that are at the same stage (phase) in a wave cycle. It is analogous to say the peak or trough of a ripple on water). In (*a*) the particle is at rest, and the frequency with which the wavefronts reach the particle, and hence the frequency and spacing (wavelength) with which the scattered wavefronts are generated, is the same as in the incident wave. In (*b*) the particle is shown moving towards the source of the incident wave. It moves forward to meet each wavefront a little earlier, and a little nearer to the source, than it would if it were stationary, resulting in an increase in the frequency with which the scattered wavefronts are generated and a reduction in their spacing along a line back to the source. Since there is no change in the speed of these back-scattered wavefronts, their closer spacing (smaller wavelength) means they arrive at the source at a frequency that is higher than the transmitted frequency. In (*c*) the particle is shown moving away from the source of the incident wave. It now meets each wavefront a little later, and a little further from the source, leading to an decrease in the rate of generation of scattered wavefronts and an increase in their spacing along a line back to the source. The frequency of arrival of scattered wavefronts back at the source is therefore lower than the transmitted frequency.

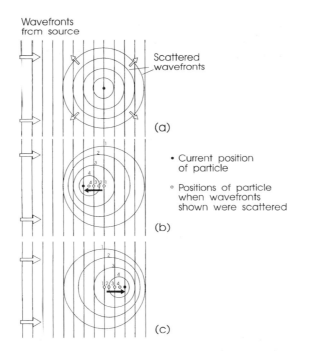

Figure 13.29. *The Doppler effect produced by a moving particle. (a) The particle is at rest. (b) The particle is moving towards the source. (c) The particle is moving away from the source. Wavefronts scattered by the particle at positions 1,2,3,... expand to become spherical wavefronts 1,2,3,... centred on these original positions.*

In medical applications, a probe on the body surface provides both a transmitter for the outgoing ultrasound wave and a receiver for detecting the scattered wave. The difference between the frequency of the transmitted wave (f_T) and that received back at the probe (f_R) is known as the **Doppler shift** (f_D).

$$f_D = f_R - f_T.$$

Note that f_D is a change in frequency and will be positive if the particle motion is towards the source, and negative if the motion is away from the source. The magnitude of the Doppler shift, expressed as a fraction of the transmitted frequency, is directly proportional to the speed of the particle, expressed as a fraction of the speed of sound (c). For the situation considered so far, where the particle is moving directly towards or away from the source (with a speed v):

$$\frac{f_D}{f_T} = 2\frac{v}{c}.$$

Note the convention, usually adopted for Doppler applications, of making v positive if the target is moving towards the source. This is different to the normal convention of making velocities away from the source positive (section 13.2.6), but it is convenient to have positive Doppler shifts associated with positive velocities. In general the direction of movement of a scattering particle will be at some angle θ to the direction of the incident wave (figure 13.30). Only motion directly towards or away from the probe produces a frequency shift, so only the component of the particle velocity that is towards or away from the source will be relevant in determining the magnitude of this shift. This component will be $v\cos\theta$. The Doppler equation, which allows particle speed v to be found from a measurement of f_D, is therefore:

$$f_D = \frac{2v \cos \theta f_T}{c}.$$

Substitution of typical values for v and f_T shows that f_D is typically a few kHz or less. For example, a blood cell moving at 0.5 m s^{-1} directly towards a probe ($\cos \theta = 1$) transmitting a frequency of 3 MHz, would produce a Doppler shift f_D equal to 2 kHz (assuming $c = 1500$ms^{-1}).

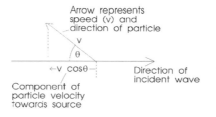

Figure 13.30. *Only the component of the target's velocity that is directly towards or away from the probe produces the Doppler effect. If the target speed is v and the angle between the ultrasound beam and the target is θ, this velocity component is $v \cos \theta$.*

13.9.   SPECTRAL DOPPLER TECHNIQUES

13.9.1.   The Doppler signal and Doppler spectrum from blood flow

Doppler systems transmit an ultrasound wave from a probe into the body and compare the power and frequency of the waves scattered back with that sent out. For each returning wave, an electrical Doppler signal is generated which has an electrical power (Doppler power) proportional to the acoustic power given to the receiving transducer by the returning wave, and a frequency (Doppler frequency) equal to the Doppler shift (f_D) of the returning wave. Since the millions of blood cells in even a small volume within a vessel will have a range of speeds, the Doppler system will, in effect, generate millions of Doppler signals with a corresponding range of Doppler frequencies. Ideally, the distribution of Doppler power versus Doppler frequency (the Doppler power spectrum) in the overall Doppler signal should match the distribution of the number of blood cells versus speed. In practice the match is only approximate, due to a number of factors discussed in section 13.9.5.

Since Doppler shifts, and hence Doppler frequencies, are at most a few kHz, the simplest way to present the Doppler signal to an operator is via a loudspeaker. The human ear–brain combination is well used to analysing complex sound signals in this frequency range, and so the operator can appreciate the relative amount of blood flowing at low and high speeds from the relative loudness of the low and high frequency content of the audible Doppler signal. Directional Doppler systems use two loudspeakers, filtering Doppler signals with negative Doppler shifts (flow away) into one channel and those with positive shifts (flow towards) into the other.

A visual representation of the distribution of power and frequency in the Doppler signal is provided by using a spectrum analyser. The Doppler spectrum is usually presented as a two dimensional plot of power (shown as a grey scale) versus frequency (vertically) and time (horizontally), as shown in figure 13.31.

13.9.2.   Continuous wave Doppler systems

The simplest type of Doppler system is a continuous wave (CW) system, shown diagrammatically in figure 13.32. The probe contains two transducers, one of which transmits continuously, while the other receives

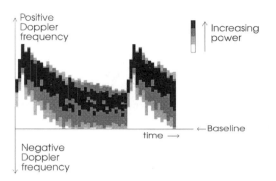

Figure 13.31. *Typical presentation of a Doppler spectrum versus time for an artery. Distance above or below the base line represents positive or negative Doppler frequency respectively. A grey scale or colour scale is used to indicate the relative power of the Doppler signal at each Doppler frequency.*

continuously. The Doppler signal generator (Doppler demodulator) compares the frequency and power of the received RF signal with a reference signal derived from the CW drive to the transmission transducer. The wall thump filter rejects Doppler signals with frequencies below about 100 Hz. This eliminates the large amplitude, but low frequency, Doppler signals produced by moving tissue, preventing them from swamping the much weaker Doppler signal from blood. A CW Doppler system is sensitive only to moving targets in the cross-over region, the region common to both the transmit and receive beams. By adjusting the sizes, shapes and relative positions and orientations of the two transducers, the manufacturer can make this region long or short, superficial or deep and narrow or wide, as appropriate to the intended application.

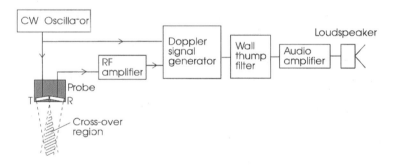

Figure 13.32. *Block diagram of a simple continuous wave (CW) Doppler system. The angle between the transmission (T) and reception (R) transducers is chosen to produce a short superficial cross-over region or one that extends deeper, as required.*

13.9.3. Pulsed Doppler and duplex systems

In many applications it is desirable to be able to measure Doppler signals from a specific depth (range) interval. For example, a cardiologist may wish to know whether blood is regurgitating back into the left atrium during systole, or a vascular surgeon may wish to know about blood flow in one particular vessel, behind or in front of other blood vessels. Pulsed Doppler allows the Doppler signal from a chosen range interval to be isolated from other Doppler signals originating along the same line.

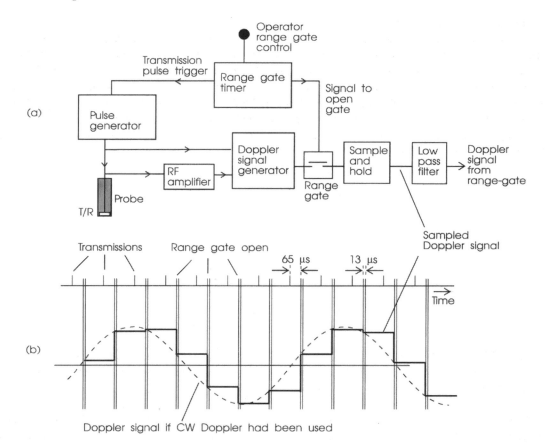

Figure 13.33. *(a) Block diagram of a pulsed Doppler system. The range gate is represented here by a switch, shown in the blocking position. (b) Assumes a single moving target, situated within a 1 cm wide range gate at a depth of 5 cm (see text). As each echo returns, the output of the Doppler signal generator momentarily becomes that which would have been produced by a CW transmission (dashed curve). Each sample is held until the next, producing an approximation to the CW Doppler signal. This is further improved by low pass filtering.*

It involves transmitting a regular sequence of ultrasonic pulses (each pulse typically consisting of 3–20 cycles) towards the target region (sample volume), using the same transducer to both transmit and receive. Samples of the Doppler signal from the chosen range interval are extracted by an electronic range gate that only opens for the short time when echoes are anticipated from that range interval (figure 13.33). For example, using the rule of 1.3 μs go and return time for every 1 mm of range (section 13.6.1), if the required sample volume extended between a range of 50 mm and 60 mm, the range gate would be opened 65 μs after the transmission of each pulse, and closed 13 μs later. Thus each transmitted pulse produces just one sample of the Doppler signal. A sequence of samples of the Doppler signal is therefore obtained, rather than a continuous Doppler signal as in CW Doppler. However, only echoes from the selected sample volume have contributed to these samples. Provided the time between samples is sufficiently small, a continuous, smooth version of the Doppler signal can be recovered by the sample and hold technique illustrated in figure 13.33(*b*), and further low pass filtering. This signal can then be input to a loudspeaker and spectrum analyser, in the same way as for a CW Doppler signal.

It is usual to combine pulsed Doppler with B-mode scanning in a duplex scanner. Duplex scanners use the same probe for producing the B-mode scan and performing the Doppler interrogation. The Doppler interrogation line and the range gate can be superimposed and adjusted on the B-mode image (figure 13.34). When the range gate has been set, the operator selects pulsed Doppler mode and the B-mode scanning action is replaced by a stationary Doppler interrogating beam at the desired position. A further advantage of duplex scanning is that it allows the machine to display blood speed rather than Doppler frequency. To do this the operator aligns a short angle cursor with the vessel or lumen axis, thereby allowing the machine to measure the angle θ needed to find v from f_D using the Doppler equation.

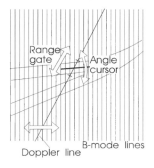

Figure 13.34. *A duplex scan display shows the sample volume in relation to the B-mode image. The operator may align an angle cursor with the blood flow, allowing the machine to measure the angle θ in the Doppler equation, and hence calculate blood speed rather than just f_D.*

A penalty of using pulsed Doppler is that there is an upper limit (**Nyquist limit**) to the Doppler frequency that can be measured. If the frequency with which the Doppler signal is sampled (i.e. the prf of the transmitted pulses) is less than twice the highest Doppler frequency present, then **aliasing** occurs (see section 10.3.3). This means that a Doppler frequency (f_D) that is outside the range $\pm\frac{1}{2}$prf is incorrectly measured as $f_D \pm N$ prf, where N is the integer value that brings the result within the range $\pm\frac{1}{2}$prf. Thus, on a spectrum analyser display, as soon as an increasing Doppler frequency exceeds the positive Nyquist limit ($+\frac{1}{2}$prf), the displayed frequency abruptly jumps to just above the negative one ($-\frac{1}{2}$prf), and then continues in the same direction (figure 13.35). The displayed frequency has thus been reduced by the prf.

Aliasing may not be a problem for moderate blood speeds in superficial blood vessels, since the go and return time is low and a high prf can be used. It may be unavoidable at larger depths (e.g. deep vessels, the heart and the foetus) since the prf is limited by the requirement that echoes from one transmission should be allowed to return before transmitting again (section 13.6.1). The solution is to use either CW Doppler or to select high prf mode. In high prf mode the above rule is broken and the pulses are transmitted at intervals equal to say one-third of the go and return time to the sample volume. The machine now has no way of knowing which of three possible pulses was responsible for any given echo, and so three sample volumes now appear on the duplex image (figure 13.36), with the possibility that the signal originated in any one of them. In practice, the operator can often eliminate all but one of these on the basis that they do not match likely sites of high speed blood flow.

13.9.4. *Interpretation of Doppler signals*

In some vessels the absolute blood speed is used to assess disease. For example, in the internal carotid artery, a peak systolic velocity of more than 1.2 m s^{-1} indicates a 50% or greater diameter stenosis, whereas the

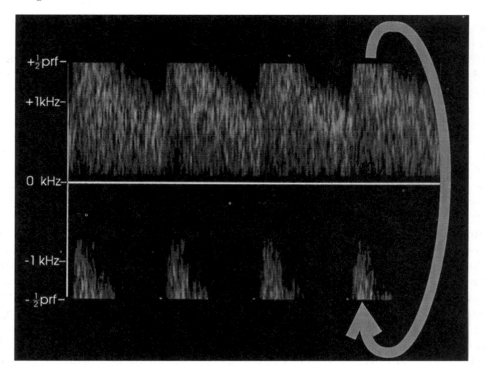

Figure 13.35. *The appearance of aliasing on a spectrum analyser. When the Doppler frequency of a signal increases beyond either the $+\frac{1}{2}$prf or the $-\frac{1}{2}$prf Nyquist limit, it jumps to the opposite limit and continues in the same direction.*

Figure 13.36. *In high prf mode, sample volumes at several different ranges are presented on the duplex display. Doppler signals originating in any or all of them will be combined in the Doppler signal presented to the operator.*

same speed at end-diastole indicates a 70% or greater stenosis. A reduced diastolic speed generally indicates increased distal vascular resistance. Alternatively, the speed at a diseased site may be compared to that at a reference site. Thus in the femoral artery a doubling of peak Doppler shift between normal and partially occluded parts of a vessel indicates a 50% stenosis. Sometimes the shape of the outline of the Doppler spectrum (the Doppler waveform) can give diagnostic information. For example a long systolic rise time indicates proximal disease, whereas a 'spiky' waveform indicates turbulence.

Often the ratio of the blood speeds at one site at two different times in the cardiac cycle can be of diagnostic value. There is then no need to know the angle of the ultrasound beam to the blood flow, since it is a constant, and the required ratio is simply that of the corresponding Doppler shifts. For example, in the lower limb, the pulsatility index (PI), defined as the difference between the maximum and the minimum Doppler shifts in a cardiac cycle divided by the mean Doppler shift over a cardiac cycle, is used to assess proximal disease on the basis of the waveform damping it produces.

13.9.5.   *Doppler artefacts*

As mentioned earlier, the Doppler power spectrum is not a perfectly accurate analogue of the distribution of number of blood cells versus speed. One reason is that, since the sensitivity within the sample volume is not uniform, those targets that happen to be in the most sensitive part will have a dominating effect on the power spectrum. Another artefact is spectral broadening, whereby targets moving with a single well defined velocity produce a range of Doppler frequencies rather than the pure single frequency one might expect. One cause of this is that ultrasound arrives at a target from all points on the transducer and therefore from a range of different directions. Referring to the Doppler equation, this means a range of values for $\cos\theta$, and hence for f_D. Another cause of spectral broadening is that the accuracy of a Doppler frequency measurement is inversely related to the short time for which a moving target is in the sample volume (or cross-over region for CW). A further cause is that blood cells tend to be arranged in irregular clusters. Just as interference between many echoes from irregularly distributed scatterers produces a speckle pattern in a B-mode image (section 13.7.10), so interference between the echoes from many blood cell clusters produces a random variation in the amplitude of the signal generated by the receiving transducer, and hence in the Doppler signal. Such Doppler speckle, like all noise, has components over a range of frequencies, which broaden the Doppler spectrum.

13.10.   DOPPLER IMAGING

By automatically and repeatedly measuring the Doppler signal at a large number of closely spaced sites within a B-mode scan plane, either the frequency or the power of the Doppler signal may be plotted as a real time colour overlay on the B-mode image (figure 13.37). The time needed to acquire the Doppler information means that only a restricted area of the B-mode field of view, lying within an operator defined 'colour box', is interrogated for Doppler information. Even so, frame rates are substantially lower than in pure B mode, and may be as low as five frames per second, or so.

Figure 13.37. *Colour Doppler imaging. Within the 'colour box' of a B-mode scan, a colour overlay shows either the mean Doppler frequency or the mean Doppler power for each 'Doppler pixel'. The scans are known as colour flow maps or images (CFM or CFI) or power Doppler scans respectively.*

The colour box is interrogated, using pulsed Doppler, in a sequence of about 20–40 Doppler scan lines, each of which is subdivided into a similar number of sample volumes. One sample volume corresponds to one 'Doppler pixel' in the display. There is a separate dedicated range gate to extract the Doppler signal from each sample volume in the Doppler line. For each Doppler line, a sequence of (say) ten pulses is transmitted, so that each range gate captures a ten sample sequence of the associated Doppler signal. The ten sample sequences from all the range gates are processed in parallel to obtain estimates of either the mean frequency or the power of the Doppler signals from the corresponding sample volumes along the Doppler scan line. These estimates are stored in a Doppler image memory, where each stored number determines the colour of the corresponding pixel. The process is repeated for all the Doppler lines in the colour box, after which a B-mode sweep is made and the whole process commences again at the first Doppler line.

If the mean Doppler frequency for each Doppler pixel is plotted, the scan is called a colour flow map (CFM) or colour flow image (CFI). A colour scale is provided to show which colour represents which Doppler frequency. There is no universal convention, but a typical scheme might be to use red through to yellow to indicate increasing positive Doppler frequencies (flow towards the probe) and dark blue through to light blue to indicate increasingly negative Doppler shifts (flow away from the probe). Zero Doppler shift is usually represented by black. It is important to remember that the colours represent Doppler frequency, not blood speed directly. Thus, remembering the $\cos\theta$ factor in the Doppler equation, the colour will change as the angle between the blood flow and the scan line changes, even though the blood speed is constant. It is also important to be aware of the possibility of aliasing. If the Doppler frequency exceeds the Nyquist limit of half the pulse repetition frequency, the displayed colour will abruptly jump from that at one end of the colour scale to that at the opposite end. Some CFM systems offer the option of showing the range of frequencies (variance) in the Doppler signal from each sample volume. This information, which can be generated at the same time as the mean frequency, indicates the degree of turbulence in the blood flow. It can be displayed by mixing an appropriate strength of another colour such as green.

Normally, CFM is used to visualize blood flow, and steps must therefore be taken by the equipment designer to filter out the low frequency, but higher power, Doppler signals (clutter) from surrounding moving soft tissue. However, in some applications, the aim is to visualize the soft tissue movement rather than the blood flow (e.g. in studying anomalous heart muscle movements, or the dynamic behaviour of tissue), and in such cases the filters are changed to preserve the soft tissue Doppler signals and reject the blood clutter signals.

If the power of the Doppler signal is plotted, the scan is called a power Doppler scan. Again, a colour scale is provided, but this is uncalibrated as it simply indicates where the power is higher or lower. Power Doppler shows where, and how much, blood is flowing, but it gives no indication of direction or speed. It is therefore not subject to any aliasing artefact. It gives a clearer indication of perfusion than does CFM, since the presence of a mixture of colours in the latter can be noiselike and difficult to interpret. Moreover, even flow at right angles to the ultrasound beam can usually be seen, since the convergence or divergence of the beam means at least part of the beam will be non-perpendicular to the flow and therefore produce a Doppler signal with measurable power. (The mean Doppler shift would be zero, however, so this situation would produce no signal in a CFM display.) A hybrid type of scan, which is sometimes called directional power Doppler, shows the power of the Doppler signal, but also indicates whether the flow is towards the probe or away from it by using say a scale of reddish colours when the Doppler frequency is positive, and bluish colours when it is negative.

13.11. THE SAFETY OF ULTRASOUND

If used prudently, diagnostic ultrasound is rightly considered to be a safe technique. However, modern machines are capable of much greater acoustic output than ever before and, if used without due care, can produce hazardous thermal and non-thermal biological effects *in vivo*. The hazard mechanisms of diagnostic ultrasound fall into the main categories of heating, cavitation, and mechanical forces.

13.11.1.   Damage mechanisms and biological effects

Tissue heating, as a result of the absorption of ultrasound energy, can cause adverse biological effects and tissue damage. Damage to dividing cells is of greatest concern, notably in the foetus during the first 8 weeks of gestation and in the foetal spine and brain up to the neonatal period. The World Federation of Ultrasound in Medicine and Biology has concluded that a temperature increase to the foetus in excess of 1.5 °C is hazardous if prolonged, while 4 °C is hazardous if maintained for 5 minutes (Barnett and Kossoff 1998). Theoretical models (e.g. Whittingham 1994) indicate that diagnostic equipment can produce worst case temperature rises of over 1 °C in B mode and up to 8 °C in pulsed Doppler mode. Pulsed Doppler mode has the greatest heating potential because of the high pulse repetition frequencies and pulse lengths used, and the fact that the beam is held stationary. The highest temperature rises are associated with bone, since this has a very high absorption coefficient. Any pre-existing temperature elevation, such as in a febrile patient, and any contribution due to self-heating of the probe must be considered as additive to the heating due to ultrasound absorption.

In order to assist operator's awareness of any thermal hazard, some machines now display a **thermal index (TI)** alongside the image on the screen (AIUM 1992). This is a number that changes in response to the settings of the controls, and is meant to provide a very approximate guide (within a factor of two) to the maximum temperature rise that would be produced, according to a simple theoretical model. Where only soft tissues are involved, or where scanned modes are employed, the soft tissue thermal index (*TIS*) is appropriate; the bone thermal index (*TIB*) applies if the exposed tissues include bone; the cranial thermal index (*TIC*) is used if bone lies very close to the probe. Thus, for example, TIS is relevant for scanning a foetus in the first trimester, while TIB would be appropriate for a pulsed Doppler scan in the second or third trimester, when bone is present.

Cavitation refers to the energetic behaviour of small gas bubbles in a liquid medium due to the pressure variations of a sound wave. In contrast to tissue heating, it is more likely to occur at low frequencies. *Stable cavitation* refers to the vigourous radial oscillations of bubbles when the frequency of the sound is close to the resonant frequency of the bubble. It is a continuous wave phenomenon, and so the only diagnostic mode of relevance is continuous wave (CW) Doppler. However, the relatively small peak negative acoustic pressures produced by CW Doppler equipment are unlikely to produce stable cavitation at diagnostic frequencies. Stable cavitation is, however, considered to be relevant to the mode of action of physiotherapy ultrasound.

Collapse (*transient* or *inertial*) *cavitation* is a more violent and damaging form of cavitation that can be produced by short pulses, provided the acoustic pressure amplitude is large enough. Here the bubble increases to many times its original size within one or two high negative pressure excursions and then violently implodes during the following high pressure excursion. During the rapid collapse the bubble temperature may rise to several thousand degrees, visible and ultra-violet light is emitted, shock waves are generated and water is dissociated into free radicals. If the bubble is near a surface, e.g. a blood vessel wall, the rapid inrush of liquid occurring during the bubble collapse can cause direct impact damage to the surface. Collapse cavitation can clearly cause tissue damage and cell death, a fact that is used to advantage in ultrasonic sterilizing tanks.

Theory (Holland and Apfel 1989) suggests that the likelihood of cavitation is related to the ratio of the peak negative pressure (MPa) to the square root of the frequency ($MHz^{1/2}$). This ratio, known as the **mechanical index (MI)**, is displayed on some machines as a continuously updated on-screen indicator of cavitation hazard, partnering the thermal index mentioned earlier.

Cavitation-related damage from diagnostic levels of ultrasound has been demonstrated in animal tissues containing gas cavities, such as the lung. These effects have been associated with MI values greater than 0.3 or peak negative pressures of more than 0.3 MPa. Such values are well within the range of up to about 5 MPa measured for diagnostic machines in all modes (Henderson *et al* 1995). There is, as yet, no direct evidence of a hazard due to cavitation from diagnostic exposures *in vivo* in other tissues, but there is ample evidence of biological effects *in vitro*. There is also evidence (Fowlkes and Hwang 1998) that the stabilized gas bubbles used as ultrasound contrast agents make cavitation more likely.

Mechanical forces. The most important direct mechanical force produced by ultrasound is the radiation force, experienced by any object that has a smaller intensity on one side than on the other. Dyson *et al* (1974) have shown that continuous waves with intensities of about 500 mW cm^{-2} can cause blood cells to agglomerate in static bands at half wavelength intervals perpendicular to the ultrasound beam. This is due to the radiation forces acting on blood cells in a standing wave, so it is relevant only to CW Doppler equipment. CW Doppler devices intended for foetal heart monitoring produce I_{SPTA} values (section 13.2.6) of less than 30 mW cm^{-2} (Duck and Martin 1992). However some devices intended for cardiac and peripheral vascular use have I_{SPTA} values up to 1000 mW cm^{-2} (Stewart *et al* 1986), so the hazard might exist if they were to be used inappropriately to monitor very low flow rates, particularly in the embryo or foetus.

Ultrasound in an absorbing liquid produces bulk motion of the liquid (streaming) due to the radiation force acting on each small region of liquid itself. Streaming can sometimes be observed in real-time ultrasound images of cysts or other liquid masses containing particulate matter, but speeds are low (several centimetres per second). Cells caught up in this motion and those at the walls of the containing structure will be subject to mechanical stresses. Streaming speeds and stress effects are likely to be higher for high frequency ultrasound exposures, since absorption by liquids is greater at higher frequencies and the total power of the beam is concentrated into narrower beams. This could be of particular significance for trans-vaginal pulsed Doppler or even M-mode exposure of an embryo.

13.11.2. *Minimizing hazard*

Keep output power as low as possible, using increased overall gain to maintain sensitivity as far as noise allows. Keep exposure times as low as possible; avoid holding the probe stationary for more than a few seconds; remove the probe from the patient when not actually examining a real-time image. Avoid scans for which there is no medical reason and ensure good training in scan technique and interpretation. Arrange for a medical physics department to measure the worst case outputs and heating potential for each machine probe and mode, and to give advice on how the controls should be set to reduce outputs. Avoid pulsed Doppler mode in early pregnancy or in the foetal or neonatal skull or spine. In scanned modes, selecting a deep write zoom box or a deep transmit focus can often mean an increase in output.

13.12. SUMMARY

Ultrasound differs from x- and gamma rays in two important ways. First, it is not an ionizing radiation in the accepted sense of the term (although it can produce free radicals by cavitation), so it does not carry the risks associated with such radiations. Secondly, it is propagated as a result of longitudinal compressions and rarefactions of the medium through which it travels. Hence the behaviour of the sound wave provides information about the density and elastic properties of the medium through which it is travelling.

Ultrasound is a wave motion and exhibits all the properties of waves such as reflection, refraction and diffraction. The compressions and rarefactions cause the particles of the medium to move back and forth about their original positions. The pressure variations are large, up to a few atmospheres, but the peak particle speeds are relatively low, up to a few centimetres per second. The ratio of the acoustic pressure to the particle speed at any instant is called the characteristic acoustic impedance of the medium, and its value is given by the product of density and speed of sound in the medium. It is important since the fraction of the incident power in a sound wave that is reflected by an interface depends primarily on the ratio of its value in the two media on either side of the interface. Ultrasound waves are significantly attenuated in tissue, the attenuation per unit distance being known as the attenuation coefficient. This quantity is usually expressed in units of dB cm^{-1} and increases with frequency, limiting the use of higher frequencies to more superficial structures.

Transducers for the generation and detection of ultrasound are constructed from thin piezoelectric discs or plates, normally of lead zirconate titanate (PZT). The transducer thickness is chosen to ensure resonance at the required frequency. Necessary additional features are a backing layer to ensure rapid damping of the transducer vibrations and an impedance matching layer to achieve maximum energy transfer into and out of the patient, as well as to provide further damping.

Lateral resolution is largely determined by the widths of the transmission and reception beams. Focusing of the reception beam is automatically controlled by the scanning system, but the operator can move the transmission focus to optimize lateral resolution at the range of interest. Focal zones as narrow as two wavelengths are possible. Axial resolution is largely determined by the pulse length, and is typically about one wavelength. Both forms of spatial resolution are improved by using higher frequencies.

The use of the Doppler effect to study movement and blood flow in the body finds widespread application. When Doppler techniques are used in conjunction with real-time imaging, useful quantitative information can be provided about the haemodynamic situation in a blood vessel or heart chamber. Doppler imaging techniques show the distribution of blood flow or soft tissue movement in a qualitative way.

To date there is no confirmed evidence of harm to a human patient or foetus from diagnostic ultrasound. However, outputs from some commercially available scanners are capable of producing hazardous temperature elevations, particularly in pulsed Doppler modes and where the target region contains bone. Furthermore harmful non-thermal biological effects have been reported when diagnostic ultrasound has been used to insonate cells *in vitro*, and animal lung tissue or intestine *in vivo*. It is therefore recommended that prudence be exercised.

REFERENCES

American Institute of Ultrasound in Medicine (AIUM) 1992 *Standard for Real-time Display of Thermal and Mechanical Output Indices on Diagnostic Ultrasound Equipment* (Rockville, MD: AIUM)

Barnett S B and Kossoff G 1998 *Safety of Diagnostic Ultrasound* (London: Parthenon)

Bass H E, Bauer H J and Evans L B 1972 Atmospheric absorption of sound: Analytical expressions *J. Acoust. Soc. Am.* **52** 821–5

Duck F A 1990 *Physical Properties of Tissue* (London: Academic)

Duck F A and Martin K 1992 Exposure values for medical devices *Ultrasonic Exposimetry* ed M C Ziskin and P A Lewin (Boca Raton, FL: Chemical Rubber Company) pp 315–44

Dyson M, Pond J, Woodward B and Broadbent J 1974 The production of cell stasis and endothelial damage in the blood vessels of chick embryos treated with ultrasound in a stationary field *Ultrasound Med. Biol.* **1** 133–48

Fowlkes J B and Hwang E Y 1998 Echo-contrast agents: what are the risks? *Safety of Diagnostic Ultrasound* ed S B Barnett and G Kossoff (London: Parthenon) pp 73–85

Fry F J and Barger J E 1978 Acoustical properties of the human skull *J. Acoust. Soc. Am.* **63** 1576–90

Henderson J, Willson K, Jago J R and Whittingham T A 1995 A survey of the acoustic outputs of diagnostic ultrasound scanners in current clinical use *Ultrasound Med. Biol.* **21** 699–705

Holland K H and Apfel R E 1998 An improved theory for the prediction of microcavitation thresholds *IEE Trans. Ultrason. Ferroelec. Freq. Control.* **36** 139

Kaye G W C and Laby T H 1995 *Tables of Physical and Chemical Constants* 16th edn (London: Longman)

Whittingham T A 1994 The safety of ultrasound *Imaging* **6** 33–51

FURTHER READING

Barnett S B and Kossoff G 1998 *Safety of Diagnostic Ultrasound* (London: Parthenon)

McDicken W N 1990 *Diagnostic Ultrasonics* 3rd edn (Edinburgh: Churchill Livingstone)

EXERCISES

1 Why is the low megahertz range used for ultrasonic imaging? Give, with reasons, the frequencies and transmission focal lengths you would choose to make ultrasound examinations of: (*a*) an eye, (*b*) a liver.

2 What factor(s) determine the reflection coefficient of an ultrasound pulse at a tissue interface? What differences would you expect in the ultrasonic appearance of specularly and diffusely reflecting interfaces?

3 Why is it necessary to use gel between the probe and the patient?

4 A source of pulsed ultrasound and a target are separated by normal soft tissue. Discuss the effect of each of the following on the amplitude of the ultrasonic pulse reflected back to the source: (*a*) the size and shape of the target, (*b*) the characteristic acoustic impedances of the target substance and intervening tissue, (*c*) the distance between source and target, (*d*) the range to which the transmission focus has been set, (e) the frequency of the ultrasound, (*f*) the acoustic power of the source.

5 What is the purpose of the backing layer, the impedance matching layer and the lens in a typical ultrasound transducer? What would be the consequences for image quality of leaving out each of these in turn?

6 Explain briefly how a focused beam of ultrasound can be obtained from the excitation of a number of transducer elements in a linear array probe.

7 Why is the active group of a linear array progressively enlarged during dynamic focusing in reception? How will the width of the effective receive beam vary with depth if the active group stops expanding when it contains about 20 elements?

8 Show that, in the scan plane, the beam of a single 1 mm wide element of a 3 MHz linear array has a near field length of only about 0.5 mm and then diverges at an angle of about $\pm 30°$ to the axis. If a transmission or receive focus were required at a range of 15 mm, would there be any point in having 20 elements in the active group?

9 Describe the different types of B-mode real-time scanner, and list their advantages and disadvantages.

10 Describe the function of the TGC facility in an ultrasonic scanner. Describe two different forms of TGC controls and suggest applications for which each might be appropriate.

11 In B mode, which of the following controls affect the lateral resolution, which affect contrast resolution, which affect sensitivity and which affect temporal resolution? transmission focus; depth of field of view; overall gain; output power; TGC; write zoom; pre-processing; dynamic range; frame averaging.

12 Describe the features seen in an ultrasound B-scan image that could be used to distinguish between a solid tumour and a cyst.

13 Describe and explain a speckle pattern, as seen in a B-mode image. Which features of such a pattern are determined by the tissue and which by the machine?

14 Describe and explain two artefacts that produce image 'echoes' that do not correspond to any real target in the scan plane, and two artefacts that cause the image echoes of genuine targets to be shown at incorrect positions.

15 Explain the Doppler effect, and outline the principles of a continuous wave Doppler device.

16 Distinguish between the terms: Doppler shift, Doppler signal, Doppler frequency, Doppler power, Doppler spectrum, Doppler wave form.

17 Describe how a pulsed Doppler device isolates the Doppler signals from a particular depth interval. What limitation is associated with Doppler frequency shift measurement by pulsed Doppler?

18 Why does CFM mode involve a lower frame rate than B mode? Why is it possible for the same blood speed to be represented by different colours on the same CFM image? What effect does aliasing have on a CFM image? Can aliasing occur in power Doppler mode?

19 Explain what is meant by the thermal index (TI) and the mechanical index (MI), as displayed on the screens of some scanners. How are these numbers meant to help the operator assess the possible hazard risk associated with a scan?

20 Which mode has the greatest potential for thermal hazard, and why? Describe the measures that should be taken to ensure the prudent use of ultrasound.

CHAPTER 14

MAGNETIC RESONANCE IMAGING

Elizabeth A Moore and Peter C Jackson

14.1.   INTRODUCTION

The phenomenon of nuclear magnetic resonance was discovered independently by two groups of workers in 1946 headed by Bloch at Stanford and Purcell at Harvard. The techniques developed were primarily used to study the structure and diffusion properties of molecules and subsequently Bloch and Purcell shared the Nobel Prize for Physics in 1952.

In later years the nuclear relaxation times and properties of tissues were exploited by Damadian and several groups in the mid-1970s, both in the United Kingdom and United States. The imaging protocols developed have been the basis of techniques now commonly known as magnetic resonance imaging (MRI) although this has subsequently incorporated the original spectroscopic techniques into a powerful investigative tool, magnetic resonance spectroscopy (MRS).

This chapter concerns itself purely with the concepts of using nuclear magnetic resonance for the purposes of diagnostic imaging. A difficulty arises in that the phenomenon can only be explained completely by using the principles of quantum mechanics which do not lend themselves to easy explanation. Most texts resort to using classical mechanics to explain the processes; however these are incomplete and in some cases inconsistent with everyday observation. Wherever possible in this chapter the use of equations and mention of quantum mechanical concepts will be kept to a minimum. It is helpful first to review the basic principles of electromagnetism.

14.2.   BASIC PRINCIPLES OF ELECTROMAGNETISM

There is a duality between electric and magnetic fields which is fundamental to many aspects of magnetic resonance imaging. The most simple observation that can be made is when an electric charge moves there will always be an associated magnetic field. This is most easily perceived when an electric current flows along a conductor and a magnetic field is generated (figure 14.1). Similarly if a magnetic field changes strength or direction in the presence of an electric conductor, movement of electric charge and therefore a current is generated in the conductor. For instance, if a simple magnet is moved through a loop of wire (or the loop of wire is moved through the magnetic field) a current will be induced (figure 14.2).

Another important concept to remember at this stage is that electric and magnetic fields can interact, and that a simple way of describing the effect of the interaction is by using vector notation. A vector has both magnitude and direction and within the concepts of MRI both will be time varying. Hence in MRI we are constantly observing the effects of a dynamic process.

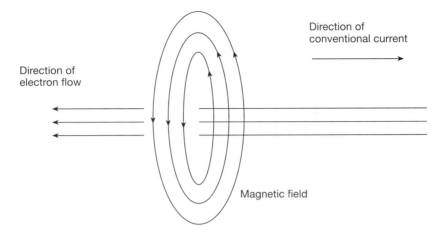

Figure 14.1. *A magnetic field is generated when electric charges move along a conductor.*

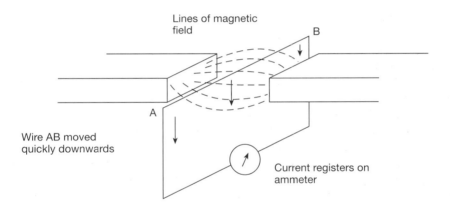

Figure 14.2. *When a conductor moves through a magnetic field (or vice versa) a current is induced.*

14.3. MAGNETIC PROPERTIES OF NUCLEI

The atom is composed of electrons associated with a nucleus containing protons and neutrons. All atomic particles are in a dynamic process and have an associated spin which can be quantized. Wherever possible similar atomic particles 'pair up' so that the effect of the spins will be cancelled: this applies to neutrons as well as protons and electrons.

From the basic understanding of electromagnetism (section 14.2) it is easy to appreciate that when an atomic particle with electric charge is moving then there is an associated magnetic field. However, the classical picture of magnetic moments being purely associated with charged particles is flawed because neutrons also have magnetic moments! This complication should not be too disturbing because from a pragmatic view point the hydrogen nucleus (^{1}H), consisting of a single unpaired proton, is the most abundant within living tissues and therefore provides the ideal nuclear species for magnetic resonance imaging.

14.4.   EFFECT OF AN EXTERNAL MAGNETIC FIELD

Fundamental to the process of MRI is the understanding of the effect of applied magnetic fields on nuclei. To assist this understanding it is helpful to draw upon a classical analogy of a spinning top (or gyroscope) in a gravitational field. The spinning top has an angular momentum and as a consequence a torque is applied to the top causing it to **precess** within the gravitational field. The rate of precession will be dependent upon the angular momentum (i.e. rotational speed and other mechanical characteristics of the top) together with the magnitude of the gravitational field (figure 14.3).

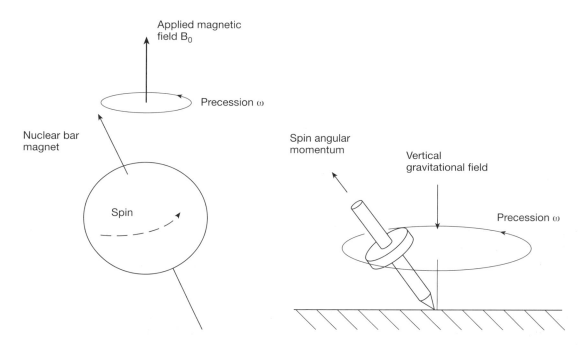

Figure 14.3. *A magnetic nucleus precesses in an external magnetic field (left). Similarly a spinning top precesses in the earth's gravitational field (right).*

A similar observation can be made with nuclei possessing both an angular and magnetic moment when placed in a static magnetic field (B_0). The nuclei exhibit a similar precessing movement that is characteristic of the nuclear species and the magnetic field strength (B_0). However, the classical observation is inconsistent since quantum mechanics allows for two possible spin orientations, which is manifest in a picture of magnetic dipoles precessing at the same frequency but in opposite orientations (figure 14.4). These two orientations correspond to a low energy state (i.e. aligned parallel with the external magnetic field B_0) and a high energy state (i.e. aligned anti-parallel to the magnetic field B_0).

These two different orientations are not equally populated. Boltzmann statistics predict that at equilibrium there should be a small excess of population in the low energy state (i.e. parallel state). This excess is dependent on the temperature but more importantly on the magnetic field strength (B_0): as the field strength is increased so the population difference of spins increases in favour of the low energy state. This population difference is the reason why magnetic field strength affects ultimately the signal-to-noise ratio (SNR) of the image formed. At approximately a field strength of 1 T there is a population difference of about 3 ppm (parts per million) which can be increased by increasing field strength (B_0) or cooling the object. At absolute zero all the spins would be perfectly aligned in a single population state; however, this is not popular with patients!

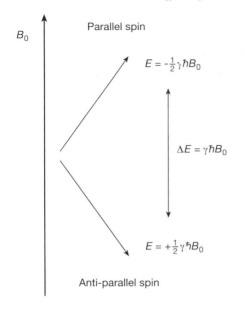

Figure 14.4. *Spin orientations in a magnetic field. The energy difference is proportional to the magnetic field strength.*

In addition to achieve a high signal intensity the abundance of suitable nuclei with a large intrinsic magnetic moment (μ) is important.

The relationship between the angular momentum (L) and the magnetic moment (μ) is given by

$$\mu = \gamma L$$

where γ is known as the **gyromagnetic ratio** which is specific for different nuclei.

The energy difference between the two population states (ΔE) depends on the gyromagnetic ratio, and the magnetic field strength and the Planck constant h

$$\Delta E = \gamma \frac{h}{2\pi} B_0. \tag{14.1}$$

It is possible to cause a transition between the two states by the emission or absorption of a photon whose energy E is given by the equation

$$E = \frac{h}{2\pi} \omega_0 \tag{14.2}$$

where ω is the frequency of radiation. Combining equations (14.1) and (14.2) yields

$$\frac{h}{2\pi} \omega_0 = \gamma \frac{h}{2\pi} B_0$$
$$\therefore \omega_0 = \gamma B_0. \tag{14.3}$$

It turns out that the frequency of radiation (ω_0) is identical to the rate of angular precession of the protons around B_0. The relationship established between ω_0 and B_0 (equation (14.3)) is known as the **Larmor equation** and states that the angular frequency (ω_0) is directly proportional to the magnetic field (B_0).

The gyromagnetic ratio (γ) for protons is 42.6 MHz T^{-1}. Hence in a magnetic field of 1 T the precessional (angular) frequency of protons is ~42 MHz, i.e. in the radiofrequency portion of the electromagnetic spectrum. In table 14.1 further information is given relating to the Larmor frequency (equation (14.3)).

Table 14.1. *The gyromagnetic properties of nuclei in different magnetic fields (B_0).*

Nucleus and gyromagnetic ratio ($/2\pi$ MHz T^{-1})	Magnetic field (T)	Larmor frequency (MHz)
^{1}H(42.6)	0.15	6.4
	0.5	21.3
	1.0	42.6
	1.5	63.9
^{31}P(17.2)	0.15	8.6
	0.5	17.2
	1.0	25.8
	1.5	34.4

At equilibrium, there is no phase coherence between spins, i.e. although the spins occupy two unequally populated energy states, they are equally distributed in space around B_0, conventionally chosen to be the z axis (figure 14.5). The individual magnetic moments (μ) of each dipole can be resolved into two components: a component along the z axis and a component located somewhere on the xy plane which is orthogonal to the z axis (figure 14.5). In considering the sum of the vector components of the magnetic moments on the xy plane it is seen that the net effect is zero magnetization within this plane. However, in considering the vector components resolved along the z axis it is apparent that there is a net excess in the parallel direction creating a net magnetization vector (M_0) in the $+z$ direction. The magnitude of the net magnetization vector is directly related to the population difference between the two energy states and consequently to the size of the magnetic field (B_0).

14.5.   RESONANCE

Although the individual magnetic moments are precessing, the resultant magnetization M_0 is fixed along the z axis and is too small to measure directly. This equilibrium has to be disturbed by the process of resonance before a signal can be obtained.

Resonance may be described as the transfer of energy (and hence interaction) by the most efficient energetic pathway: this usually occurs when the driving force matches in some way the receptor. Examples of resonant phenomena are found in the enhanced absorption of X-rays in materials (K edge absorption), i.e. the frequency (energy) of the X-rays is matched closely to the binding energy of the K shell electrons. Similarly tuning forks will vibrate if the driving force (e.g. sound) is present at a similar frequency.

The magnetic moments of the nuclei within a magnetic field (B_0) are precessing at a radiofrequency (ω_0): if a matched radio-frequency (i.e. at ω_0) is applied to the cohort of nuclei, energy will be preferentially absorbed by the system. Transitions will be stimulated from the parallel to anti-parallel, and the anti-parallel to parallel states with equal probability and the resultant net magnetization moment (M_0) will be modified according to the duration and intensity of the radio-frequency radiation. Before considering in more detail the effect of the radio-frequency driving force on the individual dipoles (magnetic moments) and net magnetization vector, it is useful to remember that a radio-frequency (RF) wave consists of both magnetic and electric field components.

14.6.   THE ROTATING FRAME

The changes which happen to the system of protons when it is excited by radio-frequency radiation and then relaxes back to equilibrium are conveniently examined by considering a rotating frame of reference. As an

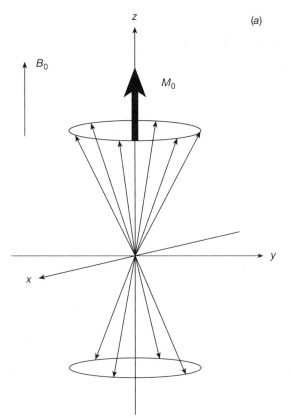

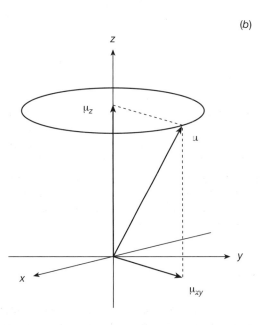

Figure 14.5. *(a) The two spin populations are randomly distributed around the z axis. (b) Each dipole moment has components along z and in the xy plane. The net magnetization (M_0) is along z.*

analogy, the motion of a bouncing ball is much easier to describe on the earth (because we are on a rotating frame of reference) than from a distant planet, which would have to include the effect of the rotation of the earth. Similarly it is easier to describe the effects of the radio-frequency pulses and subsequent recovery of the system from the perspective of a rotating frame of reference. We use a frame rotating at ω_0, the Larmor frequency, about the z' axis, and with the z' axis aligned with the z axis in the laboratory frame (defined by the magnetic field B_0). The magnetization at equilibrium thus appears as a vector of magnitude M_0 along the z (z') axis.

It is important to consider the magnetization in terms of its components in the longitudinal direction, i.e. z (z'), and in the transverse plane, i.e. $x'y'$. The vector sum of these components is equal to M_0 at equilibrium, but during the changes induced by RF excitation and with subsequent relaxation components of M_0 will exist along $z'(M_z)$ and $x'y'(M_{xy})$.

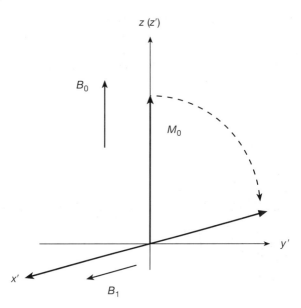

Figure 14.6. *In the rotating frame M moves around the magnetic portion of the applied RF pulse.*

14.7. EXCITATION

The matched radio-frequency radiation has the effect of forming a static magnetic field B_1, aligned along the x' axis (in the rotating frame). The net magnetization now has a magnetic field in the rotating frame, and begins to precess (in the rotating frame) about that field. Figure 14.6 shows the motion of M when the RF wave is applied.

After a certain time t_p, the magnetization will be aligned along the y' axis, i.e. it has turned through $90°$. This amount of RF energy, measured by the power of the pulse and its duration t_p, is called a $90°$ pulse. Leaving the RF on for double the time, or doubling the strength of the B_1 field, would turn the magnetization through $180°$; the pulse would then be called a $180°$ pulse. Intermediate sized pulses are easily produced, and are described by the angle through which the magnetization is turned—the so-called **flip angle** (α).

It is difficult to explain the underlying processes of how the magnetization vector is rotated using purely classical concepts. However, the absorption of RF energy will invert the two population states over a period of time whilst also achieving some phase coherence between the individual precessing magnetic moments.

Hence the degree of phase coherence at any one time will give a component of the magnetization vector in the $x'y'$ plane (M_{xy}) of varying magnitude and the shift of magnetic moments into the higher energy states will cause inversion.

14.8.   MR SIGNAL RECEPTION

When the RF wave is switched off after a 90° pulse, the magnetization vector (M) has been tipped through 90°, so that it lies along the y' axis, i.e. rotating at ω_0 in the laboratory frame. Since the RF pulse also forces all the protons to precess exactly in phase, there is now a component of magnetization along the y' axis in the rotating frame, of magnitude M_0. When the RF wave is switched off at time t_p, the system returns to equilibrium, aligned along the z (z') axis, by re-emitting the energy absorbed from the wave and phase coherence gradually being destroyed. If a pick-up coil has its axis perpendicular to B_0, it will experience a changing magnetic field as the magnetization rotates. This generates a current in the coil (section 14.2), which oscillates at a frequency of ω_0 and decays exponentially to zero (figure 14.7). This is the MR signal, known as the **free induction decay** (FID). The combination of a 90° pulse followed by a period of signal measurement is the simplest NMR experiment.

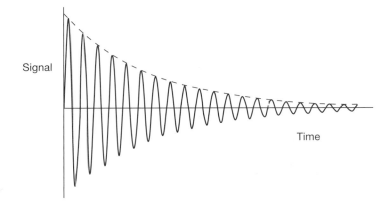

Figure 14.7. *The free induction decay (FID). The signal oscillates at ω_0 and decays exponentially to zero (dotted line).*

14.9.   RELAXATION PROCESSES

The application of an RF pulse causes a disruption to the equilibrium of the system of magnetic moments. The rate of recovery characterized by the relaxation processes conveys information on the structures in which the magnetic moments are located. There are two relaxation processes which can be used to characterize substances and influence contrast in an image.

14.9.1.   Spin–lattice relaxation

Energy may be transferred from protons to the surrounding environment, or lattice, in which they are embedded. Due to its much larger size, the lattice can absorb the energy without becoming excited. The relaxation mechanism controls the recovery of magnetization along the z axis (the longitudinal magnetization) back to

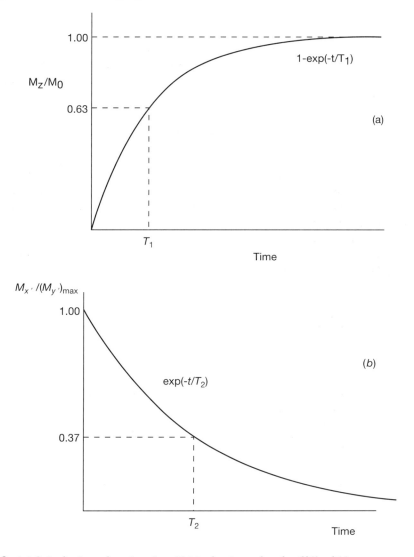

Figure 14.8. *(a) Spin–lattice relaxation time (T_1) is the time taken for 63% of M_0 to recover along the z axis. (b) Spin–spin relaxation time (T_2) describes the exponential decay of the transverse magnetization.*

its equilibrium value of M_0 (figure 14.8(*a*)). The recovery is described by the equation

$$M_z = M_0 \left[1 - \exp\left(\frac{-t}{T_1} \right) \right]$$

where t is time and T_1 is known as the spin–lattice relaxation time. The value of T_1 depends on the structure of the lattice, on temperature and on the magnetic field strength. In tissues it is typically of the order of a few hundred milliseconds (table 14.2).

14.9.2. Spin–spin relaxation

As the protons move around their environment, they experience varying magnetic fields due to their neighbours. This causes a spread of resonant frequencies, which corresponds to a dephasing of the magnetization in the (rotating) transverse plane. The relaxation in the transverse $x'y'$ plane is an exponential decay (figure 14.8(b)), governed by T_2, the spin–spin relaxation time.

$$M_{x'y'} = M_0 \exp\left(\frac{-t}{T_2}\right).$$

The value of T_2 like T_1 depends on the structure of the environment, on the mobility of protons and the irreversible dephasing of spins with respect to each other. If a tissue has a rigid structure, the movement of protons is restricted, giving rise to a very long T_1 and an extremely short T_2. The MR signal from these tissues is zero, even at very short echo times (section 14.10), and the tissue therefore appears black. For tissues which are increasingly liquid-like, T_1 times are shorter and T_2 times longer, until in pure water T_1 and T_2 become equal (about 3000 ms). Eventually, when the protons become fully dephased, the longitudinal magnetization is fully recovered and equilibrium is restored.

Table 14.2. T_1 and T_2 values at 0.5 and 1.5 T for some common tissues.

Tissue	0.5 T		1.5 T	
	T_1 (ms)	T_2 (ms)	T_1 (ms)	T_2 (ms)
Liver	400	45	500	43
Muscle	630	45	750	50
Cardiac muscle	650	75	870	45
Brain—grey matter	825	110	870	45
Brain—white matter	690	107	780	65
Fat	190	110	200	50
Kidney	765	125	760	30
Spleen	760	140	840	60

14.9.3. Inhomogeneity effects

The varying magnetic fields experienced by protons due to their neighbours may be thought of as an 'intrinsic' or microscopic magnetic field inhomogeneity. In addition, there is a macroscopic inhomogeneity in B_0 due to the technical difficulties of creating a perfectly uniform field. This 'extrinsic' inhomogeneity causes the protons to dephase in the transverse plane more quickly, as they move physically around the volume of the field. The FID seen after a 90° pulse is subject to these inhomogeneity effects and is described by the characteristic relaxation time T_2^*. T_2^* is calculated by combining the magnetic field inhomogeneity ΔB_0 with the spin–spin relaxation time (T_2):

$$\frac{1}{T_2^*} = \frac{1}{T_2} + \frac{1}{2}\gamma\Delta B_0.$$

Strictly, a term $1/T_1$ should also be added to the right-hand side of this equation, but since T_2^* is usually dominated by the effects of ΔB_0, the contribution of $1/T_1$ is negligible. The effects of inhomogeneity cannot be ignored as they produce artefacts within images which can be reduced only by careful design of pulse sequences or the improved design of magnets.

14.10. PRODUCTION OF SPIN ECHOES

In order to remove the effects of the magnetic field inhomogeneity, it is necessary to generate a spin echo using a 180° pulse. To assist in the understanding of this process it is convenient to picture a running track with athletes of different abilities competing. In the conventional 400 m race athletes run in an anticlockwise direction until they reach the finishing line: they will arrive at different times according to each individual's 'local strength' and other factors. Those other factors might include something about the local environment (i.e. the lanes) in which they run. If there are differences in the quality of the running track in the various lanes we would expect to see this reflected in the arrival times. However we cannot unmask these factors from the local strengths of the athletes.

Imagine now we start a different race in which the athletes first run in an anticlockwise direction but before they reach a complete revolution of the track another gun fires and they have to retrace their steps in a clockwise direction. What is the effect? If the track were consistent throughout all lanes the athletes would all arrive back at their original starting positions at exactly the same time irrespective of their local strengths. This is because some athletes would have run relatively faster and farther but upon retracing their steps the greater distances they would have had to travel would have nulled their 'local strength' i.e. the inhomogeneity between athletes would have been removed. If, however, there were variations in the lanes around the track that affected the frequency at which their legs moved, this would manifest itself in different arrival times at the finish.

Consider the system of protons as being split into small packets, each of which experiences a perfect magnetic field, but with inhomogeneities between the packets. After a 90° pulse which produces phase coherence, the protons are allowed to dephase for a short time TE/2 producing a fan of spin packets. This 'fan', or spreading out of vectors in the $x'y'$ plane, is caused by the variations of precessional frequency of individual 'packets' of protons experiencing slightly different magnetic fields due to the magnetic field inhomogeneities. This effect can be reversed by applying a 180° pulse along the x' axis which flips the magnetization through 180° (figure 14.9). The spin packets continue to dephase in the same sense (and direction) in the rotating frame, and after a further time TE/2 are all realigned along the Ðy' axis. The total magnetization is reduced in magnitude because of the T_2 relaxation effects within the spin packets which have also been occurring simultaneously.

Thus the spin echo is produced at time TE (time to echo) having cancelled the effect of magnetic field inhomogeneities. Each spin packet will also be slightly dephased due to the 'intrinsic' inhomogeneities of the protons, and so the magnitude of the echo will be governed by T_2. Many MR imaging sequences use spin echoes to produce signals and by varying certain timing parameters it is possible to achieve contrast based on either T_1, T_2 or proton density (section 14.13).

In summary, initially, the system is in equilibrium, with $M_z = M_0$ and $M_{xy} = 0$. A 90° RF pulse is applied, such that $M_z = 0$ and $M_{xy} = M_0$. The system returns to equilibrium, the recovery of M_z being described by the spin–lattice relaxation time T_1, and the exponential decay of M_{xy} by T_2^*. If a 180° pulse is applied after a time TE/2, a spin echo is generated at time TE whose magnitude is described by the spin–spin relaxation time T_2. Note that the effect of a 90° pulse is to sample the z component of the magnetization by flipping it into the transverse plane, where it can be detected and measured by a receiver coil. Lower flip angle (α) pulses produce $M_z \sin \alpha$ in the transverse plane, leaving the longitudinal component at $M_z \cos \alpha$; when a succession of pulses is applied the system reaches a dynamic equilibrium between excitation and relaxation. This method can be used to undertake fast imaging sequences but with reduced signal-to-noise in the image, since the magnitude of the magnetization in the $x'y'$ plane will be reduced.

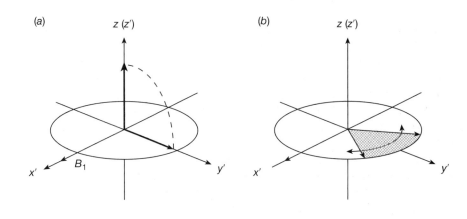

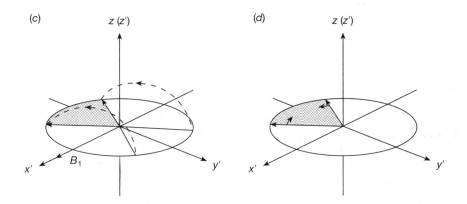

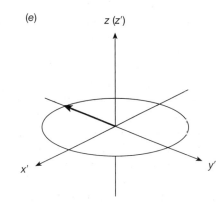

Figure 14.9. *(a) The initial* 90° *pulse flips the magnetization into the transverse plane, where (b) it begins to dephase. (c) At time TE/2 the* 180° *pulse (B₁) is applied which flips all the magnetization over. (d) The protons now begin to rephase, producing a spin echo (e) at time TE.*

14.11. MAGNETIC FIELD GRADIENTS

The signal described thus far is obtained from the system (or body) as a whole. To produce an image, some kind of **spatial** encoding is required. This is achieved using magnetic field gradients which allow selected parts of the body to be 'excited' by the RF pulse and also modify the collection of signal by 'tuning in' and measuring signal strength at various locations. Gradients are produced by passing electric currents through additional coils built in to the bore of the magnet. It is important to note that gradients modify the magnitude of B_0 in linear fashion, but that they do not change the direction of B_0 (figure 14.10). Since the resonant frequency of protons is proportional to B, which is proportional to distance in the gradient field, the result is that the MR signal frequency is directly proportional to distance.

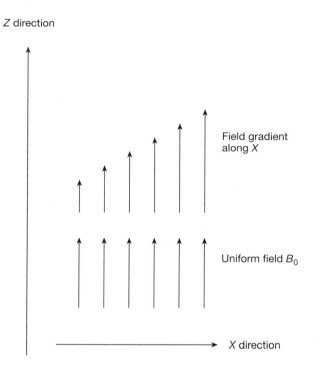

Figure 14.10. *Magnetic field gradients are superimposed on the main magnetic field, e.g. in the X direction. The strength of B varies with distance (X) but not its direction.*

Since the field at the isocentre is always B_0, the signals received have a bandwidth of frequencies centred on the Larmor frequency; the bandwidth depends on the strength of the applied gradient and the distribution of the protons. Gradients along the three principal axes are switched on and off in a repeated pattern, called a **pulse sequence**. The protons respond to changes in the magnetic field strength immediately, changing their resonant frequencies within a few picoseconds (10^{-12} s).

Throughout this section we shall use upper case (X, Y and Z) when referring to the axes of the image volume. Z is conventionally (in superconducting magnets) the direction of the magnetic field (superior–inferior in the body), Y is the vertical axis (anterior–posterior) and X is the remaining horizontal axis (right–left).

14.11.1. Fourier or reciprocal space

MRI uses a mathematical technique of image reconstruction involving Fourier analysis. In brief, the basis of Fourier analysis is that a complex waveform can be decomposed into its fundamental frequencies and amplitudes. With a suitable RF pulse and the use of magnetic field gradients it is possible to interrogate known elements of tissue because of the prior knowledge of the relationship between the magnetic field gradient and the Larmor frequency. When this is done the signal detected contains both magnitudes, ultimately representing contrast and relaxation processes, and frequencies, which represent corresponding spatial locations. The Fourier transform is able to decode and locate the different positions of inherent frequencies. As an analogy, members of an audience can locate different instruments in an orchestra using their auditory senses alone.

Thus the formation of an image requires data to be captured as spatial frequencies covering frequency or 'k' space, the intensity reflecting the magnitude of the spatial frequency and the position dictated by choice and combination of field gradients. This will be discussed further in section 14.11.5.

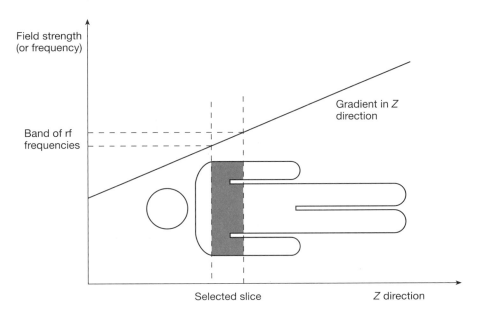

Figure 14.11. *Selective excitation of a slice of tissue, achieved by applying a narrow bandwidth RF pulse whilst a gradient is on (in the Z direction in this example).*

14.11.2. Selective excitation

The problem of spatial encoding in three dimensions can be reduced to two dimensions by only exciting protons within a slice of tissue. This is achieved by using an RF pulse containing a narrow band of frequencies, and applying a magnetic field gradient in the direction perpendicular to the plane of the slice. For example, in figure 14.11 the slice selection gradient is along Z so that transaxial slices will be generated. Slice selection gradients in the X and Y axes may be used to form sagittal and coronal cross sectional images respectively. Only protons whose resonant frequencies match those in the RF pulse will absorb energy, i.e. only those within the slice of tissue shown. The gradient is switched off after the RF pulse finishes: protons within the slice will then emit an MR signal, which may then be encoded for the two dimensions within the slice (sections 14.11.3 and 14.11.4, phase and frequency encoding respectively).

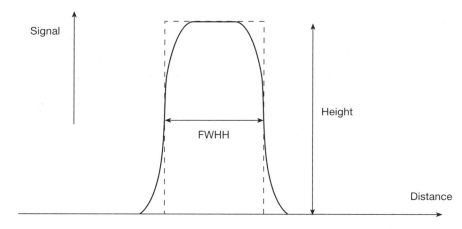

Figure 14.12. *The ideal slice profile is rectangular (dotted lines). In practice the slice is shaped according to the characteristics of the RF pulse. Slice width is defined by the full width at half height (FWHH).*

Ideally, all protons within the slice should receive a 90° pulse, while all those outside should remain unexcited. In practice however, this is impossible, and protons at the edges of the excited slice receive a smaller flip angle. Thus there is a slice profile (figure 14.12), which depends on the characteristics of the RF pulse and the slice select gradient. The slice width is defined as the full width of the slice at half the height (FWHH) of the profile (figure 14.12). Depending on the quality of the profile and the choice of slice gap, there will be signal contributions from tissue in adjacent volumes to the slice of interest ('cross talk').

14.11.3. Phase encoding gradient

Phase encoding is the most difficult technique to visualize and requires careful explanation. Consider a row of protons precessing at ω_0 in a main magnetic field B_0. There will be no phase coherence in respect of their precessional motion until a radio-frequency pulse is applied which 'drives' all protons to precess coherently (ignoring relaxation effects). A magnetic field gradient is applied along the row of precessing protons; each proton will begin to precess at a slightly different frequency in direct proportion to the local magnetic field experienced. The differences in precessional frequency between adjacent protons will cause relative phase shifts which are directly proportional to the differences in frequency, which in turn are directly related to the proton's position in the magnetic field gradient. If a linear gradient of known strength is applied for a time t_{pe}, the degree of phase shift may be predicted. When the gradient is switched off, all protons will return to the original precessional frequency, ω_0, but will retain their relative phase shift.

14.11.4. Frequency encoding gradient

The other dimension of the image slice is spatially encoded by acquiring the MR signal in the presence of a gradient in the appropriate direction. The gradient is switched on just before data collection begins, and off again when the signal has been acquired. During this time, the frequency of signals is proportional to distance in the frequency-encoding direction. The MR signal is no longer a simple oscillation, but an apparently haphazard jumble of peaks and troughs. Fourier transformation of the signal however reveals the intensity (and therefore the number of protons) at each frequency, or spatial position.

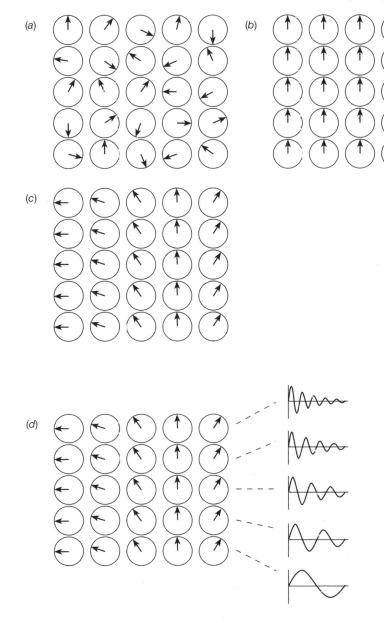

Figure 14.13. *(a) Before the RF pulse the protons have random phase (incoherent). (b) The RF pulse brings all the protons into phase with each other. (c) The phase encode gradient along X causes variations in precessional frequency and when it is switched off the columns have phase angles proportional to their position. (d) During the readout gradient the rows now have different frequencies as shown. The combination of phase and frequency information describes the position of each proton.*

14.11.5. Review of image formation

Consider an array of tissue elements (figure 14.13) in the XY plane as shown, i.e. we are potentially looking at a slice of information and the Z axis is perpendicular to the paper. There is no phase coherence in the net magnetization vectors (figure 14.13(*a*)) and hence no net signal can be measured from the slice. When an RF pulse is applied with an appropriate slice-select gradient (figure 14.13(*b*)) all vectors achieve coherence and when the RF pulse is turned off they will remain coherent (again ignoring relaxation processes), still precessing with a frequency ω_0.

Now a phase encoding gradient is applied in the X direction. This causes 'columns' of vectors to precess at slightly different frequencies according to the local magnetic field. When the phase gradient is turned off all vectors return to precessing at ω_0 but with a predefined degree of phase shift across the 'columns' (figure 14.13(*c*)). Finally a frequency encode (or read-out) gradient is used in the Y direction which causes all the 'rows' to precess at distinct frequencies (figure 14.13(*d*)).

When the signal is measured at the same time that the frequency encoding gradient is switched on, all frequencies are detected and with subsequent Fourier transformation the result is a series of peaks with different phases as well as different frequencies. However, while frequency encoding can encompass a relatively wide range of frequencies, there are only 360° of phase angle; it is impossible to distinguish between a phase angle of 10° and one of 370°. In order to overcome this the sequence of RF and gradient pulses is repeated many times, using a series of phase encoding gradients of differing size including zero. With zero phase encoding gradient, all the protons will be in phase with each other, giving the 'reference' information for measuring phase angles. Protons at a certain position within the field of view will produce a consistent pattern of phase angles; this enables their position to be properly measured. There is usually a relatively long delay between the end of the data collection and the start of the next set of pulses, to allow T_1 relaxation; the time between consecutive 90° pulses is denoted by TR (time to repetition).

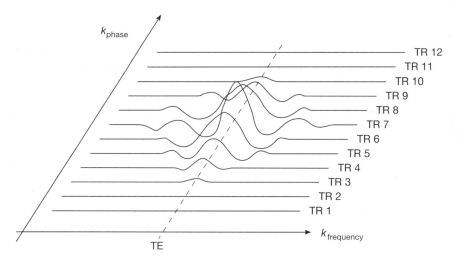

Figure 14.14. *Each set of data acquired during the readout gradient fills one line in k space. Each line has a different phase encode gradient strength.*

The result of applying different degrees of phase encoding, each time 'reading out' the signal information using the frequency encode gradient, is to build up lines of information in k or frequency space (figure 14.14). These data are then subjected to a two dimensional Fourier transform which produces a recognizable image. Due to the nature of the Fourier transform process, different parts of k space do not correspond to the same parts of the resulting image. At the centre of k space, where the phase encoding gradients are very small, the data contain all the signal-to-noise and contrast information. At the edges, at high phase encoding gradients, the data contain high spatial frequencies which correspond to the fine detail or resolution of the image.

An understanding of the nature of the information contained in k space enables different image sequences to be developed that allow minimum sets of information to be acquired (thus reducing acquisition times or computer memory constraints), or sampling k space in more efficient ways (e.g. by not using linear gradients).

14.11.6. Dephasing effects of gradients

Magnetic field gradients may be considered as large magnetic field inhomogeneities, which cause more rapid dephasing of the magnetization in the transverse plane. If this were not compensated for, the MR signal would quickly decay to zero. However, by reversing the current in the gradient coils, an inhomogeneity is created in the opposite sense, which allows the signal to be rephased. The signal will be exactly rephased when

$$\varphi_{\text{rephase}} = -\varphi_{\text{dephase}}$$

i.e. when the time integrals of the gradient pulses are equal and opposite. On the pulse sequence diagram, this is when the areas of the gradient pulses are equal.

As can be seen (figure 14.15), equal areas can be achieved with short, high gradient pulses, or long and low pulses, or anywhere in between. Since we are usually trying to perform the sequence as fast as possible it is best to use the shortest gradient pulse within the limits of the gradient set. However there are technical as well as safety considerations that limit the frequency and speed of application of magnetic gradients. Because it is the area which is important, it is even possible to split a gradient pulse and move part of it to a different time in the sequence (figure 14.15(*d*)). This is essential for flow compensation (section 14.15), since for moving protons balanced gradients are needed to rephase the signal. In the case of the phase encoding gradient, rephasing is unnecessary (the phase angles give positional information). For the frequency encoding gradient, the maximum signal (the spin echo) should occur in the middle of the data capture window; it is important that the net phase effect is zero at this time. Thus, a 'frequency dephase' gradient pulse is applied before the frequency encoding pulse, with area equal to half that of the frequency encoding pulse. The dephase lobe comes before the 180° pulse which effectively inverts the phases: therefore in a spin echo sequence the frequency dephase gradient is positive.

The situation for the slice selection gradient is slightly more complex, because the RF pulse has a changing shape. By convention however, the shaped RF pulse is considered to act as a sharp spike at the middle of the pulse; i.e. during the first half of the pulse there is no excitation, then all protons in the slice receive a 90° pulse instantaneously. Whilst there are no excited protons, the gradient can have no dephasing effect, so the period up the middle of the RF pulse is ignored. The remaining area of the slice selection gradient will cause dephasing of the newly excited protons, which may be remedied by applying a negative gradient with approximately equal area (the 'slice rephase' gradient). It is necessary slightly to adjust the size of the rephase gradient since the RF pulse does not act instantaneously.

14.12. PRODUCTION OF GRADIENT ECHOES

From the above discussion, it should be clear that it is possible to create a signal echo without using a 180° pulse, but simply using the gradients. All imaging sequences use either a spin echo or a gradient echo to read the signal.

A gradient echo sequence uses slice selection and phase encoding gradients as usual. However, since there is no 180° RF pulse, a negative frequency dephase gradient is used. When the frequency encoding gradient is switched on to start data capture, the effect of the original dephasing is gradually reversed, until at the centre of the data capture window it is zero, and the signal is a maximum (figure 14.16). The continuing gradient then dephases the signal in the opposite direction decreasing the signal again; thus an echo has been formed.

Due to the absence of the 180° pulse, gradient echo signals are described by T_1 and T_2^*, not T_2. By varying parameters, images may be formed which have T_1 or T_2^* contrast. In practice, T_2^*-weighted images

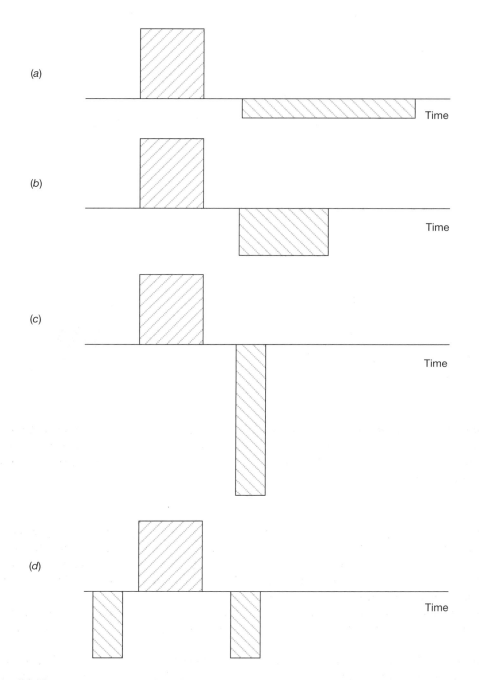

Figure 14.15. *(a) to (c) A range of gradient heights and durations is possible to rephase the same initial gradient. (d) The rephase gradient can be split and still perform its function. In these examples the time integral (i.e. shaded areas above and below the time axis) are balanced and therefore equal zero.*

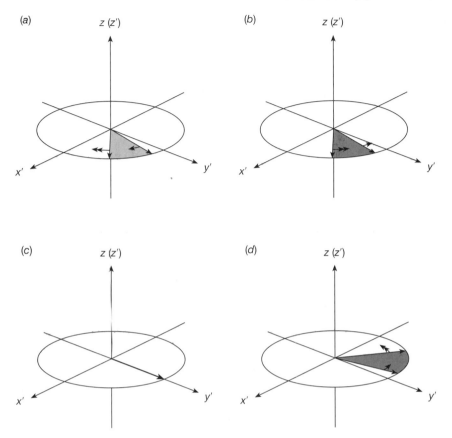

Figure 14.16. *(a) The dephase lobe of the gradient accelerates the dephasing of the protons after the RF pulse. (b) When the gradient is reversed (at the start of the readout lobe) the protons begin to rephase. (c) The protons come back into phase, forming the gradient echo. (d) The continuing gradient now dephases them in the opposite direction.*

have similar contrast to T_2 weighted images, and are often described as 'gradient-echo T_2 weighted'; this is strictly incorrect and it should be remembered that the magnetic field inhomogeneity usually dominates the T_2^* decay. However magnetic field inhomogeneities have been greatly reduced during the past decade by improvements in the design and construction of magnets: if this were not the case, inferior image quality would render gradient echo images unacceptable.

Whilst all imaging sequences use either a spin echo or gradient echo to acquire data various additional pulses may be used before the imaging sequence to 'prepare' tissues, so that different contrast effects may be produced.

Whichever pulse sequence is chosen for imaging, it is important to understand the T_1 and T_2 relaxation curves in order to optimize the SNR in the final image. Figure 14.17 shows the relaxation curves for two tissues. The TR chosen for the sequence determines the T_1 weighting, while the TE affects the T_2 weighting. It can be seen that at a long TR, both tissues are able to fully relax and, since they have similar proton densities, will have similar signal intensities on the image. Reducing the TR still allows tissue 1 to relax fully, giving high signal, but tissue 2 is only able to partially relax. Consequently the signal will be reduced in the final image, reducing the SNR but improving the tissue contrast. Similar analysis shows that maximum SNR is

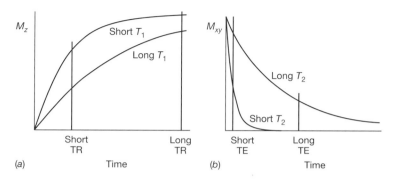

Figure 14.17. *(a) T_1 and (b) T_2 relaxation curves for two tissues. TR affects the amount of T_1 relaxation in the sequence, while TE controls the T_2 weighting.*

obtained with minimum TE, but that contrast is improved with longer echo times.

Several other factors affect the SNR and are common to all pulse sequences. For example, the signal intensity of a pixel is also proportional to the amount of tissue contained in that volume. SNR can be improved by increasing the pixel size (i.e. field of view) and the slice thickness; however, this is at the expense of image resolution. Another method to increase SNR is to use extra signal acquisitions, i.e. repeat the pulse sequence: the disadvantage of this is the increased scan time.

14.13.   IMAGING SEQUENCES

14.13.1.   *Spin echo*

As previously described (section 14.10), this sequence uses two RF pulses (90° and 180°) which creates a spin echo at time TE. Varying TE and TR (the sequence repetition time) allows images to be formed with different contrast (figure 14.17). The echo time, TE, controls the amount of T_2 weighting. T_1 weighting is controlled by the TR; shorter repetition times have less signal from long T_1 tissues. This loss of signal is due to **saturation** of the protons. In quantum mechanics terms, complete saturation occurs when the populations of the two energy levels are equal and no signal can be obtained. 'Partial saturation' describes the intermediate situation when the difference in populations is reduced from the equilibrium, and there is a partial loss of signal. Maximum SNR is achieved by having a short TE and a very long TR; such images are proton density (PD) weighted. Decreasing the TR or increasing the TE will give a lower overall SNR, but improves the contrast between different tissues. (Combining a short TR with a long TE would give both T_1 and T_2 weighting, but with very little signal—not a useful sequence!)

Since both PD- and T_2-weighted images have long TRs, it is possible to use repeated 180° pulses to produce more than one spin echo. With two spin echoes, one at a short TE and the other with long TE, both PD and T_2 images are produced within the same TR (and therefore the same scan time). Such 'dual echo' sequences are much more common than only T_2 or PD images.

With technical improvements, the minimum echo times have reduced, and a new type of sequence has been developed to take advantage of this. Variously known as **fast or turbo spin echo** (FSE or TSE, respectively), these sequences use a repeated series of 180° pulses after the initial 90°, which produce a train of spin echoes. Each echo has a different phase encoding gradient, so that data are acquired many times faster than conventional sequences, where a single phase encoding gradient is used in each TR. However, the echo time is no longer a single value: the middle of the echo train is taken to be the 'effective TE', but there is some averaging of signals over the T_2 decay curve, and the image contrast is not exactly the same as for traditional spin echo sequences.

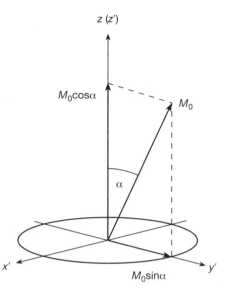

Figure 14.18. *Reduced flip angles split the magnetization. With a 30° angle for example, the signal in the transverse plane is 50% of the maximum. However, because the longitudinal magnetization is only reduced to 87% instead of zero, T_1 relaxation is complete in a much shorter time than with a 90° pulse.*

14.13.2. Gradient echo

Gradient echo (GE) sequences are generally much faster than those using spin echoes (SE) and there are several variations on the basic sequence. The most important difference between SE and GE is that the flip angle is reduced from 90°. In these sequences, the TR is often less than 100 ms, which allows very little recovery of longitudinal magnetization. If the flip angle α is reduced, the transverse magnetization becomes $M_0 \sin \alpha$, and the longitudinal component is left with $M_0 \cos \alpha$ (figure 14.18). M_z recovers back towards M_0, controlled by T_1 as usual, and as successive RF pulses are applied, an equilibrium is reached. Thus there are three parameters that have an effect on image contrast—TE, TR and α. Generally, TE and TR are an order of magnitude smaller than SE parameters; unlike spin echo however, multiple gradient echoes are rarely formed. T_2-weighted (strictly T_2^*-weighted) images are produced using a relatively long TR, very low α and a long TE. As with SE T_2 weighting, signals from tissues with a long T_2 are relatively enhanced. Images which are T_1 weighted use a short TR and very short TE, combined with a large α (although usually not 90°). Proton density contrast can be achieved using a short TE, very low α and relatively long TR; however, this is rarely used, since the contrast is not true PD, and T_2 weighted images have to be acquired separately.

14.13.3. Inversion recovery

Inversion recovery sequences use a 180° pulse before the imaging sequence to prepare the signal. The initial magnetization is turned through 180°, so that M_z becomes $-M_0$; M_{xy} remains at zero. M_z recovers as usual, controlled by T_1 relaxation, until a time TI (the inversion time). Then the spin echo imaging sequence is applied, to produce an image which is heavily T_1-weighted. If TI is such that the signal from a particular tissue has $M_z = 0$ (the 'null point') the tissue will appear black on the resulting image. This is often used to remove fat signals from images; fat has a short T_1, and a TI of 100–110 ms is at the null point (for a 0.5 T system). This particular sequence is called **STIR**—short TI inversion recovery.

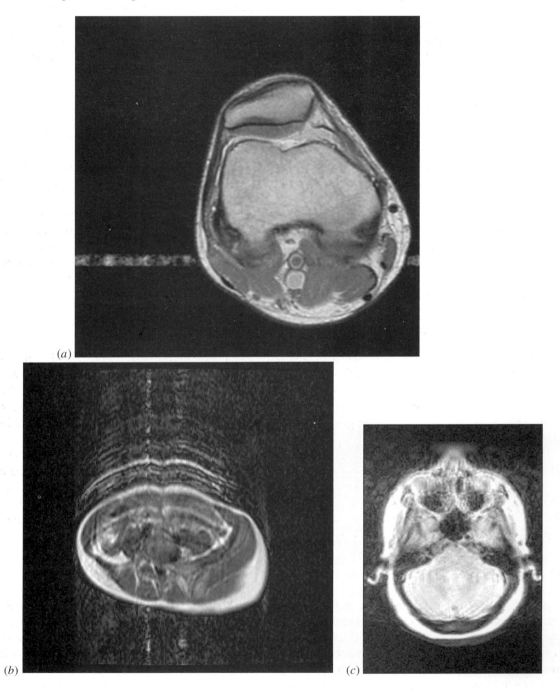

(a)

(b)   *(c)*

Figure 14.19. *Motion artefacts on images, always in the phase encoding direction. (a) Blood vessel artefact, (b) respiratory artefact in the abdomen and (c) gross patient motion.*

14.14.   GATED SEQUENCES

The effect of gradients on protons described in section 14.11 depends on the protons remaining stationary. If there is movement of spins during the imaging sequence, artefacts are produced on the image (figure 14.19). In the body there are many moving protons due to the cardiovascular system, respiration and peristalsis. Fortunately, because they occur in a regular cycle, the effects of cardiac and respiratory motion can be removed by using gated sequences.

14.14.1.   Cardiac gating

To 'freeze' the motion of the heart, the pulse sequence is triggered using the ECG waveform instead of a regular TR. Electrodes are attached to the patient in the usual way, and the waveform is electronically monitored. As each R wave is detected the imaging sequence is started, so that the effective TR is the RR interval which depends on the heart rate of the patient. Each line of data for a particular image is acquired at the same time relative to the cardiac cycle and thus motion artefacts do not appear.

ECG gating is most commonly used when imaging the thorax for cardiac or pulmonary disease. Gated imaging can also be useful in situations where a major artery is in the field of view, or to remove artefacts due to CSF pulsation in a spinal imaging. In the latter cases it is more convenient to use a photoplethysmographic detector on a toe or finger, which detects the arrival of arterial blood in the digit and thus provides an R-wave trigger. Since the 'effective TR' is determined by the RR interval the contrast and SNR in an image will to some extent be dependent on pulse rate, and images are usually T_1 weighted.

14.14.2.   Respiratory gating

The rise and fall of the diaphragm and chest wall during respiration has a much slower cycle. It can be measured by connecting expandable bellows that contain a fixed volume of air around the chest; the scanner is able to detect the expansion and collapse of the bellows by measuring the pressure in the bellows. Some scanners use respiratory triggering to remove motion artefacts, which works in much the same way as cardiac gating. However, the effective TR is lengthened to several seconds because of the slow cycle, and this produces an extremely long scan time.

An alternative method of compensating for respiratory motion is called phase re-ordering which relies on the fact that the largest phase encode gradients are least sensitive to motion of protons. Respiration is measured for two or three cycles to determine the breathing pattern and repetition time. Instead of varying the phase encode gradient from positive to negative in a regular way, the scanner now re-orders the sequence so that the largest gradients occur during the periods of most respiratory motion. This is a much more time-efficient method, but its success does rely on the patient's breathing remaining regular.

14.15.   MR ANGIOGRAPHY

Moving blood within arteries and veins in the field of view of the image causes distinctive artefacts, always in the phase encoding direction. On spin echo images, the lumen of the blood vessels appears black, because the blood receives either the 90° or the 180° pulse but not both, and therefore cannot produce an echo. Gradient echo sequences produce the opposite appearance ('bright blood') because the echo is rephased by the gradients not an RF pulse. However, these rules only apply in the simplest of cases; any form of turbulence or reduction in flow rate changes the appearance of moving fluid in the image, usually by a reduction of intensity.

In order to remove the flow artefact, it is necessary to use balanced gradients as mentioned in section 14.11.6. Known as 'gradient moment nulling', this technique rephases the signal from protons even though they are moving during the image acquisition.

Instead of trying to compensate for the motion of protons in flowing blood, it is possible to exploit this property to produce MR angiograms, which show only the blood vessels. There are two types of MRA sequence, **time-of-flight** (TOF) techniques (also called in-flow methods) and **phase contrast** (PC).

TOF angiography uses a gradient echo sequence with a very short TR, which saturates the signal from static tissues within the slice. Blood flowing perpendicular to the slice however has 'fresh' magnetization for each excitation and produces a high signal. The contrast between this bright signal and the low signal from static tissues produces an angiographic appearance.

Slowly flowing blood, or blood in vessels that run across the slice, experiences several RF excitations and the magnetization becomes saturated. Turbulent flow also produces a low signal, due to rapid dephasing of the spins. These effects provide many traps for the unwary in the interpretation of TOF appearances. In addition, short T_1 tissues such as fat leave residual signals on the image, which can obscure the anatomy of blood vessels.

Phase-contrast MRA is a quite different sequence, which uses special gradient waveforms to achieve gradient moment nulling. This allows the velocity of the protons to be encoded in the phase angle of the signal (note that this is separate from phase encoding for position). To produce an angiogram, two images are acquired with and without gradient moment nulling. Subtraction of one from the other gives the phase angle, and thus the velocity, which is combined with the magnitude image for display purposes. The advantages of PC-MRA are that static protons have zero phase, and thus produce no signal on the final image, and that flow may be quantified. The gradient strengths may be adjusted to make the sequence sensitive to slow or fast flows, allowing arterial blood to be easily distinguished from venous return. The disadvantages are that to visualize flow in all directions, gradient moment nulling must be applied to all three gradients, which extends the imaging time. In addition, to fully characterize the flow in a given vessel, a series of images with different velocity encoding should be acquired since each shows only a range of velocities.

14.16.   CONTRAST AGENTS

MRI has a large amount of intrinsic contrast, due to the combinations of proton density, relaxation times and sequence timings available. Despite this, certain types of pathology are not easily visible and such examinations benefit from the use of contrast agents. There are two groups of agents, those which affect relaxation times (gadolinium- and manganese-based compounds) and those which affect magnetic susceptibility (iron-oxide-based compounds).

By far the most common contrast agent is gadolinium (Gd) administered intravenously to enhance a range of tumours. Since gadolinium is extremely toxic it is complexed to a large molecule such as DTPA which does not pass through the healthy blood–brain barrier, but passes through a disrupted barrier. Because gadolinium is strongly paramagnetic, it reduces the T_1 relaxation time of the tumour tissue, causing a relative enhancement on T_1-weighted imaging.

Recently iron-oxide-based contrast agents have become available, which work by causing local field inhomogeneities. The T_2 and T_2^* relaxation times are reduced in these regions, so that tissues which take up the contrast agent lose signal on T_2- or T_2^*-weighted images. Administration may be as an oral suspension for abdominal imaging, or intravenous for liver and spleen imaging.

14.17.   TECHNICAL CONSIDERATIONS

MR systems comprise a large number of complex components, the specifications of which change very rapidly. Central to the system is a large bore, high field magnet with high homogeneity of the magnetic field. There are three types of magnet, with advantages and disadvantages as shown in table 14.3, but the most common are superconducting magnets. These rely on a bath of liquid helium at 10 K ($-263\,°$C) to cool the current carrying wires so that they have almost zero resistance. Once the current is set to provide the correct magnetic field, no electrical power is necessary and the major running cost is that of topping up the liquid helium. Very good homogeneity is achievable, often to less than 0.2 ppm over the volume of the head coil.

Table 14.3. *Comparison of various magnet types.*

Magnet type	Properties	Advantages	Disadvantages
Permanent	Made of iron alloy blocks permanently magnetized	• Cheap; zero running costs • Open design	• Relatively low field (<0.3 T) • Weight >40 tonnes
Electromagnetic	Water-cooled coils of copper	• Field can be turned off in emergency • Open design • Mid-cost	• High power requirements • Poor stability
Superconducting	Liquid-helium-cooled wires	• High field • Good homogeneity	• High capital cost • Cost of helium • Relatively enclosed bore

The gradient set contains three sets of coils, one for each orthogonal direction, which produce linear magnetic field gradients when current is passed through them. Gradient sets are usually characterized by a maximum gradient strength, expressed in mT m^{-1}, and the rise time from zero to that maximum in milliseconds. The maximum gradient has implications for resolution, while both strength and speed are important for fast imaging.

Modern MR systems are equipped with a variety of RF coils, each designed to optimize SNR from a particular part of the body. The body coil is a large cylindrical coil, usually manufactured as part of the gradient set. Like the head coil (which is simply a smaller version), the body coil is used both to excite the protons with the RF pulses and to receive the signal. Such coils are termed 'transmit-and-receive' coils. Many of the other coils are 'receive-only' or 'surface' coils. When these coils are used, the RF pulses are transmitted on the body coil to achieve a uniform excitation of protons. The signal is picked up by the surface coil, which has much higher sensitivity because it is adjacent to the area of interest. The sensitivity of such a coil depends on distance, with a fall-off proportional to $1/r^2$ as shown in figure 14.20. This can be exploited to avoid signals from unwanted structures. The increased sensitivity of surface coils, together with the decreased field of view, allows high resolution images to be acquired with good SNR.

RF coils are also described as quadrature, linear or phased array. The simplest 'linear' coils are single loops of copper, rarely circular but shaped appropriately for part of the body. Quadrature coils have two loops, usually overlapping, and are able to improve the SNR by reducing the amount of noise detected. Phased array coils are a relatively recent development, using several coils to pick up signal. Each coil forms an image, and after reconstruction all the images are joined together to form a single high SNR image. Such coils are able to extend the field of view compared to simple surface coils, but maintain high sensitivity over the whole region.

The whole system is controlled by a central computer, with dedicated processors to do certain tasks. For example, the sequence of gradient and RF pulses is defined by a pulse generator processor, and the final Fourier transforms are processed by an array processor capable of handling large numbers of data at high speed.

In order to prevent the small RF signal being swamped by the ambient RF noise, the magnet is enclosed in a screened room, or 'Faraday cage'. Sheets of copper built into the walls, floor and ceiling form a totally closed metal box around the system, and unwanted RF noise cannot pass through. It does not block the magnetic field however; containment of the magnet's fringe fields is achieved with iron or steel shielding, usually close to the magnet. An alternative means of containing the magnetic field is the development of

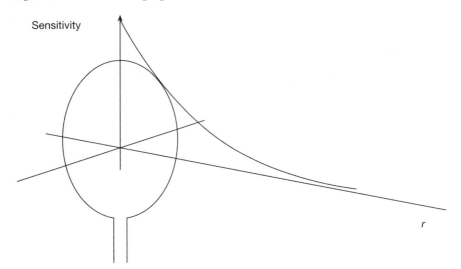

Figure 14.20. *The sensitivity of a circular surface coil falls off with distance from the plane of the coil.*

'active shields'; these consist of an additional set of coils in the magnet generating an opposing magnetic field to reduce the fringe magnetic field.

14.18.   HAZARDS OF MRI

MRI is often described as a safe imaging technique, and indeed it is much less hazardous than methods involving ionizing radiation. There are three areas to be considered for MRI safety: the large main magnetic field, the gradient fields which are switched at low frequencies and the radio-frequency radiation.

The Food and Drug Administration (FDA) in the USA and the National Radiological Protection Board (NRPB) in the UK have issued guidelines for the safe use of diagnostic MRI; although these are not statutory limits for practical purposes standard clinical scanners do comply with them. The Medical Devices Agency (MDA) of the Department of Health and the International Electrotechnical Commission (IEC) have produced extended safety guidelines for best practice, and in addition every MR unit should have local safety rules. It is strongly recommended that all radiologists and radiographers working in MRI should be familiar with all these documents.

14.18.1.   Main magnetic field

The field strength of the main magnetic field (B_0) ranges from 0.2 T to 3 T in most clinical systems. It is generally accepted that fields of up to 2 T produce no harmful bio-effects, including no chromosomal effects. In higher fields, effects are related to the induction of currents in the body (which is a conductor) when it moves through the field. These may cause visual sensations (magnetophosphenes), nausea, vertigo and a metallic taste. However, these effects are short term, usually disappearing when the body is no longer in the magnetic field.

14.18.2.   Projectile effect of B_0

Although the main field is not intrinsically hazardous, its powerful effect on ferromagnetic objects constitutes a major problem. Magnetic bodies within the 0.1 mT fringe field experience an attractive force towards the centre of the magnet. If they are free to move, they can acquire high velocities and cause significant damage to equipment and persons in their path.

The strength of the force and thus the resulting acceleration depend on several factors, including distance from the field centre, mass of the object and its magnetic properties. It is essential therefore that all potential projectiles are kept outside the magnetic field. The list of such objects is extensive, and includes keys, scissors, paper clips, stethoscopes, gas cylinders, drip stands, wheelchairs etc. Apart from being a potential hazard and causing personal injury or damage to the equipment, small metal objects within the bore of the magnet will also cause image artefacts. Medically or accidentally implanted objects in the patient's body may also be ferromagnetic. The degree of hazard depends both on the type of implant and its location. Of particular concern are intra-orbital foreign bodies, which are generally not fixed by scar tissue, and aneurysm clips. In these cases movement of the implant can cause blindness or fatal haemorrhage respectively.

Other items may not become missiles but will be damaged by the magnetic field and should be removed. Examples include analogue watches and cards with magnetic strips. Various magnetically activated devices also fall into this category such as cochlear implants and cardiac pacemakers. Persons with these implanted devices may not enter the 0.5 mT fringe field and therefore cannot be scanned.

14.18.3. Gradient fields

When gradient fields are switched on and off, they may induce currents in the body. The size of the current will depend, among other factors, on the maximum gradient field strength and on the switching time; it is important to remember that the gradient fields increase away from the centre of the magnet, so the area for concern is not necessarily the part being scanned. Induced currents may be large enough to stimulate nerves, muscle fibres or cardiac tissue; effects may include magnetophosphenes, muscle twitches, tingling or pain or in the worst case ventricular fibrillation. Theoretical calculations and experimental evidence indicate that such effects will be avoided if the gradient switching is kept below 20 T s^{-1}. Most existing clinical scanners are well below this limit, but as scans become faster and technology improves the capabilities of the gradient sets, it is important that this hazard is carefully monitored.

14.18.4. RF fields

Radiofrequency waves contain both electric and magnetic fields oscillating at MHz frequencies. At these rates, the induction of circulating current in the body is minimal, as there are high resistive losses. Most of the RF power is therefore converted to heat in the body, and bio-effects and safety limits are considered accordingly. In healthy tissues, a local temperature rise caused by RF power deposition will trigger the thermoregulatory mechanism of increased perfusion to dissipate the heat around the rest of the body. If the rate of power deposition is very high, or the thermoregulation system is impaired in some way, heat will accumulate locally, eventually causing tissue damage. Some areas of the body are particularly heat sensitive, for example the eyes, the testes and the embryo, and extra care should be taken. The safety limits are somewhat complex, as they depend on the duration of the exposure and the area under consideration, but basically they are designed to limit the temperature rise of the body to $1\,^{\circ}\text{C}$. For the whole body in a 30 minute examination, the specific energy absorption rate (SAR) limit is 1 W kg^{-1} (NRPB guidelines; 0.4 W kg^{-1} according to FDA guidelines), while for a head scan of similar duration it is 2 W kg^{-1}.

Particular care must be taken when metal objects are in the imaging field, e.g. ECG leads or non-removable metallic implants. Such objects absorb RF energy very efficiently and may become hot enough to burn the skin if in direct contact. It is worth noting that RF burns to patients form the majority of MRI-related accidents reported to either the FDA or MDA.

14.18.5. *MRI in pregnancy*

There is some evidence that exposure to switched magnetic fields may cause abnormal development of the embryo. The effects of excess heat on the foetus are well established and evidence that exposure to RF radiation may cause heat-induced damage is available. However, there are also studies showing no harmful effect with either gradients or RF. Overall, the evidence is contradictory and neither proves nor disproves the safety of MR imaging during pregnancy, although there has been no indication that clinical MRI produces adverse effects. The NRPB recommends that scans are not performed in the first trimester. However if other non-ionizing imaging methods are inadequate or the alternative diagnostic method involves ionizing radiation, and if the information is regarded as clinically important, a scan may take place at any stage of pregnancy.

14.19. CONCLUSIONS AND FUTURE DEVELOPMENTS

It is impossible for a single chapter in a book such as this to explain all the difficult concepts involved in MRI. However, there are many other textbooks, and each offers a slightly different view of the topic. Readers are encouraged to explore as many books as possible to improve their understanding.

MRI is a valuable addition to the range of radiological methods, and has proved superior to other techniques in many applications. The complexity of the underlying science is balanced by the flexibility of the information which may be deduced, from high resolution anatomy to flow, diffusion and other physiological processes. Advanced research in recent years suggests that MRI is even able to detect metabolic changes in the brain during various activities, making it possible to measure functionality as well as anatomy. The pace of development in MRI has not slowed over the last 15 years, with major research areas expanding all the time. The future of MRI in clinical radiology can only expand as technology improves and new techniques gain acceptance in the medical community.

FURTHER READING

Brown M and Semelka R 1995 *Magnetic Resonance: Basic Principles and Applications* (New York: Wiley–Liss)
Elster A D 1994 *Questions and Answers in MRI* (Mosby)
Foster M A and Hutchison J M S 1987 *Practical NMR Imaging* (IRL Press)
Hashemi R H and Bradley W G 1997 *MRI: The Basics* (Baltimore, MD: Williams and Wilkins)
Hovowitz A L 1992 *MRI Physics for Radiologists* 2nd edn (Berlin: Springer)
NRPB 1991 Principles for the protection of patients and volunteers during clinical magnetic resonance diagnostic proce-
     dures *Documents of the NRPB* vol 2, No 1 (London: HMSO)
Rinck P A 1993 *Magnetic Resonance in Medicine* 3rd edn (Oxford: Blackwell Scientific Publications)
Shellock F G and Kanal E 1994 *Magnetic Resonance: Bioeffects, Safety and Patient Management* (Raven)

EXERCISES

1   What is meant by the term 'resonance' in relation to MRI?
2   Describe what is meant by the Larmor frequency and how is it related to the applied magnetic field.
3   What are the Larmor frequencies for water nuclei in magnetic fields of 0.5 T, 1.0 T and 1.5 T?
4   The tesla (T) is a unit used to measure the magnetic field strength in MRI. What is the strength of the earth's magnetic field?
5   What is meant by 'net magnetization'?
6   Classically the spin population states are divided into 'parallel' and 'antiparallel': what is the approximate population difference between these two states at 1.5 T?
7   What is the purpose of the radio-frequency transmit and receive coils?
8   What determines the frequency of the rotating frame of reference?
9   What is meant by flip angle (α)?
10 Describe what is meant by free induction decay (FID) of the MR signal.
11 Describe what is meant by spin–lattice relaxation. How is this characterized mathematically?
12 Describe what is meant by spin–spin relaxation. How is this characterized mathematically?
13 At 0.5 T what are the approximate T_1 and T_2 for fat and muscle?
14 Describe the spin echo sequence ignoring the imaging process. How might the timing parameters be adjusted to reflect T_1, T_2 and proton density in the image?
15 What is the difference between a spin echo and a gradient echo sequence and how are these differences useful?
16 Describe how magnetic field gradients are used in the imaging process.
17 What is meant by frequency and phase encoding?
18 Describe how you would recognize motion and chemical shift artefacts. How are they related to the magnetic field gradients?
19 Describe the inversion recovery sequence. How might a modification of this sequence be used to remove the fat signal from the image?
20 What is the purpose of the Faraday cage and where might it be located?
21 If the Faraday cage were ineffective how might this manifest itself in an image?
22 What is the fringe field?
23 What is the maximum value in the fringe field that is generally regarded as safe for a person with a cardiac pacemaker to stand?
24 List the main contraindications for MRI.
25 What precautions should be taken for staff or patients who may be pregnant in relation to magnetic fields?
26 What is the biological effect of applying radio-frequencies to tissues?
27 What is meant by SAR? Define the units used to measure this parameter. What is the whole body limit and which organs might be of particular concern to radio-frequency effects?
28 List documents which are relevant to MRI safety considerations.
29 Describe what quality assurance should be undertaken on an MRI system.
30 Briefly describe the different magnet technologies and the associated benefits of each.

APPENDIX

The following multiple choice questions have been constructed so as to test your knowledge of the subject. Most answers are either given in the book or may be obtained by deductive reasoning. A few may require some additional reading, particularly for a fuller explanation.

The questions have been grouped in the same order as the chapter headings. There is no constraint on the number of parts of any question that may be right or wrong.

Unless otherwise stated, assume in all practical radiographic situations that any change in conditions is accompanied by an adjustment of mA s to give a similar optical density on the film. Note that without this constraint many of the questions would be ambiguous.

1.1 The nucleus of an atom may contain one or more of the following:

(a) Photons.
(b) Protons.
(c) Neutrons.
(d) Positrons.
(e) Electron traps.

1.2 The atomic number of a nuclide:

(a) Is the number of neutrons in the nucleus of an atom.
(b) Determines the chemical identity of the atom.
(c) Will affect the attenuation properties of the material at diagnostic X-ray energies.
(d) Is the same for 123-I as for 131-I.
(e) Is increased when a radioactive atom decays by negative beta emission.

1.3 Orbital electrons in an atom:

(a) Contribute a negligible fraction of the mass of the atom.
(b) Are equal in number to the number of neutrons in the nucleus.
(c) Can be raised to higher energy levels when the atom is excited.
(d) Are sometimes ejected when atoms decay by electron capture.
(e) Play an important part in the absorption of X-rays.

1.4 All isotopes of an element:

(a) Have the same mass number.
(b) Have the same physical properties.
(c) Emit radiation spontaneously.
(d) Have the same number of extra nuclear electrons as the neutral atom.
(e) When excited emit characteristic X-rays of the same energy.

1.5   With respect to radioactive decay processes:

(a) The activity of a radionuclide is 1 MBq if there are 1000 nuclear disintegrations per second.
(b) The half-life is 0.693 multiplied by the decay constant.
(c) After 10 half-lives the activity of a radionuclide will have decreased by a little more than 1000-fold.
(d) Different radioisotopes of the same element always have different half-lives.
(e) The physical half-life is independent of the biological half-life in the body.

1.6   The following are properties of *all* ionizing radiations:

(a) Decrease in intensity exponentially with distance.
(b) Have energy of at least 8 keV.
(c) Produce a heating effect in tissue.
(d) Produce free electrons in tissue.
(e) Act as reducing agents.

1.7   The following statements are correct in respect of radionuclides:

(a) They have energy levels in the nucleus.
(b) There is no change in mass number when gamma rays are emitted.
(c) There is a change in mass number when negative beta particles are emitted.
(d) Metastable states have half-lives of less than a minute.
(e) Orbital electron capture results in the emission of characteristic X-rays.

1.8   The exponential process:

(a) Applies to both radioactive decay and absorption of monoenergetic gamma rays.
(b) Is expressed mathematically by the inverse square law.
(c) When applied to radioactive materials means constant fractional reduction in activity in equal intervals of time.
(d) When applied to absorption of monoenergetic gamma rays means the half value thickness will be the same at any depth in the absorber.
(e) Will never reduce the intensity of a beam of gamma rays to zero.

2.1   The anode angle of a diagnostic X-ray tube:

(a) Is the angle the face of the target makes with the direction of the electron beam.
(b) Is usually about 25°.
(c) Influences the size of the effective focal spot.
(d) Determines the largest X-ray film that can be adequately covered at a given distance from the focus.
(e) Affects the amount of heat produced in the target for a given kV and mA s.

2.2   Tungsten is used as the target material in an X-ray tube because it has the following desirable properties:

(a) A high atomic number.
(b) A high density.
(c) A high melting point.
(d) A relatively low tendency to vaporize.
(e) A high thermal conductivity.

2.3   The effective focal spot size of an X-ray tube depends on:

(a) The size of the filament.
(b) The anode material.
(c) The anode angle.
(d) The kV and mA s selected.
(e) The amount of inherent filtration.

2.4   Heat produced in the target of a rotating anode tube:

(a) Is a potential source of damage to the tube.
(b) Helps the process of X-ray emission.
(c) Accounts for a very large percentage of the incident electron energy.
(d) Is dissipated mainly by radiation.
(e) Reduces the speed of rotation of the anode.

2.5   The energy of the characteristic radiation emitted from an X-ray tube:

(a) Is independent of the type of generator used.
(b) Depends on the mA.
(c) Decreases as the atomic number of the target material increases.
(d) Must be less than the maximum energy of the electrons striking the target.
(e) Depends on the tube shielding.

2.6   The quality of the X-ray spectrum emerging from an X-ray tube depends on:

(a) The peak voltage applied to the tube.
(b) The waveform of the applied voltage.
(c) The atomic number of the anode.
(d) The angle of the anode.
(e) The atomic number of the tube window.

2.7   The following statements relate to the continuous part of an X-ray spectrum:

(a) The maximum energy of X-rays generated at a fixed kV is independent of the voltage profile applied to the tube.
(b) The intensity of X-rays is influenced by the anode–cathode distance.
(c) The intensity of X-rays is influenced by the atomic number of the anode.
(d) There are two maxima.
(e) It is explained by considering deceleration of electrons in the target.

2.8   In carrying out quality assurance checks on X-ray equipment:

(a) An ionization chamber may be used to detect leakage radiation.
(b) Light beam alignment may be checked by imaging a rectangular metal frame.
(c) Total filtration may be estimated by measuring the half value thickness of aluminium for the beam.
(d) Intensifying screens may be checked using visible light.
(e) A resolution of 2.0 line pairs per millimetre would be acceptable for a fluoroscopy system.

2.9   The following statements are true regarding the rating of an X-ray tube:

(a) It specifies maximum safe electrical and thermal operating conditions for a long tube lifetime.
(b) Full wave and half wave rectified tubes have the same rating curves.
(c) The maximum tube current is inversely proportional to the area of the focal spot target.
(d) The limiting mA s increases with time of exposure because of cooling.
(e) During fluoroscopy, a radiographic exposure causes a sharp increase in the temperature of the anode.

2.10   In carrying out quality assurance checks on X-ray equipment:

(a) The output of the X-ray set would be measured with a Geiger–Müller monitor.
(b) kV can be measured with a film sensitometer.
(c) Exposure times as short as 0.05 s may be checked using an electronic timer.
(d) A higher output would be expected with the broad focus spot than with the fine focus.
(e) The input dose rate to the image intensifier should be checked.

2.11 Advantages of a three phase generator compared to a single phase generator are the following:

  (a) A higher tube rating for short exposures.
  (b) Near maximum loading may be applied to the tube throughout the exposures.
  (c) Shorter exposure times.
  (d) Higher repetition rate.
  (e) A higher maximum energy of photons.

2.12 The following statements are true concerning X-ray beam filtration:

  (a) The inherent filtration of a conventional X-ray tube is of the order of 0.7 mm aluminium equivalent.
  (b) High inherent filtration is essential for mammography.
  (c) Added filtration becomes unnecessary above 125 kV.
  (d) It reduces the skin dose.
  (e) It reduces the integral absorbed dose.

3.1 The photoelectric attenuation process:

  (a) Occurs when photons collide with atomic nuclei.
  (b) Decreases continuously with increasing radiation energy.
  (c) Is negligible in water at 120 keV.
  (d) Gives rise to scattered radiation outside the body.
  (e) Is the main reason for lead being such a good protective material for diagnostic X-rays.

3.2 If the half value thickness in aluminium for a beam of X-rays from an X-ray tube operating at 80 kVp is 2 mm:

  (a) The intensity of the beam will be halved if 2 mm of aluminium is placed in the beam.
  (b) The intensity of the beam will be reduced by 75% if 4 mm of aluminium is placed in the beam.
  (c) The half value thickness would be greater if measured in copper.
  (d) The half value thickness in aluminium would be less if the tube were operated at 100 kVp.
  (e) The half value thickness in aluminium would be less if the tube current (mA) were reduced.

3.3 When X-rays interact with matter, the following statements are true:

  (a) Absorption of X-radiation by the photoelectric effect results in annihilation radiation.
  (b) The photoelectric effect usually predominates over the Compton effect in X-ray computed tomography.
  (c) In the Compton effect the energy of the incident photon is shared between an electron and a scattered photon.
  (d) In the Compton effect the majority of photons are scattered backwards at the scattering centre at diagnostic energies.
  (e) Absorption of X-radiation by pair production is always accompanied by the release of 0.51 MeV radiation.

3.4 The first half value thickness for 30 keV gamma rays in water is 20 mm. The following statements are true.

  (a) The second half value thickness is also 20 mm.
  (b) 60 mm of water will stop 95% of 30 keV X-rays.
  (c) The half value thickness in healthy lung for 30 keV gamma rays is more than 20 mm.
  (d) The half value thickness of water for X-rays from a tube operating at 30 kVp is 20 mm.
  (e) The second half value thickness of water for X-rays from a tube operating at 30 kVp is greater than the first half value thickness.

3.5   When an X-ray photon in the diagnostic energy range passes through matter it may:

(a) Be deflected through more than 90° and lose energy.
(b) Be deflected without any change of energy.
(c) Be transmitted without any change in energy or direction.
(d) Cause the production of characteristic radiation.
(e) Undergo an interaction resulting in the creation of two particles with the same kinetic energy.

3.6   The tenth value layer of a diagnostic X-ray beam is:

(a) The depth at which the beam intensity has fallen to one-tenth.
(b) The depth at which the maximum X-ray energy has fallen to one-tenth.
(c) Approximately five times the half value layer in soft tissue.
(d) Greater in materials of higher atomic number.
(e) Independent of the voltage supply.

3.7   When 60 keV photons of gamma ray energy are absorbed in a crystal of sodium iodide:

(a) All the photons undergo a photoelectric interaction.
(b) Each gamma ray photon releases one visible light photon.
(c) The number of visible light photons may be increased by adding certain impurities to the crystal.
(d) Visible light is released from many different depths in the crystal.
(e) The wavelength of light produced is directly proportional to the wavelength of the absorbed gamma photon.

4.1   The following statements relate to the characteristic curve of an X-ray film:

(a) The curve is plotted on axes of optical density against exposure.
(b) Film speed is the exposure to cause an optical density of 1.0.
(c) The maximum slope of the curve is the film gamma.
(d) The film latitude is the X-ray exposure range that will produce useful optical densities.
(e) At very high exposures the optical density may start to decrease.

4.2   When X-rays interact with photographic film the process depends on:

(a) Photoelectric interactions with the gelatine.
(b) Impurities in the crystal structure.
(c) Heat generation by the X-rays.
(d) Formation of clusters of neutral silver atoms.
(e) Removal of excess silver by the developer.

4.3   Intensifying screens:

(a) Emit electrons when bombarded with X-rays.
(b) Contain high atomic number nuclides to improve X-ray stopping efficiency.
(c) Reduce the patient dose by at least a factor of 10.
(d) Increase the sharpness of a radiograph.
(e) When used in combination with film have a higher speed than the film itself.

4.4   In an image intensifier:

(a) Caesium iodide is chosen for the input phosphor because its light emission is well matched to the spectral response of the eye.
(b) Sensitivity is improved by increasing the input phosphor thickness.
(c) Electrons emitted by the input phosphor are focused onto the output phosphor.
(d) Electrons are accelerated by potential differences of about 25 kV.
(e) Image brightness is increased by minification by a factor of about 10.

4.5   Film fog is increased by the following:

   (a) Storage of film at high temperature.
   (b) Prolonged storage of film before use.
   (c) Use of rare earth screens.
   (d) Excessively strong fixation.
   (e) Poor viewing conditions.

4.6   The use of intensifying screens:

   (a) Increases the fraction of radiation emerging from the patient that is used to form the image.
   (b) Improves the spatial resolution in the final image.
   (c) Reduces the amount of quantum mottle in the image.
   (d) Increases film latitude.
   (e) Reduces the patient dose.

5.1   The following statements are true regarding the radiographic image:

   (a) Increasing the field size increases markedly the geometric unsharpness.
   (b) The heel effect is most pronounced along the anode side of the X-ray beam.
   (c) The visible contrast on a fluorescent screen between two substances of equal thickness depends on the differences between their mass attenuation coefficients and densities.
   (d) For an X-ray generator operating above 150 kV bone/soft tissue contrast is less than soft tissue/air contrast.
   (e) The limiting spatial resolution between different parts of the image depends on the contrast.

5.2   The exposure time for taking a radiograph:

   (a) Is decreased if the kV is increased.
   (b) Is increased if the mA is increased.
   (c) Is decreased by increasing the added filtration.
   (d) Depends on the thickness of the patient.
   (e) Is decreased if a grid is used.

5.3   When a radiograph of a section of the body is to be taken with the film cassette kept as close to the patient as possible, the skin dose to the patient in the beam will be increased if:

   (a) The focus–film distance is increased.
   (b) The kV used is increased.
   (c) More sensitive screens are used in the cassette.
   (d) A grid is inserted between the cassette and the patient.
   (e) The mA is increased while keeping the mA s constant.

5.4   In a normal radiographic set-up:

   (a) A shorter focus–film distance reduces the magnification.
   (b) The smaller the object the greater the magnification.
   (c) The image of an object increases in size as its distance from the film increases.
   (d) The image of an object increases in sharpness as its distance from the film decreases.
   (e) The magnification depends on the focal spot size.

5.5   The following statements relate to the correct use of linear grids:

   (a) An increased grid ratio will reduce the scattered radiation reaching the film.
   (b) An increased grid ratio will reduce the field of view.
   (c) The width of the lead strips in the grid must be less than the required resolution.
   (d) The density of the radiograph will be reduced towards the edges.
   (e) The dose to the patient will increase by the ratio lead strip thickness/interspace distance.

5.6   In abdominal radiography the amount of scattered radiation reaching the film may be reduced, relative to the primary beam by:

(a) Increasing the kV.
(b) Increasing beam filtration.
(c) Increasing the tube–patient distance.
(d) Increasing the patient–film distance.
(e) Abdominal compression.

5.7   Contrast on a radiograph is increased by:

(a) Increasing the distance from the patient to the film.
(b) Decreasing the focal spot size.
(c) Increasing the filtration.
(d) Decreasing the field size.
(e) Using a film of higher gamma.

6.1   The exit dose from a patient in the primary X-ray beam:

(a) Is caused primarily by backscattered electrons.
(b) Depends on X-ray field size.
(c) Depends on X-ray focal spot size.
(d) Decreases if tube voltage is increased with mA s constant.
(e) Increases if tube filtration is increased but other factors remain constant.

6.2   Absorbed dose:

(a) Is directly proportional to the mA s.
(b) Is typically a few μGy at the skin surface in a diagnostic examination of the abdomen.
(c) Is less in fat than in muscle for a given exposure of 30 kV X-rays.
(d) Is about the same in bone and muscle for a given exposure of 30 kV X-rays.
(e) Is the only factor determining biological effect.

6.3   If the gonads are outside the primary beam during the taking of a radiograph, the dose to them will be reduced if:

(a) A grid is used.
(b) A more sensitive film screen combination is used.
(c) The field size is reduced.
(d) A smaller focal spot size is selected.
(e) The patient film distance is increased.

6.4   A typical P–A chest radiograph will:

(a) Include the female gonads in the direct beam.
(b) Give a dose to the male gonads of about 10 μGy.
(c) Give a skin dose to the patient in the direct beam of about 2 mGy.
(d) Use X-rays generated above 60 kV.
(e) Give a lower effective dose than an A–P chest radiograph.

6.5   Absorbed dose can be specified in

(a) joule/kilogram.
(b) millisieverts.
(c) gray.
(d) coulomb/kilogram.
(e) electron volts.

6.6 If the dose rate in air at 100 cm from an X-ray machine when operating at 100 kV and 100 mA is 600 mGy min^{-1} then:

    (a) The dose rate at 50 cm will be 2.4 Gy min^{-1}.
    (b) The dose rate at 25 cm will be 4.2 Gy min^{-1}.
    (c) The dose rate at 100 cm will be halved if the machine is operated at 100 kV and 25 mA.
    (d) The dose rate will remain unchanged if the kV is altered.
    (e) Placing an extra filter in the beam will decrease the dose rate.

6.7 A free air ionization chamber is more sensitive to 100 kV X-rays, if totally irradiated, i.e. produces a larger current, for a given exposure rate, if:

    (a) Its air volume is increased.
    (b) The ambient temperature is greater.
    (c) The ambient pressure is greater.
    (d) The effective atomic number of the wall material is less than that of air.
    (e) The voltage across the electrodes is less than that necessary to saturate the chamber.

6.8 In diagnostic radiology using film/screens the entrance dose to the patient on the axis of the beam is reduced by:

    (a) Increasing the kV.
    (b) Using a grid.
    (c) Patient compression.
    (d) Using a small focal spot.
    (e) An air gap technique.

6.9 The effective dose to a patient from a single plane film radiograph is:

    (a) Decreased at lower mA.
    (b) Decreased by the use of grids.
    (c) Decreased by using a faster film.
    (d) Decreased by decreasing the field size.
    (e) Always less than 5 mSv.

6.10 Absorbed dose:

    (a) Is a measure of the energy absorbed per unit mass of material.
    (b) Is the energy released in tissue per ion pair formed.
    (c) May be determined from an ionization chamber measurement.
    (d) Is different in different materials exposed to the same X-ray beam.
    (e) Decreases exponentially as the X-ray beam from an X-ray tube passes through soft tissue.

6.11 The following are features of a Geiger–Müller counter used for measuring radioactivity:

    (a) It has a rather poor efficiency for detecting gamma rays.
    (b) A thin window is desirable to detect beta radiation.
    (c) The collection volume is open to the atmosphere.
    (d) The operating voltage may be similar to that of an ionization chamber.
    (e) It is better than a scintillation crystal monitor for checking surface contamination in the nuclear medicine department.

6.12 In a primary diagnostic X-ray beam, the radiation dose at a depth of 20 cm in a thick region of the patient:

   (a) Is independent of mA s.
   (b) Decreases if tube voltage is increased but film density is kept constant.
   (c) Decreases if the tube to patient distance is increased but film density is kept constant.
   (d) Is about half the dose at a depth of 10 cm.
   (e) Is independent of the X-ray field size.

6.13 The following dosimetric quantities are approximately correct:

   (a) The effective dose from a barium enema is 7 mSv.
   (b) The annual whole body equivalent dose limit for members of the public is 15 mSv.
   (c) A chest X-ray entrance skin dose is 150 μGy.
   (d) The effective dose to a patient from a lung perfusion scan with 100 MBq Tc-99m is 1 mSv.
   (e) The average annual effective dose to the UK population from medical exposures is 350 μSv.

6.14 In comparison with a Geiger–Müller counter, a scintillation counter:

   (a) Is more sensitive to gamma rays.
   (b) Is more portable.
   (c) Can be adapted more readily for counting radiation from one radionuclide in the presence of another.
   (d) Is better for dealing with high count rates.
   (e) Requires less complex associated apparatus.

7.1 Technetium-99m is a suitable radionuclide for imaging because:

   (a) Its half-life is such that it can be kept in stock in the department for long periods.
   (b) Its gamma ray energy is about 140 keV.
   (c) It emits beta rays that can contribute to the image.
   (d) It can be firmly bound to several different pharmaceuticals.
   (e) A high proportion of disintegrations produce gamma rays.

7.2 A molybdenum–technetium generator contains 3.7 GBq of molybdenum-99 at 0900 hours on a Monday. The activity of technetium-99m in equilibrium with the molybdenum-99:

   (a) At any time depends on the half-life of technetium-99m.
   (b) Is 3.7 GBq at 0900 hours on the Monday.
   (c) Depends on the temperature.
   (d) Will be about 370 MBq at 0900 hours on the following Friday.
   (e) At any time will be reduced if the generator has been eluted.

7.3 Desirable properties of a radionuclide to be used for skeletal imaging are:

   (a) Adequate beta ray emission.
   (b) Gamma ray emissions of two distinct energies.
   (c) A non-radioactive daughter.
   (d) A medium atomic number.
   (e) A half-life of less than 10 minutes.

7.4 The effective dose to an adult patient from a radionuclide bone scan:

   (a) Is affected by radiochemical purity.
   (b) Increases with increasing body weight.
   (c) Is reduced if the patient empties their bladder before the scan.
   (d) Is typically about 15 mSv per examination.
   (e) Is high enough for the patient to be designated as a classified person.

7.5   When using a collimator with a gamma camera:

(a) The main purpose of the collimator is to remove scattered radiation.
(b) A high resolution collimator is required for dynamic studies.
(c) High energy collimators give poorer resolution than low energy collimators.
(d) With a parallel hole collimator resolution is independent of depth.
(e) A diverging collimator will increase the field of view.

7.6   Diagnostic doses of radiopharmaceuticals are chosen so that the effective dose does not exceed:

(a) 5 mSv.
(b) The dose limit for occupational exposure.
(c) The dose limit for members of the public.
(d) The variation of natural background doses.
(e) The effective dose from an X-ray CT scan.

7.7   In a gamma camera:

(a) Resolution increases with increasing thickness of the crystal.
(b) Non-linearity is caused more by the scintillation detector than by the photomultiplier tubes.
(c) The pulse height analyser eliminates all scattered photons.
(d) The pulse height analyser enables two radionuclides to be detected at the same time.
(e) Monoenergetic gamma rays produce light flashes of fixed intensity in the detector.

7.8   The resolution of a gamma camera image is affected by:

(a) The septal size on the collimator.
(b) The energy of gamma rays.
(c) The count rate.
(d) The number of photomultiplier tubes.
(e) The window width on the pulse height analyser.

7.9   The following properties are desirable in a radionuclide used for organ imaging:

(a) A photon energy below 80 keV.
(b) A single gamma ray energy.
(c) Absence of beta activity.
(d) A low tumour/background ratio.
(e) A high critical organ affinity.

8.1   In the analysis of the spatial frequency response of an imaging system which uses an image intensifier as the receptor:

(a) The modulation transfer function (MTF) is normalized to one at zero spatial frequency.
(b) The resultant MTF is the sum of the MTFs of the components.
(c) MTF and line spread function (LSF) contain the same information.
(d) The MTF always decreases with increasing spatial frequency.
(e) The resultant MTF is independent of the design of the X-ray tube.

8.2   The amount of quantum noise in the image produced by an image intensifier and television system for fluoroscopy can be reduced by:

(a) Increasing the screening time.
(b) Increasing the exposure rate.
(c) Use of thicker intensifier input screens.
(d) Increasing the brightness gain on the image intensifier.
(e) Reducing electronic noise in the television chain.

8.3   In relation to the radiographic image:

(a) Contrast between two parts of the film is the ratio of their optical densities.
(b) Full width half maximum is a measure of resolution.
(c) Modulation transfer function is a method of measuring contrast.
(d) Noise increases with increasing patient dose.
(e) Fine detail corresponds to high spatial frequencies.

8.4   The following statements correctly relate to the radiographic image:

(a) Noise depends on the brightness of the film viewing box.
(b) Contrast between two areas can be defined as the difference in optical density on the two parts of the film.
(c) Definition can be related to the response of a sharp edge.
(d) Total unsharpness is the sum of geometric, screen and movement unsharpness.
(e) Resolution can be derived from the modulation transfer function measured on a low contrast object.

8.5   In the assessment of a diagnostic imaging procedure:

(a) The line spread function is an effective measure of resolution.
(b) The modulation transfer function measures the performance of the complete imaging system.
(c) The sum of the true positive and false positive images will be constant.
(d) A strict criterion for a positive abnormal image is a better way to discriminate between two imaging techniques than a lax criterion.
(e) Bayes theorem provides a way to allow for the prevalence of disease.

8.6   Receiver operator characteristic (ROC) curves:

(a) Provide a means to compare the effectiveness of different imaging procedures.
(b) Are obtained when contrast is varied in a controlled manner.
(c) Normally plot the false positives on the X axis against the false negatives on the Y axis.
(d) Require the observer to adopt at least five different visual thresholds.
(e) Reduce to a straight line at $45°$ when the observer guesses.

9.1   In asymptomatic mammographic screening:

(a) X-rays generated below 20 kV must be used.
(b) The X-ray tube may have a molybdenum target.
(c) A focus–film distance of at least 1 metre is necessary.
(d) Doses to the breasts must be less than 2 mGy.
(e) Rare earth screens must be used.

9.2   Xerography:

(a) Produces more contrast than conventional radiography between images of structures of different densities.
(b) Depends on the phenomenon of photoconductivity.
(c) Makes use of plates that can be re-used repeatedly.
(d) Has been recommended for routine breast screening as the radiation dose required is lower than for any other technique.
(e) Needs a special X-ray generator for its successful application.

9.3 If the focus–film distance for a particular examination is increased from 80 cm to 120 cm with the film cassette kept as close to the patient as possible:

(a) The image is magnified more.
(b) Geometric unsharpness is reduced.
(c) A larger film should be used.
(d) The entrance skin dose to the patient is reduced.
(e) The exit skin dose to the patient is unchanged.

9.4 Requirements for good macro-radiography include:

(a) A tube–patient distance of at least 1 m.
(b) A fine focus X-ray tube.
(c) A higher mA s than conventional radiography of the same part at the same kV.
(d) Stationary grids.
(e) Slow screens to counteract magnified grain size.

9.5 The spectrum of X-rays emitted from an X-ray tube being used for mammography:

(a) Will contain characteristic X-rays if a molybdenum anode is being used.
(b) Is likely to contain X-rays with a maximum energy of 50 keV.
(c) At fixed kilovoltage, has a total intensity that depends on the atomic number of the anode.
(d) Is independent of the tube window material.
(e) Will pass through additional filtration before the X-rays fall on the patient.

9.6 The following statements relate to digital radiography:

(a) Digital information is any information presented in discrete units.
(b) A variable window width facility is a means of manipulating contrast.
(c) If a uniform source is examined with an ideal imager, variations in grey scale across the image may be reduced by increasing the radiation dose.
(d) If N frames from a digitized set of images are added, the fractional variation in statistical noise is reduced by a factor of N.
(e) Digital data must be converted to analogue form prior to storage.

10.1 In X-ray computerized tomography:

(a) Compton interactions predominate in the patient.
(b) A cadmium tungstate detector is more sensitive than a xenon detector.
(c) Filtered backprojection is a method of data reconstruction.
(d) The Hounsfield number for water is zero.
(e) Small differences in contrast are more easily seen by decreasing the window width.

10.2 In X-ray computerized tomography:

(a) X-ray output must be more stable than in conventional radiology.
(b) The detectors are normally photomultiplier tubes.
(c) The spatial resolution is limited by the number of pixels.
(d) A pixel may have a range of attenuation coefficients if it contains several different tissues.
(e) Hounsfield numbers may be changed by altering the window level and window height.

10.3 The thickness of the slice seen in focus on a conventional tomogram may be reduced by:

(a) Increasing the angle of traverse.
(b) Increasing the mA s.
(c) Reducing the kV.
(d) Using a more sensitive film–screen combination.
(e) Increasing the fixation time.

10.4 X-ray computed tomography:

   (a) Is more sensitive than conventional radiography to differences of atomic number in tissues.
   (b) Can only resolve objects greater than 100 mm in diameter.
   (c) Depends on the different attenuation of X-rays in different tissues.
   (d) Is usually performed with X-rays generated below 70 kV.
   (e) Is used to obtain pictures of transaxial body sections.

10.5 In X-ray computed tomography:

   (a) Some CT scanners have more than 1000 detectors.
   (b) The partial volume effect occurs when the X-ray beam does not pass through part of the patient cross section.
   (c) Filtered backprojection provides additional attenuation between the patient and the detector.
   (d) If a water phantom is imaged, all the computed Hounsfield numbers will be close to zero.
   (e) Decreasing the window width is a mechanism for visualizing small differences in contrast.

10.6 In X-ray computed tomography:

   (a) The majority of X-ray interactions in the patient are inelastic collisions.
   (b) If xenon gas detectors are used, the gas is above atmospheric pressure.
   (c) Each detector has its own collimator.
   (d) Resolution is improved by increasing the pixel size.
   (e) The amount of noise in the image can be estimated by imaging a water phantom.

11.1 Equivalent dose is:

   (a) Affected by radiation weighting factors.
   (b) The absorbed dose averaged over the whole body.
   (c) For X-rays numerically equal to the absorbed dose in grays.
   (d) Used to specify some dose limits.
   (e) Used to specify doses to patients from X-ray examinations.

11.2 The following statements refer to the biological effects of radiations on cells and tissues:

   (a) The risk of fatal cancer is higher than the risk of severe hereditary disease.
   (b) A deterministic effect increases steadily in severity from zero dose.
   (c) Stochastic effects may be caused by background radiation.
   (d) The period of maximum risk for severe mental retardation is between the 16th and 24th weeks after conception.
   (e) Equal equivalent doses of different radiations cause equal risk to different tissues.

11.3 With respect to the responses of cells to ionizing radiation:

   (a) Stem cells are generally more sensitive than differentiated cells.
   (b) Cellular recovery is apparent by 6 h after irradiation.
   (c) Dose rate effects are evidence of recovery.
   (d) The RBE of neutrons (*in vitro*) will increase with increasing dose.
   (e) Cells show more evidence of recovery after X-ray irradiation than after neutron irradiation.

11.4 The following statements relate to the long term effects of radiation:

   (a) Stochastic risks are additive when radiation doses are fractionated.
   (b) Mutagenic effects have been demonstrated statistically following radiation exposure of humans.
   (c) Early transient erythema is a deterministic effect.
   (d) There is no evidence of radiation-induced malignancy from the diagnostic use of X-rays.
   (e) Radiation-induced solid tumours may not appear until 40 years after exposure.

11.5 The risk from injected radioactivity:

    (a) Is independent of the biological half-life in the body.
    (b) Depends on both the radionuclide and on the pharmaceutical form.
    (c) Decreases steadily with time after injection.
    (d) May be primarily due to the dose to a single organ.
    (e) Is a factor limiting the quality of radionuclide images.

11.6 The following statements relate to the use of diagnostic X-rays during pregnancy:

    (a) The dose to the uterus is usually calculated to obtain an estimate of the dose to the foetus.
    (b) Severe mental retardation is the most serious risk during the first few weeks of pregnancy.
    (c) Hereditary effects may be no higher than after birth.
    (d) The effective dose to the foetus cannot be higher than the effective dose to the mother.
    (e) The risk of fatal cancer to age 15 will be about 1 in 30 000 for each 1 mGy of absorbed dose to the foetus.

12.1 Thermoluminescent dosimeters:

    (a) May be made of lithium chloride.
    (b) Depend for their action on the emission of light when the dosimeter is heated after irradiation.
    (c) Can only be used for measuring doses over 10 mGy.
    (d) Can be used for *in vivo* measurements in body cavities.
    (e) Have a response which is practically independent of radiation energy over a wide range of energies.

12.2 Photographic film as a dosimeter:

    (a) Measures dose rather than dose rate.
    (b) Is commonly used for personal monitoring.
    (c) Can cover a wide range of dose.
    (d) Has a response which is independent of the radiation quality.
    (e) Requires carefully controlled development conditions.

12.3 Thermoluminescent dosimetry:

    (a) Depends on the emission of light from an irradiated material after heating.
    (b) Can only be used in a very restricted dose range.
    (c) Is particularly useful for *in vivo* dosimetry because the detector is tissue equivalent.
    (d) Is an absolute method of determining absorbed dose.
    (e) Is used for routine personal monitoring.

12.4 The lead equivalent of a protective barrier:

    (a) Is the amount of lead the barrier contains.
    (b) Depends on the density of the barrier material.
    (c) Increases as radiation energy increases in the range up to 5 MeV if the atomic number of the barrier material is less than that of lead.
    (d) Must be at least 2 mm of lead in diagnostic radiology.
    (e) Varies with the amount of filtration in a diagnostic X-ray beam.

12.5 In a film badge:

    (a) The film is single coated.
    (b) The open window enables beta ray doses to be assessed.
    (c) The lead–tin filter is to protect the film from fogging.
    (d) Doses in air of 100 mGy diagnostic X-rays can be measured.
    (e) α-particles will not be detected.

12.6 The following statements concerning the Ionising Radiations Regulations (1985) are true:

   (a) Local rules are required for any work notifiable to the Health and Safety Executive under the Regulations.
   (b) Staff must be designated as classified radiation workers if the dose to their hands is likely to exceed 150 mSv per annum.
   (c) Radiation protection supervisors are appointed to assist the radiation protection adviser with their work.
   (d) All radiation workers must have their personal dose monitored at regular intervals.
   (e) Boundaries to controlled areas must correspond with physical barriers such as walls, doors etc.

13.1 The following statements concerning ultrasound are true:

   (a) It is electromagnetic radiation in the range 1 to 15 MHz.
   (b) It can be focused.
   (c) The speed of sound in tissue is independent of temperature.
   (d) The absorption of ultrasound in tissues is frequency dependent.
   (e) Swept gain is used to compensate for tissue attenuation.

13.2 In Doppler ultrasound:

   (a) The ultrasound must be used in continuous wave mode.
   (b) The frequency of the Doppler-shifted ultrasound is above the audible range in medical applications.
   (c) The Doppler-frequency shift depends on the acoustic properties of the medium between the probe and the moving target.
   (d) No Doppler signal will be recorded unless there is a difference in characteristic impedance across the moving boundary.
   (e) The Doppler-frequency shift decreases as the angle between the beam and the normal to the moving boundary decreases.

13.3 Two similar-sized patients are to have consecutive liver ultrasound scans. The following parameters may need to be adjusted between patients:

   (a) Ultrasound frequency.
   (b) Focusing.
   (c) Dynamic range.
   (d) Swept gain.
   (e) Input power.

13.4 Decreasing the pulse length from a conventional B scanner has the effect of:

   (a) Increasing the dynamic range.
   (b) Increasing the frequency bandwidth.
   (c) Increasing the axial resolution.
   (d) Increasing the lateral resolution.
   (e) Increasing the penetration.

13.5 Attenuation of an ultrasound beam in soft tissue is:

   (a) Independent of frequency.
   (b) Affected by 'swept gain' settings.
   (c) Principally due to reflection and scattering at boundaries.
   (d) Affected by the amount of energy involved in relaxation processes.
   (e) Exponential with range in the far field.

13.6 The following statements are correct for ultrasonic radiation used in medical diagnosis.

    (a) It is generated by transverse vibration of air molecules in the frequency range 1 to 15 MHz.
    (b) It has a wavelength in soft tissue of between 0.1 and 1.5 mm.
    (c) At a fixed frequency it has a longer wavelength in bone than in soft tissue.
    (d) It can be focused using plastic lenses.
    (e) It is more heavily absorbed in tissue at high frequencies than at low frequencies.

13.7 When ultrasound is used for conventional B-mode imaging:

    (a) The higher the frequency the greater the depth that can be scanned.
    (b) Resolution along the direction of propagation of the beam depends on pulse length.
    (c) Lateral resolution in the near field depends on the frequency of the ultrasound.
    (d) The transducer is in the receiving mode for over 99% of the time.
    (e) Each echo on the display corresponds to reflecting or scattering boundary in the tissue.

13.8 Swept gain (time gain compensation) is used in amplifiers of ultrasound signals to:

    (a) Improve axial resolution.
    (b) Compensate for differences in attenuation between different tissues.
    (c) Compensate for the time difference between the front and back of a single pulse.
    (d) Obtain signals of similar intensity from all similar boundaries.
    (e) Widen the field of view seen by the detector.

13.9 In order to detect multiple tissue boundaries within the patient using conventional ultrasound:

    (a) An ultrasound frequency of at least 1 MHz is required.
    (b) The densities of the media on either side of each boundary must be different.
    (c) Pulsed ultrasound must always be used.
    (d) At least 10% of the incident energy must be reflected back to the transducer by each boundary.
    (e) The reflecting surface must be stationery.

14.1 Imaging by magnetic resonance imaging:

    (a) Requires at least a one tesla static magnetic field.
    (b) Depends on excitation of nuclei by a time varying RF magnetic field.
    (c) Can demonstrate blood flow without injection of contrast medium.
    (d) Produces a digital image.
    (e) Can use diamagnetic materials to enhance contrast.

14.2 In magnetic resonance imaging:

    (a) Only nuclides with an odd number of protons can be investigated.
    (b) The strength of the signal increases as the strength of the static magnetic field increases.
    (c) Short T_1 values are associated with highly structured tissues.
    (d) Field gradients must be applied to obtain spatial information.
    (e) The major hazard to health limiting the static magnetic field is the associated temperature rise in the tissues.

14.3 In nuclear magnetic resonance imaging of protons:

    (a) The SI unit of magnetic induction (magnetic flux density) is teslas per metre.
    (b) A magnetic field varying with the Larmor frequency is used to define the slice to be imaged.
    (c) Magnetic field gradients may be applied in more than one direction.
    (d) The strength of the static magnetic field is a factor in determining the duration of the time varying field.
    (e) Local variations in the static magnetic field affect T_2^* values.

14.4 In nuclear magnetic resonance imaging:

(a) Theoretically, the static magnetic field can act in any direction.
(b) The magnitude of the magnetic field gradient determines the thickness of the slice to be imaged.
(c) A superconducting magnet has almost no electrical resistance.
(d) The gyro-magnetic ratio is one of the factors that determines the strength of the signal from a given volume.
(e) The gradient magnetic field can cause a depolarizing potential across circulating blood cells.

14.5 In magnetic resonance imaging, the Larmor frequency:

(a) Measures the gyroscopic rotation of the nuclei.
(b) Is the radiowave frequency at which the magnet system can absorb energy.
(c) Is one of the factors contributing to the strength of the net magnetic moment.
(d) Is directly proportional to the Boltzmann constant.
(e) Is the frequency of the detected radio signal.

14.6 In magnetic resonance imaging, for a fixed pixel matrix, the signal intensity within a pixel is affected by the following:

(a) The gyromagnetic ratio.
(b) The amplitude of the applied radiofrequency magnetic field.
(c) The repetition time.
(d) Flow.
(e) The field of view.

14.7 Magnetic resonance imaging

(a) May utilize the carbon content of the body to generate the image.
(b) Uses signals which depend on the magnitude of the magnetic vector.
(c) Needs a radio-frequency wave to induce a signal.
(d) Is contra-indicated for patients with aneurism clips.
(e) Produces stray fields proportional to the static field strength.

14.8 In magnetic resonance imaging

(a) The net spin angular momentum for protons should be zero.
(b) The ratio of nuclei in the low/high energy state depends on the magnetic field.
(c) Precession occurs when the resultant magnetization vector of the nuclei is parallel to the external field.
(d) T_2 is always shorter than T_1 in biological tissues.
(e) The STIR sequence is used to suppress the high signal from fat.

14.9 In magnetic resonance imaging the magnetic field gradient

(a) Is generated by a superconducting magnet.
(b) Is applied together with a 90° pulse to form the image slice.
(c) Is applied perpendicular to the slice plane in selective excitation.
(d) Inverts the nuclear magnetization.
(e) Produces a linear variation of resonant frequency with position.

ANSWERS

Question	(a)	(b)	(c)	(d)	(e)	Question	(a)	(b)	(c)	(d)	(e)
1.1	F	T	T	F	F	6.5	T	F	T	F	F
1.2	F	T	T	T	T	6.6	T	F	F	F	T
1.3	T	F	T	T	T	6.7	T	F	T	F	F
1.4	F	F	F	T	T	6.8	T	F	T	F	F
1.5	F	F	T	T	T	6.9	F	F	T	T	T
1.6	F	F	T	T	F	6.10	T	F	T	T	F
1.7	T	T	F	F	T	6.11	T	T	F	T	F
1.8	T	F	T	T	T	6.12	F	T	T	F	F
2.1	F	F	T	T	F	6.13	T	F	T	T	T
2.2	T	T	T	T	F	6.14	T	F	T	T	F
2.3	T	F	T	T	F	7.1	F	T	F	T	T
2.4	T	F	T	T	F	7.2	F	T	F	F	F
2.5	T	F	F	T	F	7.3	F	F	T	F	F
2.6	T	T	T	T	T	7.4	T	F	T	F	F
2.7	T	F	T	F	T	7.5	F	F	T	F	T
2.8	T	T	T	F	T	7.6	F	F	F	F	F
2.9	T	F	F	T	T	7.7	F	F	F	T	F
2.10	F	F	T	F	T	7.8	T	T	T	T	T
2.11	T	T	T	T	F	7.9	F	T	T	F	T
2.12	T	F	F	T	T	8.1	T	F	T	T	F
3.1	F	F	T	F	T	8.2	F	T	T	F	F
3.2	T	F	F	F	F	8.3	F	T	F	F	T
3.3	F	F	T	F	T	8.4	F	T	T	F	F
3.4	T	F	T	F	T	8.5	T	T	F	F	T
3.5	T	T	T	T	F	8.6	T	F	F	F	T
3.6	T	F	F	F	F	9.1	F	T	F	T	T
3.7	F	F	T	T	F	9.2	F	T	T	F	F
4.1	F	F	T	T	T	9.3	F	T	F	T	T
4.2	F	T	F	T	F	9.4	F	T	T	F	F
4.3	F	T	T	F	T	9.5	T	F	T	F	T
4.4	F	T	F	T	F	9.6	T	T	T	F	F
4.5	T	T	F	F	F	10.1	T	T	T	T	T
4.6	T	F	F	F	T	10.2	T	F	T	F	F
5.1	F	T	T	T	T	10.3	T	F	F	F	F
5.2	T	F	F	T	F	10.4	F	F	T	F	T
5.3	F	F	F	T	F	10.5	T	F	F	T	T
5.4	F	F	T	T	T	10.6	T	T	T	F	T
5.5	T	T	F	T	F	11.1	T	F	T	T	F
5.6	F	F	F	T	T	11.2	T	F	T	F	F
5.7	T	F	F	T	T	11.3	T	T	T	F	T
6.1	F	T	F	F	F	11.4	T	F	T	F	T
6.2	T	F	T	F	F	11.5	F	T	F	T	T
6.3	F	T	T	F	F	11.6	T	F	T	F	T
6.4	F	F	F	T	T	12.1	F	T	F	T	T

Question	(a)	(b)	(c)	(d)	(e)	Question	(a)	(b)	(c)	(d)	(e)
12.2	T	T	T	F	T	13.8	F	T	F	T	F
12.3	T	F	T	F	T	13.9	F	F	T	F	F
12.4	F	T	T	F	T	14.1	F	T	T	T	F
12.5	F	T	F	T	T	14.2	F	T	T	T	F
12.6	T	T	F	F	F	14.3	F	F	T	F	T
13.1	F	T	F	T	T	14.4	T	T	T	T	T
13.2	F	T	F	T	F	14.5	T	T	F	F	T
13.3	F	T	T	T	T	14.6	T	F	T	T	T
13.4	F	T	T	F	F	14.7	F	T	T	T	T
13.5	F	F	F	T	F	14.8	F	T	F	T	T
13.6	F	T	T	T	T	14.9	F	T	T	F	T
13.7	F	T	F	T	F						

Footnotes:

1.3(d) As Auger electrons (see Persson L 1994 The Auger effect in radiation dosimetry *Health Phys.* **67** 471–6).

1.6(e) Ionizing radiations are oxidizing agents, e.g. ferrous ion $\rightarrow$ ferric ion.

1.8(d) This is frequently used as a practical definition of a diagnostic X-ray beam when it has been heavily filtered.

2.2(b) The high density prevents too much electron penetration into the anode.

2.3(d) kV and mA s will affect the effective spot size because they will influence the performance of the cathode focusing cup in forming a small target spot on the anode surface.

2.6(d) We know there is substantial self-absorption of X-rays in the anode and this will be affected by the angle.

3.1(b) Because of absorption edges

3.1(d) Any secondary radiation, not strictly scattered, will be of such low energy that it is absorbed inside the body.

3.2(b) The second half value thickness will be greater because of beam hardening.

3.5(a) There will still be a small amount of elastic scattering

3.5(d) Characteristic radiation will be produced if a K shell (or higher shell vacancy) is created. If this occurs in the body the radiation may be of too low energy to escape.

3.7(a) A few photons will not interact at all and a few will be Compton scattered.

4.3(e) The slope of the linear portion of the characteristic curve becomes steeper.

4.6(c) Although the intensifying screen stops a higher fraction of photons than film, the number of incident photons will be greatly reduced.

5.3(b) Assuming film density remains the same—see general instructions at the beginning of the questions.

5.4(e) If the penumbra are excluded the larger focal spot will give a smaller image; if penumbra are included it will give a larger image.

5.6(c) This assumes the field size is set at the cassette, if it were set on the patient surface there might be less scatter because of less beam divergence.

5.7(a) and (d) both reduce the scatter reaching the film.

6.1(e) Filtration cannot increase any component of the spectrum so the exit dose cannot increase with the specified conditions.

6.8(c) See general instructions at the beginning of the questions. Compression will actually reduce the amount of tissue in the beam.

6.9(a) See general instructions.

6.10(b) Some energy will be deposited as excitation.

6.11(d) Although the electric field will be much higher than for an ionization chamber, the operating voltage may be similar.

6.12(d) For this to be true, the half value thickness would have to be 10 cm—it is much less.

6.13(c) A reasonable mean from quite a wide range of quoted values.

7.2(d) 4 days is less than two half-lives for Mo-99.

7.2(e) Note the question says 'equilibrium', by the time the generator has returned to equilibrium, it will not matter whether the generator has been eluted or not.

9.6(c) A major contribution to the variations in grey scale will be quantum mottle.

12.1(e) The mean atomic number is similar to that of air and soft tissue.

13.2(b) It is the beat frequency that is audible.

13.2(c) Neglecting small changes in the speed of sound in the medium.

13.5(e) The beam is also diverging so the intensity of radiation from a point source decreases by both attenuation (exponential) and inverse square law effects.

13.7(e) False because signals will also arise from noise artefacts.

14.7(a) The amount of C-11 in the body is too small to give an adequate signal.

INDEX